Bioactive Compounds from Natural Products: Separation, Characterization and Applications

Bioactive Compounds from Natural Products: Separation, Characterization and Applications

Editor

Emanuel Vamanu

MDPI • Basel • Beijing • Wuhan • Barcelona • Belgrade • Manchester • Tokyo • Cluj • Tianjin

Editor
Emanuel Vamanu
University of Agricultural Sciences and Veterinary Medicine
Romania

Editorial Office
MDPI
St. Alban-Anlage 66
4052 Basel, Switzerland

This is a reprint of articles from the Special Issue published online in the open access journal *Applied Sciences* (ISSN 2076-3417) (available at: https://www.mdpi.com/journal/applsci/special_issues/Bioactive_Products).

For citation purposes, cite each article independently as indicated on the article page online and as indicated below:

LastName, A.A.; LastName, B.B.; LastName, C.C. Article Title. *Journal Name* **Year**, *Volume Number*, Page Range.

ISBN 978-3-0365-4197-6 (Hbk)
ISBN 978-3-0365-4198-3 (PDF)

Contents

About the Editor . . . . **vii**

Emanuel Vamanu
Bioactive Compounds from Natural Products: Separation, Characterization, and Applications
Reprinted from: *Appl. Sci.* **2022**, *12*, 3922, doi:10.3390/app12083922 . . . . **1**

Adriana Mirela Tache, Laura Dorina Dinu and Emanuel Vamanu
Novel Insights on Plant Extracts to Prevent and Treat Recurrent Urinary Tract Infections
Reprinted from: *Appl. Sci.* **2022**, *12*, 2635, doi:10.3390/app12052635 . . . . **5**

Miguel Cascais, Pedro Monteiro, Diana Pacheco, João Cotas, Leonel Pereira,
João Carlos Marques and Ana M. M. Gonçalves
Effects of Heat Treatment Processes: Health Benefits and Risks to the Consumer
Reprinted from: *Appl. Sci.* **2021**, *11*, 8740, doi:10.3390/app11188740 . . . . **23**

Sofia Pappou, Maria Myrto Dardavila, Maria G. Savvidou, Vasiliki Louli, Kostis Magoulas
and Epaminondas Voutsas
Extraction of Bioactive Compounds from *Ulva lactuca*
Reprinted from: *Appl. Sci.* **2022**, *12*, 2117, doi:10.3390/app12042117 . . . . **35**

Lucienne Gatt, David G. Saliba, Pierre Schembri-Wismayer and Marion Zammit-Mangion
Tyrosol, at the Concentration Found in Maltese Extra Virgin Olive Oil, Induces HL-60
Differentiation towards the Monocyte lineage
Reprinted from: *Appl. Sci.* **2021**, *11*, 10199, doi:10.3390/app112110199 . . . . **53**

Eduardo Padilla-Camberos, Omar Ricardo Torres-Gonzalez,
Ivan Moises Sanchez-Hernandez, Nestor Emmanuel Diaz-Martinez,
Oscar Rene Hernandez-Perez and Jose Miguel Flores-Fernandez
Anti-Inflammatory Activity of *Cnidoscolus aconitifolius* (Mill.) Ethyl Acetate Extract on Croton
Oil-Induced Mouse Ear Edema
Reprinted from: *Appl. Sci.* **2021**, *11*, 9697, doi:10.3390/app11209697 . . . . **73**

Nicoleta Mirela Blebea, Dan Rambu, Teodor Costache and Simona Negreș
Very Fast RP–UHPLC–PDA Method for Identification and Quantification of the Cannabinoids
from Hemp Oil
Reprinted from: *Appl. Sci.* **2021**, *11*, 9414, doi:10.3390/app11209414 . . . . **81**

Emilia Janiszewska-Turak, Weronika Kołakowska, Katarzyna Pobiega
and Anna Gramza-Michałowska
Influence of Drying Type of Selected Fermented Vegetables Pomace on the Natural Colorants
and Concentration of Lactic Acid Bacteria
Reprinted from: *Appl. Sci.* **2021**, *11*, 7864, doi:10.3390/app11177864 . . . . **95**

Chunjian Zhao, Shen Li, Chunying Li, Tingting Wang, Yao Tian and Xin Li
Flavonoids from Fig (*Ficus carica* Linn.) Leaves: The Development of a New Extraction Method
and Identification by UPLC-QTOF-MS/MS
Reprinted from: *Appl. Sci.* **2021**, *11*, 7718, doi:10.3390/app11167718 . . . . **109**

Ştefania Adelina Milea, Oana Crăciunescu, Gabriela Râpeanu, Anca Oancea, Elena Enachi, Gabriela Elena Bahrim and Nicoleta Stănciuc
Multifunctional Ingredient from Aqueous Flavonoidic Extract of Yellow Onion Skins with Cytocompatibility and Cell Proliferation Properties
Reprinted from: *Appl. Sci.* **2021**, *11*, 7243, doi:10.3390/app11167243 . . . . **127**

Emanuel Vamanu, Laura Dorina Dinu, Cristina Mihaela Luntraru and Alexandru Suciu
In Vitro Coliform Resistance to Bioactive Compounds in Urinary Infection, Assessed in a Lab Catheterization Model
Reprinted from: *Appl. Sci.* **2021**, *11*, 4315, doi:10.3390/app11094315 . . . . **139**

Cristina Florentina Pelcaru, Mihaela Ene, Alina-Maria Petrache and Daniel Constantin Neguț
Low Doses of Gamma Irradiation Stimulate Synthesis of Bioactive Compounds with Antioxidant Activity in *Fomes fomentarius* Living Mycelium
Reprinted from: *Appl. Sci.* **2021**, *11*, 4236, doi:10.3390/app11094236 . . . . **151**

Alexandru Petre, Mihaela Ene and Emanuel Vamanu
Submerged Cultivation of *Inonotus obliquus* Mycelium Using Statistical Design of Experiments and Mathematical Modeling to Increase Biomass Yield
Reprinted from: *Appl. Sci.* **2021**, *11*, 4104, doi:10.3390/app11094104 . . . . **163**

Lyudmila Asyakina, Svetlana Ivanova, Alexander Prosekov, Lyubov Dyshlyuk, Evgeny Chupakhin, Elena Ulrikh, Olga Babich and Stanislav Sukhikh
Determination of the Qualitative Composition of Biologically Active Substances of Extracts of In Vitro Callus, Cell Suspension, and Root Cultures of the Medicinal Plant *Rhaponticum carthamoides*
Reprinted from: *Appl. Sci.* **2021**, *11*, 2555, doi:10.3390/app11062555 . . . . **175**

Yannik K. Schneider, Solveig M. Jørgensen, Jeanette Hammer Andersen and Espen H. Hansen
Qualitative and Quantitative Comparison of Liquid–Liquid Phase Extraction Using Ethyl Acetate and Liquid–Solid Phase Extraction Using Poly-Benzyl-Resin for Natural Products
Reprinted from: *Appl. Sci.* **2021**, *11*, 10241, doi:10.3390/app112110241 . . . . **185**

About the Editor

Emanuel Vamanu

Emanuel Vamanu, Ph.D., is the coordinator of a team specialized in studying human microbiota at the Faculty of Biotechnology Bucharest, Romania. At this time, he coordinates the only research laboratory in Romania that aims to study microbiota, testing the modulation process and the differences between metabolomic and microbiota patterns from different target groups. In addition, the lab started the base for the first microbiota pattern collection—www.gissystems.ro. In 2020, he started a collaboration with the Centre of Biotechnology, University of Allahabad, Prayagraj, India, and published several papers related to neurodegenerative pathologies. He is a Topic Editor at the *Applied Sciences* journal (IF 2.679, Q2 JCR), a Guest Editor *at the Applied* Sciences (IF 2.679, Q2 JCR), *Biomedicines* (IF 4.717, Q1 JCR), and *Evidence-Based Complementary and Alternative Medicine* (IF 2.629, Q1 JCR) journals, an *Associate Editorial Board Member at Current Pharmaceutical Biotechnology* (IF 2.837, Q2 JCR), a member of the Editorial Board at the *International Journal of Food Science and Biotechnology*, and a member of an Editorial Advisory Board at the *Revista de Chimie*. He was an invited speaker in 2016 at the international conference "Bonding tradition with innovation, successful strategies in food chain value", Olsztyn, Poland, and in 2018 at the 12th World Congress on Polyphenols Applications. The Project Manager is a member of the Romanian Inventors Forum from 2019 and a member of IFIA, WIIPA. He is a member of other international associations, such as FEBS, IUBMB, FEMS, SRBBM, and SRM.

MDPI

Editorial

Bioactive Compounds from Natural Products: Separation, Characterization, and Applications

Emanuel Vamanu

Faculty of Biotechnology, University of Agricultural Sciences and Veterinary Medicine, 011464 Bucharest, Romania; email@emanuelvamanu.ro

1. Introduction

The study of bioactive compounds represents a new and innovative section of biotechnology, with applications to the well-being of humans. The therapeutic potential of natural health products could be demonstrated by in vitro and/or in vivo biological activities, because the primary aim is to prevent or alleviate different chronic/degenerative diseases. Natural extracts or functional foods (including byproducts) could be valuable substrates through which new products could be obtained. These products display low correlations in in vitro and in vivo activities (antioxidant, anti-inflammatory, or antimicrobial) due to their bioavailability, which is one of their significant limiting factors. The fermentative activities of human microbiota could modulate microbial fingerprints and metabolomic profiles. They alleviate chronic diseases, and their administration could have potent effects (for type 2 diabetes, cardiovascular dysfunctions, or neurodegenerative diseases).

2. Issue Contents

This volume includes eleven research articles and three reviews that vary in their content but all cover research that aimed to characterize the biological effects of natural compounds and the results of quantitative analysis. All of the results displayed are supported by in vitro and in vivo studies or innovative fermentation/extractive processes.

In [1], the authors used *Ulva lactuca* microalgae to extract bioactive compounds with high biological activity. Four solvents were used, of which an ethanol/water mixture (70:30 v/v) was proven to be the most effective in terms of total carotenoid and phenolic contents. Quantitative analysis was performed using the high-performance liquid chromatography (HPLC) technique. Gallic acid was the predominant phenolic compound, and all-trans-violaxanthin and all-trans-lutein were predominantly identified among the carotenoid compounds. Following in vitro determinations, it was concluded that *U. lactuca* extracts could possibly be used as ingredients in the cosmetic, pharmaceutical, and food industries.

In [2], the authors determined the induction of HL-60 (acute myeloid leukemia cell line) differentiation using biochemical and morphological methods. The data showed that tyrosol (in Maltese Extra Virgin Olive Oil) induces cytotoxicity in HL-60 cells. These data demonstrated that tyrosol is helpful as an adjuvant for differentiating AML models. Upregulated NFkB2 and RELB was displayed after tyrosol treatment, showing this compound's efficacy in reducing inflammatory processes and treating cancer by inducing HL-60 cell apoptosis.

In [3], the anti-inflammatory activity of *Cnidoscolus aconitifolius* was presented using an ethyl acetate extract, and the role of flavonoid content was determined in this study. The anti-inflammatory effect of *C. aconitifolius* extract was determined using the mouse ear edema test induced by the activities of croton oil. The presence of flavonoids was determined using thin-layer chromatography and a spectrophotometric method. The extract resulted in a reduction in induced ear edema at 25–50 mg/kg doses. The results suggest that its use in the treatment of inflammation would be beneficial.

Citation: Vamanu, E. Bioactive Compounds from Natural Products: Separation, Characterization, and Applications. *Appl. Sci.* **2022**, *12*, 3922. https://doi.org/10.3390/app12083922

Received: 8 April 2022
Accepted: 12 April 2022
Published: 13 April 2022

Publisher's Note: MDPI stays neutral with regard to jurisdictional claims in published maps and institutional affiliations.

In [4], hemp oils, isolated from *Cannabis sativa* L., were analyzed using the reverse-phase–high-performance liquid chromatography–photodiode matrix system (RP–UHPLC–PDA) method. A rapid method was developed that can identify and quantify 10 of the most common cannabinoids in *C. sativa* oils: cannabivarin, cannabidiolic acid, cannabigerol acid, cannabigerol, cannabidiol, cannabinol, 9-tetrahydrocannabinol, 8-tetrahydrocannabinol, cannabichromene, and tetrahydrocann. Phytocannabinoid analysis showed which compounds can support the biopharmaceutical use of these products. Such a rapid method has improved the screening of these compounds and helped obtain controlled composition products.

In [5], the authors presented the influence of the drying process on the physicochemical properties of fermented vegetables (beetroot, red pepper, and carrot) in a ratio of 1:1:1 with *Levilactobacillus brevis*, *Lactiplantibacillus plantarum*, and *Limosilactobacillus fermentum*. Two convection drying methods or freeze-drying were used, and the dry matter, color, active substances (betalains and carotenoids), and the bacterial load were determined. Experimental data showed that pomace from fermented vegetables has a vital role in increasing the functional properties of food. Products from dried vegetables can be used in various foods, leading to improved color and an additional supply of probiotic bacteria that ensures the high stability of the products they are incorporated in.

In [6], an ultrasonic enzyme-assisted ATPE (UEAATPE) method was developed to extract flavonoids from the leaves of the *Ficus carica* L. plant which are usually considered waste. UPLC-QTOF-MS/MS was used to identify the flavonoid component, and the significant component identified was quercetin 3-O-hexobioside. The study showed that the recovery of these plant residues could be an essential source of biologically active compounds. In this way, the impact of the accumulation of plant waste on the environment is minimal.

In [7], the use of by-products from yellow onion skins (*Allium cepa* L.) was detailed, mainly by capitalizing on the flavonoid component. Extraction in hot water at 70 °C was performed, and the resulting extract was freeze-dried. Several formulas were extracted before being dried, and an in vitro digestion study demonstrated stability in gastric transit and a controlled release of flavonoid bioactive components during the intestinal phase. The study also showed thermal stability and a controlled pH, indicating that they can be used as functional food ingredients.

In [8], the use of a functional product to inhibit the proliferation of uropathogenic strains of *Escherichia coli* was reported. The product consisted of an ethanolic extract of cranberry fruit (*Vaccinium vitis idaea* L.), St. John's wort aerial parts (*Hypericum perforatum*), thyme aerial parts (*Thymus vulgaris* L.), propolis tincture with 27% dry mass, thyme essential oil, and rosemary essential oil. The tests were performed in a new in vitro bladder simulation system. The in vitro study showed the crucial inhibitory capacity of the proposed product. This initial in vitro study intended to investigate recurrent infections. New functional products were also tested, considering the inhibition of biofilm formation, which was an innovative in vitro research perspective. The system allowed various uropathogenic strains, which are associated with the use of a urinary catheter and participate in the formation of biofilm and the proliferation of inflammatory processes, to be tested.

In [9], the effect of gamma radiation on the mycelium of the fungus *Fomes fomentarius* was presented. A ^{60}Co source was used to induce oxidative stress. The impact on antioxidant stability and total phenol content was determined after 24 h. It was found that 300 Gy was the optimal dose for the extracts in the functional products industry.

In [10], a process for use in the submerged cultivation of the mycelium of the fungus *Inonotus obliquus* was presented to optimize the culture medium to obtain biomass. By analyzing six environmental components and pH, batch bioreactor cultures were performed to confirm the data that was gained from mathematical modeling. This optimized medium had a composition that contained (g/L): malt extract 2.15299, yeast extract 3.99296, fructose 11.0041, soluble starch 17.4, $MgSO_4$ 0.1 g, and $CaCl_2$ 0.05.

In [11], chromatographic and NMR methods described the chemical composition of extracts from callus, cell suspension, and *Rhaponticum carthamoides* roots. Flavonoids, such as ecdysteroids and anthocyanins, can inhibit the proliferation of tumor cells and have other biological effects, which can be used in the future to produce biopharmaceuticals.

In [12], the authors addressed a current health problem: recurrent urinary tract infections and the increasing presence of antibiotic-resistant strains. The paper considered the effect that some less-used plant species have on managing recurrent urinary tract infections. These were pomegranate (*Punica granatum* L.), chokeberry (*Aronia melanocarpa* Michx.), and cornelian cherry (*Cornus mas* L.), and evidence of their potential in the prevention of these infections or as an adjunct to antibiotic therapy was provided.

In [13], the effect of heat treatment on bioactive compounds contained in microalgae was presented as an alternative to increasing the consumption of functional foods. The study showed that there was an increase in oxidative processes, even in the case of short heat treatment times. Phenolic compounds are among the most affected molecules. The effect on the pigments causes discoloration, and the amount of iodine is significantly reduced. The conclusion of this study was that the effect varies from one species to another, and the protein level is also strongly disturbed.

In the last paper [14], aspects related to obtaining an extract from a culture of *Flavobacterium* sp. were presented. The study showed the differences in yield between liquid–liquid extraction using ethyl acetate and liquid–solid extraction using polybenzyl resin. The culture of *Flavobacterium* sp. was isolated from natural products of microbial and plant origin. The effect of pH on the extraction yield was determined. Although both methods isolated the same compounds, differences were identified between the protocols used. These details were considered necessary for various practical aspects.

These articles have demonstrated the potential of using bioactive compounds from natural products in applications associated with human well-being. This Special Issue outlines a broad picture of the study of functional compounds, with many new details on separation, characterization, and biopharmaceutical applications.

These studies have addressed alternative possibilities for preventing and reducing the occurrence of chronic and associated diseases. A new in vitro model has been proposed for the study of recurrent urinary tract infections. The data presented will be a starting point in developing alternative technologies for the extraction of natural compounds but will also serve as an innovative possibility to characterize strategies for the early prevention of infections with antibiotic-resistant strains. The proposal of new sources of functional compounds has been an attractive aspect of capitalizing on lesser-known natural sources.

Funding: This research received no external funding.

Conflicts of Interest: The author declares no conflict of interest.

References

1. Pappou, S.; Dardavila, M.M.; Savvidou, M.G.; Louli, V.; Magoulas, K.; Voutsas, E. Extraction of Bioactive Compounds from *Ulva lactuca*. *Appl. Sci.* **2022**, *12*, 2117. [CrossRef]
2. Gatt, L.; Saliba, D.G.; Schembri-Wismayer, P.; Zammit-Mangion, M. Tyrosol, at the Concentration Found in Maltese Extra Virgin Olive Oil, Induces HL-60 Differentiation towards the Monocyte lineage. *Appl. Sci.* **2021**, *11*, 10199. [CrossRef]
3. Padilla-Camberos, E.; Torres-Gonzalez, O.R.; Sanchez-Hernandez, I.M.; Diaz-Martinez, N.E.; Hernandez-Perez, O.R.; Flores-Fernandez, J.M. Anti-Inflammatory Activity of *Cnidoscolus aconitifolius* (Mill.) Ethyl Acetate Extract on Croton Oil-Induced Mouse Ear Edema. *Appl. Sci.* **2021**, *11*, 9697. [CrossRef]
4. Blebea, N.M.; Rambu, D.; Costache, T.; Negreș, S. Very Fast RP–UHPLC–PDA Method for Identification and Quantification of the Cannabinoids from Hemp Oil. *Appl. Sci.* **2021**, *11*, 9414. [CrossRef]
5. Janiszewska-Turak, E.; Kołakowska, W.; Pobiega, K.; Gramza-Michałowska, A. Influence of Drying Type of Selected Fermented Vegetables Pomace on the Natural Colorants and Concentration of Lactic Acid Bacteria. *Appl. Sci.* **2021**, *11*, 7864. [CrossRef]
6. Zhao, C.; Li, S.; Li, C.; Wang, T.; Tian, Y.; Li, X. Flavonoids from Fig (*Ficus carica* Linn.) Leaves: The Development of a New Extraction Method and Identification by UPLC-QTOF-MS/MS. *Appl. Sci.* **2021**, *11*, 7718. [CrossRef]

7. Milea, Ș.A.; Crăciunescu, O.; Râpeanu, G.; Oancea, A.; Enachi, E.; Bahrim, G.E.; Stănciuc, N. Multifunctional Ingredient from Aqueous Flavonoidic Extract of Yellow Onion Skins with Cytocompatibility and Cell Proliferation Properties. *Appl. Sci.* **2021**, *11*, 7243. [CrossRef]
8. Vamanu, E.; Dinu, L.D.; Luntraru, C.M.; Suciu, A. In Vitro Coliform Resistance to Bioactive Compounds in Urinary Infection, Assessed in a Lab Catheterization Model. *Appl. Sci.* **2021**, *11*, 4315. [CrossRef]
9. Pelcaru, C.F.; Ene, M.; Petrache, A.-M.; Neguț, D.C. Low Doses of Gamma Irradiation Stimulate Synthesis of Bioactive Compounds with Antioxidant Activity in *Fomes fomentarius* Living Mycelium. *Appl. Sci.* **2021**, *11*, 4236. [CrossRef]
10. Petre, A.; Ene, M.; Vamanu, E. Submerged Cultivation of *Inonotus obliquus* Mycelium Using Statistical Design of Experiments and Mathematical Modeling to Increase Biomass Yield. *Appl. Sci.* **2021**, *11*, 4104. [CrossRef]
11. Asyakina, L.; Ivanova, S.; Prosekov, A.; Dyshlyuk, L.; Chupakhin, E.; Ulrikh, E.; Babich, O.; Sukhikh, S. Determination of the Qualitative Composition of Biologically Active Substances of Extracts of In Vitro Callus, Cell Suspension, and Root Cultures of the Medicinal Plant *Rhaponticum carthamoides*. *Appl. Sci.* **2021**, *11*, 2555. [CrossRef]
12. Tache, A.M.; Dinu, L.D.; Vamanu, E. Novel Insights on Plant Extracts to Prevent and Treat Recurrent Urinary Tract Infections. *Appl. Sci.* **2022**, *12*, 2635. [CrossRef]
13. Cascais, M.; Monteiro, P.; Pacheco, D.; Cotas, J.; Pereira, L.; Marques, J.C.; Gonçalves, A.M.M. Effects of Heat Treatment Processes: Health Benefits and Risks to the Consumer. *Appl. Sci.* **2021**, *11*, 8740. [CrossRef]
14. Schneider, Y.K.; Jørgensen, S.M.; Andersen, J.H.; Hansen, E.H. Qualitative and Quantitative Comparison of Liquid–Liquid Phase Extraction Using Ethyl Acetate and Liquid–Solid Phase Extraction Using Poly-Benzyl-Resin for Natural Products. *Appl. Sci.* **2021**, *11*, 10241. [CrossRef]

applied sciences

MDPI

Review

Novel Insights on Plant Extracts to Prevent and Treat Recurrent Urinary Tract Infections

Adriana Mirela Tache, Laura Dorina Dinu * and Emanuel Vamanu *

Faculty of Biotechnology, University of Agricultural Sciences and Veterinary Medicine, 011464 Bucharest, Romania; tacheadriana95@yahoo.com
* Correspondence: laura.dinu@biotehnologii.usamv.ro (L.D.D.); email@emanuelvamanu.ro (E.V.)

Abstract: Urinary tract infections (UTI) represent one of the most widespread infections, and frequent recurrent episodes, induced mostly by uropathogenic *Escherichia coli*, make them increasingly difficult to treat. Long-term antibiotic therapy is an effective approach to treat recurrent UTI but generates adverse effects, including the emergence of pathogenic strains resistant to the vast majority of antibiotics. These drawbacks have enhanced the interest toward new alternatives based on plant extracts to prevent and treat recurrent UTI, especially in a synergistic antibiotic approach. Therefore, this review highlights the potential of some medicinal plants to be used in the management of recurrent UTI, including plants that have been approved for the treatment of urinary infections and promising, but less studied, plant candidates with proven anti-uropathogenic activity. Pomegranate (*Punica granatum* L.), black chokeberry (*Aronia melanocarpa* Michx.), and cornelian cherry (*Cornus mas* L.) have great potential to be used for prevention or in a combined antibiotic therapy to cure UTI, but more studies and clinical trials in specific population groups are required. Further progress in developing plant-based products to cure rUTI will be supported by advances in UTI pathogenesis and human-based models for a better understanding of their pharmacological activities.

Keywords: uropathogenic; herbal medicines; plant extract; bioactive molecules

Citation: Tache, A.M.; Dinu, L.D.; Vamanu, E. Novel Insights on Plant Extracts to Prevent and Treat Recurrent Urinary Tract Infections. *Appl. Sci.* **2022**, *12*, 2635. https://doi.org/10.3390/app12052635

Academic Editor: Akikazu Sakudo

Received: 13 January 2022
Accepted: 2 March 2022
Published: 3 March 2022

Publisher's Note: MDPI stays neutral with regard to jurisdictional claims in published maps and institutional affiliations.

1. Introduction

Urinary tract infections, abbreviated UTI, frequently occur in the urinary tract, affecting up to 150 million people/year worldwide [1,2]. Recurrent UTI (rUTI) is more frequent in young females, but it has been estimated that up to 50% of women experience a UTI at least once in their lives [3,4]. A notable proportion of patients who develop UTI have no discernible causes. However, UTI is mainly associated with patients with detected urinary tract abnormalities, suppressed immune systems, long-term catheter use, and recent urinary procedures. A lot of recent studies suggest a genetic predisposition to the recurrent symptomatic UTI. Numerous genes seem to contribute and have been strongly associated with UTI-prone patients [2,5].

rUTI is defined as two or more episodes of symptomatic lower UTI in the last six months or three or more episodes in the past 12 months, based on the European Union guidelines for rUTI in adults. Uropathogenic *E. coli* (UPEC) is the most common bacterial agent for all types of UTI and is responsible for 70–90% of urinary infections, followed by *K. pneumoniae* in uncomplicated UTI and *E.* spp. in complicated UTI [6]. *E. coli* belongs to the *Enterobacteriaceae* family of bacteria. Despite being one of the most well documented and studied bacterial species globally, it continues to pose a constant threat to human health, particularly in medical settings [7,8].

Prophylactic low-dose antimicrobial therapy represents the current gold standard practice. First-line antibiotics, such as trimethoprim, nitrofurantoin, amoxicillin, and cefalexin are recommended, while second-line quinolones should be considered for more severe infections (NHS Guidelines). The continuous antibiotic prophylaxis reduces the rate of UTI but induces adverse effects, including vaginal and oral candidiasis, and gastrointestinal symptoms, and the spread of antibiotic resistance. Following an antibiotic resistance index assessment, these infections have become increasingly difficult to treat since the mid-2000s [9]. This emphasizes the need for new treatment regimens against urinary infections with innovation by introducing alternative therapies based on herbal extracts, nanoparticles, and dietary supplements [10]. Plant remedies are a natural option for the long-term treatment of rUTI, especially in a synergistic antibiotic approach, as they have antimicrobial effects, can reduce different symptoms (e.g., anti-inflammatory, antipyretic, and analgesic effects), and help decrease adverse antibiotic effects. Moreover, they can be a prophylactic treatment to prevent rUTI for patients that have experienced urinary infection for years and in precision medicine.

This review aims to present the current knowledge about promising plant extracts, including candidate plant-based products that can be applied in the management of the rUTI as prophylactic or treatment therapy, and future directions for research are highlighted.

2. Pathogenesis of rUTI

The most common risk factors for UTI are female anatomy, age, sexual, birth control activity, and menopause (Figure 1). However, other risk factors, such as abnormalities and blockages in the urinary tract, catheter, or recent urinary procedures should be considered [11]. The studies on hereditary UTI started when it was noticed that the infection tends to be clustered in specific individuals and is relevant among females with relatives that suffer from recurring UTI. The candidate genes determining human/animal UTI susceptibility have been identified starting from the early 1980s using various genetic and immunological tools [12]. These studies concluded that increased susceptibility to UTI is a complex heritable trait with numerous genes that seem to contribute to the UTI-prone phenotype. These genes are involved in antimicrobial defense [13]. The host resistance to recurrent UTI relies strongly on the pathways that signal the innate immune system and the genes that control those pathways. In contrast, different immune variants can exacerbate acute and chronic infection or be protective [12,14].

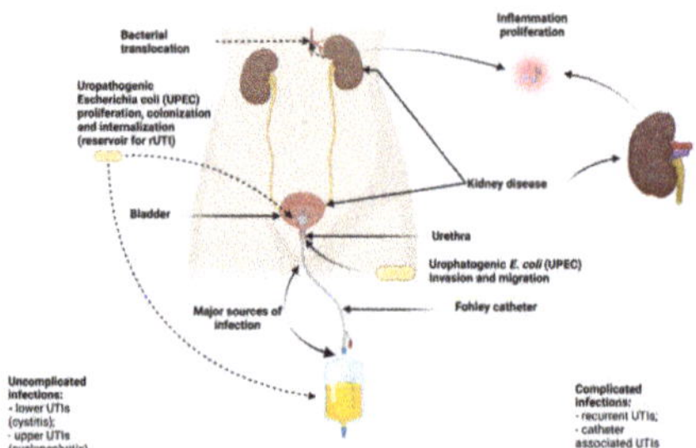

Figure 1. The pathogenesis of rUTI and their effects on human urinary system homeostasis. The major sources of infection are intestinal, perineal, vaginal, and preputial microflora, then *E. coli* (UPEC) migrate and invade the bladder tissue where they form biofilm and internalize. The migration of *E. coli* (UPEC) affects the kidney's normal functions as an effect of inflammation proliferation. The persistence of the inflammatory process at this level could determine bacterial translocation. The picture was created with the Biorender app.

Based on the current knowledge about the pathogenesis of rUTI, there are three main components involved: pathogen virulence, the host's defense mechanism, and the environment [15]. The etiology of UTI includes Gram-positive and Gram-negative bacteria and fungi, but UPEC accounts for a significant proportion of rUTI [6]. Uropathogenic *E. coli* strains from reservoirs (intestinal, perineal, vagina, and preputial microflora) invade the urinary tract. Bacterial virulence factors (e.g., flagella/pili, extracellular polysaccharides, toxins, and enzymes) may contribute to UTI recurrence [2,14]. The bacterial urogenital mucosa attachment activates a cascade of signaling molecules in the host from innate immune responses that stimulate cytokine release resulting in an inflammatory response and symptoms. In response, bacteria produce molecules that specifically inhibit the signaling pathways and use toxins to kill the mucosal barrier cells and enter the bloodstream in the acute phase of infection. Thus, the severity of a UTI depends on the innate defensive response of the host and the virulence of the infecting agents. In the case of rUTI, bacteria form biofilm. Moreover, they can persist in quiescent communities, surviving the immune response and resisting antibiotic attacks. Recurrent urinary tract infections are based on the same pathogenic microorganism that underwent the initial condition, *E. coli* strains that use the reservoirs associated with the host in the gastrointestinal tract and bladder tissue to repopulate/reinforce the host [9]. In human anatomy, *E. coli* is primarily found in the gastrointestinal tract. The proximity of the urethra to the anus, particularly in female patients, makes *E. coli* a significant contributor or activator in most catheter-associated urinary tract infections (CAUTI) for intermittent catheter users [7]. Another cause of rUTI is bladder dysbiosis, an alteration of the urinary microbiota by strains from reservoirs that compete with the natural microbial community, and uropathogens can become dominant. Antibiotic prophylactic treatment is another critical factor affecting rUTI because it favors the development of multidrug-resistant (MDR) uropathogenic bacteria, which are more virulent and persistent [2,14].

3. Classification of Causes in Cases of UTI

The following stages of the evolution of a urinary tract infection were highlighted due to the morphology of the patient's urinary tract and past incidences of the disease [16]:

- Uncomplicated Infection: It can be defined as the infection of a relatively healthy urinary tract patient who responds to antibiotic treatment;
- Complicated Infection: It occurs mainly in people with an abnormal urinary tract, which are often obstructed by stones or bladder-ureter reflux;
- Isolated Infections: Represent the mother infection or the infection that appears at an interval of six months, without any connection between the two episodes;
- Unresolved Infection: It is the opposite of uncomplicated infection; it does not respond to antibiotic treatment;
- Reinfection/Recurrence: Reinfection is the last stage of the evolution of urinary tract infections, described as a persistent bacterial infection. At this stage, the individual is reinfected with the same pathogen two weeks after treating a urinary tract infection.

4. In Vitro Urinary Tract Models

To develop new plant-based treatment strategies, it is imperious to improve our knowledge about UTI pathogenesis using different model systems that will provide practical information about the host–pathogen interaction and unique aspects of the urinary tract environment [17]. The mechanism of rUTI is complex, and over the last decades, several hypotheses have emerged based on animal models or advanced human cell-based in vitro models. A recent exhaustive review discussed the importance of model systems in revealing the crucial insights about UTI pathogenesis, including their strengths and limitations [2]. Further progress in improving the plant-based strategies to cure rUTI will be supported by advances in human-based models that are more accurate and controllable than animal models and can reveal bottleneck points of UTI pathogenesis [18,19].

In general, in vitro urinary tract infection models are based on the cells in static culture; however, the intricacy and antibiotic resistance of these infections have necessitated the need to expand these studies to include the characteristics of the bladder's physiological conditions (e.g., urine and fluid flow). Advanced in vitro human cell-based models that provide a complementary platform (along with animals) for drug detection and alternative treatment strategies are directly proportional to studies on the complexity of recurrent urinary tract infections [2].

Some models were based on a virulent bacterium colonizing a bladder in vitro at the laboratory level. These models were created based on research about preventing symptomatic urinary tract infection caused by aggressive bacteria by colonizing the bladder with strains that cause asymptomatic bacteriuria (ABU), which are less virulent bacteria. These asymptomatic *E. coli* strains were created by comparing isolated ABUs from diabetic patients in terms of their virulence-associated phenotypes and virulence in an in vitro model with murine sepsis (from laboratory rats) and the content of their genome [20,21].

Generalized Lotka–Volterra modeling (gLVM) has revealed the protective role of two taxonomic classes of bacteria, *Actinobacteria* and *Bacteroides*, which appear to suppress pathogen growth. For longitudinal data, the original BEEM approach (biomass estimation and model inference with a maximizing of expectancies) was utilized to simulate the microbiota shift after 24 h of antibiotic therapy [22].

These models are highly desirable because of their advantages compared to animal models and their potential to study a wide variety of conditions, such as UTI in bladder cancer, tissue regeneration/transplantation, and the effect of xenobiotics that come into contact with the urine, in vitro in the human bladder environment and reconstitution of the urothelium (or even organ) [2,23].

While the procedure can preserve a precise 3D cell architecture, layering, and differentiation, it is time-consuming, requires quick access to fresh tissue, and may show variability between samples [20]. There is also a growing chance that stromal and urothelial cells will come into contact with undesired combinations during cultivation, with stromal cells outgrowing the urothelium [24,25].

Because the existing experimental models have flaws and only provide partial results, new tactics have been created, including new experimental 3D bladder models that strive to reproduce/imitate the human urothelium employing stem cells or stem cells that have been propagated, expanded, and stratified/differentiated. Some are even protected by patents [26].

4.1. In Vitro Catheterization Model (IVCM)

The research was carried out in the Laboratory of Pharmaceutical Biotechnologies of the Faculty of Biotechnology using an in vitro catheterization model (IVCM) to analyze the effect on UTI for a variety of plant extracts and compared the results with data obtained using known antibiotics (Figure 2). In vitro testing of sample viability, antibacterial activity, and characterization of functional extracts were performed to use them as adjuvants in treating urinary tract infections.

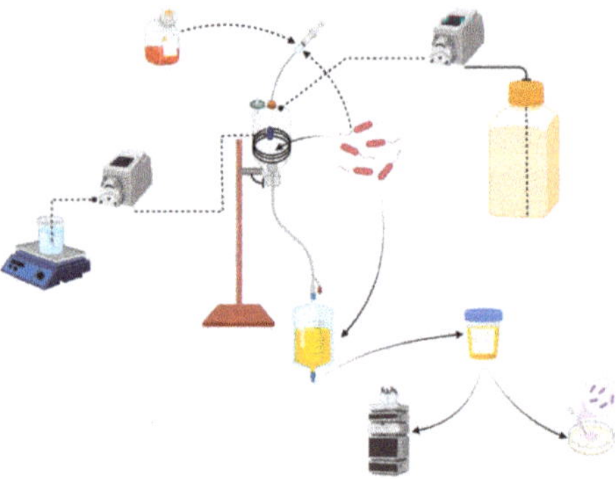

Figure 2. After the update of the constructive model, the redrawn schematic version of the in vitro catheterization model from the Faculty of Biotechnology, UASVM Bucharest. The figure contains the full details of the system, with a heating system made by the authors (left part of the picture). Samples are analyzed by microbiologic and chromatographic methods to fully understand the in vitro simulation [27]. The figure was created with the Biorender app.

The general operating principle is presented below [23]:

- The peristaltic pump ensured the continuous supply of sterilized artificial urine at a rate of 1 L per day. The bladder temperature was maintained at 35–37 °C by a circulatory system that included a peristaltic pump and a ceramic heating plate;
- A urinary catheter was used to drain the bladder. The leftover urine contributed to the inflated balloon's physiological stagnation in a catheterized bladder. A urine catheter was attached to a 'drainage bag'-like effluent collection vessel;
- The bladder had rubber septa ports that facilitated the aseptic inoculation of bacteria of interest. The catheter tip was placed into the bladder and secured by inflating the balloon with 15 mL of sterile phosphate buffered saline (PBS). Urine flow was started to fill the bladder to the necessary capacity (80–100 mL) so that the balloon could be covered;
- After that, the flow of urine from the tank was turned off to allow bacteria to be inoculated into the septum port's 'bladder.' To simulate the introduction of contaminating bacteria into the bladder during catheter insertion at the clinical site, approximately 5×10^7 CFU of *E. coli* were injected into the bladder. This would result in a contaminating bacteria concentration of 5×10^5 CFU/mL;
- To prevent bacteria from being removed immediately after inoculation, the urine flow was interrupted for one hour and then resumed [28].

4.2. In Vivo Vision of In Vitro Urinary Tract Models

Human urothelial models represent a chapter of interest in the current research. In vivo, cells are exposed to urine and create a barrier supplied by nearby cells/tissues. Although urine may be required to differentiate the cells in vitro, it is a harsh environment, and, even in the short term, it may not be the only fluid used. Only part of the urothelium would be exposed to urine in an ideal in vitro model, whereas an appropriate differentiation medium would support the basolateral compartment. This is why advanced urothelial models will benefit from platforms that have porous substrates. The most straightforward implementation would be a transwell system or similar permeable membrane systems [29]. The volume of urine added, the possibility to change urine concentrations over time, and donor characteristics are considered factors (sex, age, and any diseases or disorders that could alter the urine's composition). Fluid flow platforms, which are still being extensively investigated, can be implemented in more advanced models [2].

5. Management of Recurrent UTI by Non-Antibiotic Plant-Based Prevention and Treatments

Prophylactic antibiotics therapy is the current standard of care to prevent recurrent UTI in many worldwide guidelines [30]. However, recent UK, European, and US guidelines advise prudent antibiotic prescribing to reduce antimicrobial resistance [31]. The reason is that many studies have confirmed the emergence of antibiotic-resistant UTI pathogens after a few weeks of prophylactic therapy, including multiple drug resistance (MDR). It has been reported that most *E. coli* urinary isolates have resistance to amoxicillin, while colistin and carbapenem resistance are emerging in *Enterobacteriaceae* [1]. Moreover, the prophylactic antibiotic treatment remains of doubtful usefulness, and some of the MDR uropathogenic strains will soon be untreatable with the current antibiotics on which UTI-prone patients rely [1,2].

The problem of antimicrobial resistance is recognized as a global public health concern that provokes high financial costs to healthcare systems and is fueled by the overuse of antibiotics. Since 2014 World Health Organization has highlighted that antibiotic resistance in uropathogens is a key pressure point in the global antimicrobial resistance crisis. Therefore, there is an urgent need to understand UTI pathophysiology better and develop innovative products that use non-antibiotic strategies to prevent and treat UTI. A different non-antibiotic approach might provide the optimal treatment to reduce rUTI. A recently published comprehensive article discusses the future of non-antibiotic prevention and management of recurrent urinary tract infections [1].

Phytotherapy, alternative medicine based on plant extracts [32], is an essential non-antibiotic strategy that has been used for centuries to prevent or treat different diseases, including UTI [33]. Medicinal plants have long been an excellent place to search for innovative ways to treat human health issues. The biologically active substances present in medicinal plants have evolved to protect the plant from pathogens and, therefore, could prevent/treat human infections [34].

Complementary and alternative plant medicine has been an effective practical approach for treating rUTI, especially as a prophylactic therapy of antibiotics, as both have a synergistic effect. Plant medicines are a realistic option for the long-term prevention of rUTI. They are cost-effective, readily available, safe to use, with fewer reported side effects, and do not induce bacterial resistance. The innovation in prophylactic plant-based treatments, especially for UTI-prone patients with high recurrence rates, should target the bottleneck points of UTI pathogenesis. The bioactive plant molecules used in rUTI treatment must prevent uropathogen adherence to urothelial cells, internalization or biofilm formation, and the production of inhibitory molecules against defensive immune response and modulate the natural microbial community, restoring the microbiome dysbiosis. Moreover, phytotherapy is critical in providing a way to reduce rUTI symptoms without using antibiotics. However, aggressive acute UTI treatment requires antibiotic therapy associated with plant-based therapies that increase the antibiotic effect and the host's immune response.

However, the mechanism of plant medicines used for the treatment of rUTI is still poorly understood. Further studies are required to disclose the potential of all active compounds responsible for pharmaceutical activities and their potential adverse effects, including mutagenic and carcinogenic activities. The antimicrobial effects of plant extracts are attributed to multiple mechanisms: directly killing microbes, interfering with microbial adhesion to epithelial cells and biofilm formation, stopping the multiplication of pathogens, restoring dysbiosis, improving the host protection acting on natural barriers, serving as immunomodulators, or boosting body oxidant status [35]. Therefore, using whole plant extracts is recommended, as the biologically active compounds work synergistically and have different mechanisms of action [36]. In the last years, some reports have enlisted botanicals helpful in preventing and curing UTI and made synopses with the therapeutical potential of medicinal plants for the management of UTI [1–6,36–39].

The present review covers plants with recognized anti-uropathogenic activities and promising medicinal plants that can treat infection recurrence but are less studied [39,40].

6. Plants Commonly Used in the Treatment of rUTI

A vast majority of the medicinal plants known in traditional medicine in different areas worldwide have been studied in vitro and in clinical trials for the prophylaxis of UTI or combined treatment with antibiotics. Currently, cranberry is considered the most effective plant extract to prevent rUTI. At the same time, plants with high berberine contents, such as uva-ursi or different Chinese herbs, are frequently prescribed for acute UTI. The review selected only a few well-known plants that have been approved for the treatment of rUTI and discussed studies that confirmed the active compounds responsible for their therapeutic potential.

6.1. Herbal Chinese Medicine

The herbal products used in traditional Chinese medicine have been proven efficient to treat UTI symptoms for over 2000 years because they have significant diuretic, antimicrobial, anti-inflammatory, and immune-enhancing activities. Different medicinal herbs have been studied in vitro and in clinical trials for the prophylaxis of active and recurrent UTI or are used in combination with antibiotic treatment. Extensively studied Chinese herbs named Huang Liam (*Coptis chinensis* Franch) have shown in vitro inhibitory and anti-inflammatory effects against a broad spectrum of uropathogenic strains, including UPEC. These herbs contain highly active berberine alkaloid compounds (berberine, coptisine, and palmatine) with specific inhibitory activity against *E. coli* [13]. Antimicrobial substances from *Salvia plebeia*, a native herb from Asia, have been shown to inhibit the growth of *E. coli* at a minimal inhibitory concentration of 0.25 g/mL [29]. Moreover, several clinical trials havee proven that adding Chinese herbs to antibiotic prophylaxis is more effective in reducing rUTI rates, especially in postmenopausal women, but more extensive randomized trials are required. Although different Chinese plant remedies are routinely prescribed in rUTI treatment, Sihra et al., 2018 concluded that high-quality and well-conducted clinical studies should be undertaken [1].

6.1.1. *Arctostaphylos uva-ursi*

Leaf extracts from *A. uva-ursi*, an herb found in Europe and northern America and named "grape of the bear", obtained Germany's governmental approval for treatment of UTI [6]. The leaves contain flavonoids, tannins, terpenoids, iridoids, and arbutin, a glycoside with recognized antibacterial activity [26]. *Uvacin* and *Uva-E* are commercial products based on uva-ursi extracts from leaves and have been reported to be effective in preventing rUTI. However, other herbal formulae with this plant have dose-dependent activities against uropathogens and may serve as an essential adjuvant treatment to antibiotic prophylaxis in reducing rUTI.

6.1.2. *Vaccinium macrocarpon*—Cranberry

Currently, no definitive mechanism of action has been clarified to explain the prophylactic treatment of UTI, but there is much clinical evidence that the berries and cranberry-containing products can be used as prophylaxis for rUTI [1]. Cranberry is a plant from the family Ericaceae, known as *V. macrocarpon*. It was used by native Americans to treat infections [41]. Other varieties, including *V. oxycoccos*, a small/bog cranberry, and *V. erythrocarpum*, a southern mountain cranberry, have also been used to treat UTI because these berries contain many biologically active compounds that are supposed to be responsible for anti-uropathogenic activities [42]. Cranberries contain fructose and A-type proanthocyanidins that have been found to prevent the adhesion of *E. coli* to uroepithelial cells, and prolonged exposure to cranberry induced a change in the morphology of *E. coli* to the spherical shape that affects adherence [36,43]. Ursolic acid inhibits *E. coli* biofilm formation, and quinic acid is responsible for the excretion of hippuric acid and acidifies the urine. In contrast, sialic acid has an anti-inflammatory and painkilling effect, but other cranberry components have a complementary or synergistic role that contributes to the antibacterial development [1,40].

Ellura, a proanthocyanidin-rich commercial product from cranberry, is used for its preventive role in rUTI. Recent reviews analyzed the data obtained during clinical trials that have been conducted to assess the efficiency of cranberry extracts/juice/syrup/tablets/capsules in UTI treatment. They concluded that they are mixed results and more significant well-designed trials are required [1,40]. However, in the case of rUTI, several randomized controlled trials with a large number of participants suggested that cranberry products might be efficient in specific populations (women with rUTI and children). Still, some side effects could occur [1]. An interesting result was reported in a clinical trial when cranberry extract (500 mg/kg) administered for six months decreased the UTI to the same level as the trimethoprim (100 mg) [44].

6.2. Promising Medicinal Plants to Treat rUTI

6.2.1. *P. granatum* L.—Pomegranate

Pomegranate (*P. granatum* L.) is a fruit-bearing shrub in the *Punicacea* family, described initially throughout the Mediterranean region but widely cultivated in the present [45]. Pomegranate has been known for centuries for its multiple health benefits, including antimicrobial activities, due to various biologically active substances from all plant parts, including the fruit juice, peel, arils, flowers, and bark [45,46]. Different parts of the pomegranate (fruit, horns, skin, and seed) have been extracted utilizing an extraction procedure based on maceration with various extraction solvents (water, ethanol, methanol, ethyl acetate, etc.). Aqueous and organic extracts from the fruits and by-products have antibacterial constituents, such as hydrolyzable tannins (punicalagin and penicillins, ellagic acid, and gallic acid) synergized with bioactive flavonols, i.e., myricetin, quercetin, and anthocyanins, e.g., cyanidin-3-glucose and pelargonidin-3-galactose [47,48]. Seed extracts have urobactericidal activity against different *E. coli* clinically isolated from patients suffering from urinary tract infections [49]. The researchers suggested that therapeutic antioxidants and natural glycosides are supposed to act as molecular decoys to prevent the adhesion of pathogenic bacteria from hosting cells, thereby inhibiting future pathogenesis, but further studies are required to confirm it [49]. Pomegranate peel is a waste obtained through juice processing. It contains significant amounts of bioactive compounds (total phenols, flavonoids, anthocyanins, tannins, and proanthocyanidins) with variations between different cultivars and drying processes [50]. The primary polyphenols in pomegranate that have been shown to have antimicrobial, antioxidant, and anti-inflammatory bioactivities are ellagitannins and anthocyanins, which are concentrated in the fruit's peel and kernels. Aqueous pomegranate peel extract's antimicrobial and antioxidant effect against uropathogenic *E. coli* has recently been reported [51]. The inhibitory activity was dose- and pH-dependent, with a minimum inhibitory concentration value of 0.6 mg/mL that produced a reduction of up to 80% of the adhesion index, accompanied by a decrease in motility and ornithine decarboxylation. In comparison, total inhibition was reported at a 1.2 mg/mL value. Thus, the peel extract inhibits the biofilm formation and cellular adherence of *E. coli* and reduces bacterial motility and the polyamine production that helps bacteria to survive during the oxidative stress produced by some antibiotics (fluoroquinolones, aminoglycosides, and cephalosporins) [51].

Different parts of *P. granatum* (pericarp, leaves, flowers, and seeds) have antibacterial activities against antibiotic-resistant *E. coli* strains, but the ethanol extract of pericarp has a significant effect on animal *E. coli* isolates [52]. Another significant result showed by this study is that water extracts were not as active as organic extracts against tested *E. coli*. In another study, a methanolic section of pomegranate fruit pericarp tested with ciprofloxacin against extended-spectrum β-lactamase (ESBL)-producing *E. coli* showed an enhancement of ciprofloxacin activity, probably due to the bacterial efflux pump inhibitor activity of the polyphenolic constituents from the extract [53]. However, the isolates exhibiting a high level of ciprofloxacin resistance did not respond to ciprofloxacin–extract combinations, and the authors consider that this result is due to target site modification that is not influenced by plant inhibitor molecules [38].

An aqueous extract of pomegranate leathery exocarp showed antifungal and antibacterial effects, including on *E. coli*, related to an anti-acetylcholinesterase inhibitory effect and antioxidant activities. This promising antimicrobial effect was compared to the synthetic drug activity. At an extract concentration of 10 mg/mL, the result against *E. coli* was equal to kanamycin. In contrast, the lower concentration of the extract (5 mg/mL) showed higher bactericidal activity than tetracycline [54].

Pomegranate aril extracts (30–90 μg/mL) from six different varieties from Turkey (Mediterranean region) have proven antimicrobial effects on other microorganisms [55], including *E. coli* DM. The authors demonstrated that the antimicrobial effect of extracts is dependent on the plant cultivar, not on the phenolic content. The most acidic plant variety with the second-highest phenolic content from all varieties tested had the most significant inhibitory effect on *E. coli* [55].

Ellagic acid is a bioactive tannin from the methanolic extract of pomegranate that inhibits biofilm formation by bacteria *Staphylococcus aureus* and *E. coli* [56]. Moreover, it has been proven that pomegranate polyphenols or the interactive effect of different compounds present in the extract may indirectly inhibit quorum-sensing, the signaling mechanism used by bacteria to communicate and form a biofilm [43].

Because of the health benefits linked to its polyphenol content, pomegranate extract pills/juice are becoming increasingly popular. They have been acknowledged as an effective plant extract for urinary tract infections [57]. Still, clinical trials that use only pomegranate extracts to prove their efficiency in UTI prevention or treatment are missing. Polyphenols, ellagic acid, anthocyanins, and glycosides are the main bioactive substances that support the use of pomegranate in rUTI prophylaxis, alone or in combination with antibiotic therapy.

6.2.2. *A. melanocarpa* (Michx.) Elliott—Black Chokeberry

Sometimes known as scores, chokeberry fruits are members of the *Rosaceae family's Aronia genus*. *A. melanocarpa* (Michx.) Elliott—black chokeberry and *A. arbutifolia* L.—red chokeberry are the two shrubs species utilized as fruits in the genus (red scab) [58]. The fruit of *A. melanocarpa* (Figure 3) is known for its high concentration of bioactive compounds, such as polyphenols with significant antioxidant and antimicrobial effects, which have been studied intensively for the last 15 years [59]. The proanthocyanidin concentration of *A. melanocarpa* is four times higher than in cranberry, and aronia berries contain a higher dose of quinic acid; both compounds help fight and prevent UTI [60]. Moreover, 100 g of raw cranberries were scaled at 9584 on the USDA ORAC (oxygen radical absorbance capacity) value scale, with the highest antioxidant capacity from common fruits, while aronia berry was 16,862 [61].

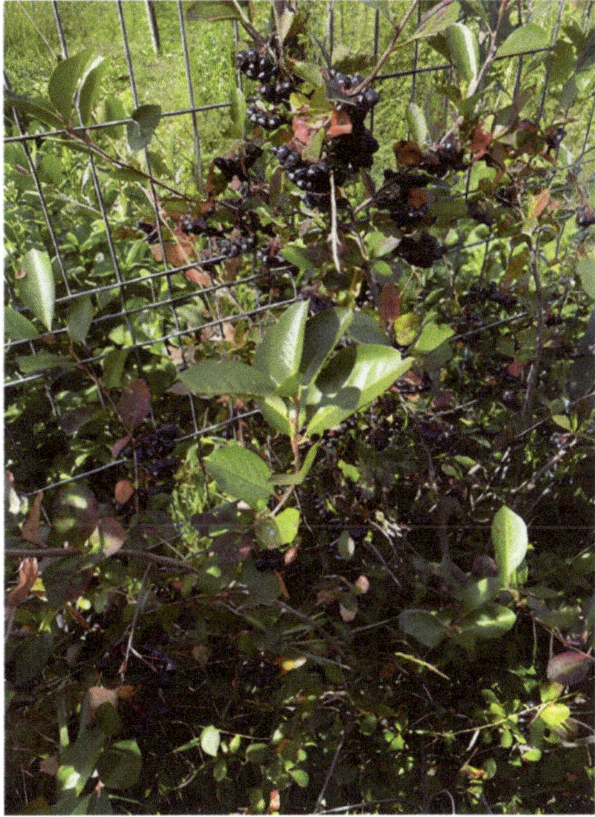

Figure 3. The fruit of *A. melanocarpa* growing in Ionășeni village, Vârfu Câmpului, Botoșani county, Romania.

The mechanism of action for anthocyanins from *A. melanocarpa* proved that at a minimum inhibitory concentration (MIC) of 0.625 mg/mL and minimum bactericidal concentration of 1.25 mg/mL, anthocyanins compromise the integrity of the *E. coli* cell wall and membrane, bind directly to the bacterial DNA, and interfere with protein homeostasis [62]. Epicatechin, a proanthocyanidin from *Aronia* extracts, has been involved in the biofilm formation effect in the case of uropathogenic *E. coli* CFT073 (ATCC 700928). It supposes that the active substance interferes with quorum-sensing chemotaxis/or mobility genes [63]. Bacteria in a biofilm are often responsible for recurrence and antibiotic treatment failure. Therefore, finding new plants rich in bioactive molecules with anti-biofilm activity, such as epicatechin from *A. melanocarpa*, ellagic acid from *P. granatum*, and ursolic acid from cranberry, is a good non-antitoxic inhibition strategy with a reduced potential for resistance development.

The results of the antibacterial activities of black chokeberry fruit extract are mixed. In a recent study, the leaf aqueous extract had a bacteriostatic effect against *E. coli* and *L. monocytogenes*, decreasing the growth rate and extending the lag phase [64]. An HPLC analysis showed that *Aronia* extract contained 15 phenolic compounds and 11 flavonoids, mainly flavonols and hydroxycinnamic acid. The total concentration of phenolic acids (114.66 µg/mL) was higher than that of flavonoids (93.40 µg/mL). However, no clear relationship was found between the influence of the aronia extracts and whether the tested bacteria belonged to the Gram-positive or Gram-negative groups. In another study, a common inhibition effect was noticed for 20 µL of black chokeberry leaf extract that reduced the growth of the tested *E. coli* strain by 23% [64,65]. Similarly, it has been shown that aqueous and ethanolic extracts of fresh, dried, or frozen fruits have efficacy against *Bacillus cereus*, *S. aureus*, and *Pseudomonas aeruginosa* but did not influence *E. coli* 332 [66]. In a study, *E. coli* strains were resistant to extracts from leaves, berries, and stems obtained by subcritical water extraction [66,67]. These contradictory results could be explained by differences in the extract preparation method and the tested *E. coli* strains.

Anthocyanidin-rich extracts obtained as lyophilized powders from mature fruits of *A. melanocarpa* L., cultivated in the ecological environment in the Eastern Carpathians in Romania, were used to test the synergism-based antibiotic treatment on pathogens from the urinary tract [68]. The extract contained anthocyanins and epicatechin units, and the extraction method had an essential impact on the quantity of each component of the plant extract. The antioxidant activity of the extract was in close relation to the anthocyanidin concentration, and it was noticed that the ethanolic extracts proved to be active against 10 out of the 18 tested strains at doses of 2.5–5 mg/mL. Moreover, a significant inhibition of monoculture biofilm formation of 11 out of all tested strains was reported. Antibiotic disks (amikacin, tetracycline, nitrofurantoin, imipenem, and norfloxacin) supplemented with the extract slowly increased the growth inhibition zone in the case of *P. aeruginosa*, *E. coli*, and *M. morganii* [69]. The synergistic antibiotic extract effect that has been shown for pomegranate and black chokeberry could be explained by the presence in the extract of other secondary metabolites with antibacterial activities or the increase in antibiotic activity due to the existence of active biomolecules from the plant extract (polyphenols, tannins, etc.). In this way, the efficacy of an existing drug could be improved with the help of an inexpensive alternative therapy.

Accurate and reliable clinical evidence about the efficiency of black chokeberry in UTI treatment is missing; only one clinical trial has been found that reported the reduction in UTI frequency after three months of juice administration in a specific population group (residents from six nursing homes) [70]. Polyphenols, anthocyanidins, epicatechin, and quinic acid, with anti-biofilm formation and antioxidant activities, are bioactive substances that raised attention to aronia berries and their role in fighting rUTI. Moreover, *A. melanocarpa* fruits appear to have more excellent health and wellness benefits, higher antioxidant levels, and fewer side effects than cranberry.

6.2.3. *C. mas* L.—Cornelian Cherry

The cornelian cherry (*C. mas L.*) is a member of the *Cornaceae* family (*Cornaceae*), found in the wild throughout Europe's central and south-eastern areas and Asia, where fruits have been used for centuries in traditional medicine [71]. The considerable number of anthocyanins in cornelian cherry is linked to the fruits' main pro-health benefits; however, the chemical makeup of *C. mas* is varied and primarily depends on variety, culture, and environmental and climatic circumstances [72]. Extracts from different parts of the plant revealed broad pharmacological activities. Various studies identified 101 compounds among active biological anthocyanins, flavonoids, iridoids, and vitamin C, especially in the fruits [73]. The highest antimicrobial activity of the *C. mas* leaf extract was correlated with iridoids, ellagic acid, and ellagitannins, substances that were not detected in *A. melanocarpa* leaf extract [64]. This study noted the more substantial effect of cornelian cherry leaf extract (1%) on Gram-negative bacteria compared to Gram-positive bacteria and explained that some interactions that may occur between polyphenols, iridoids, and cell walls, but the results are not in agreement with other data reported in the literature. Biologically active substances from cornelian cherry leaf extract produced morphological changes in bacterial aggregate formation-induced reactive oxygen species with various adverse intracellular effects, including suppressing the DNA gyrase activity that inhibits DNA synthesis [64]. Second-line antibiotics recommended in the therapy of rUTI, such as quinolones, especially ciprofloxacin, have a similar mechanism of action and target DNA gyrase. However, the *C. mas* bark extract showed moderate antibacterial activity on some bacteria but no effect on others, including *E. coli*, or on the yeasts *C. albicans* and *R. rubra* [74].

Cornelian cherry fruits are non-toxic and safe foods, based on toxicity studies in rat and human models. They can mainly treat diabetes, obesity, atherosclerosis, skin diseases, and gastrointestinal and rheumatic problems [75]. The data in the literature have suggested that polyphenols and iridoids are responsible for the therapeutic benefits of cornelian cherry fruits on many physiological parameters [76]. Moreover, the leaves from *C. mas* have shown significant anti-inflammatory effects connected with inhibiting aldose reductase, an overexpressed enzyme in many inflammation processes and cancers [77].

The recent trends in herbal medicine open new insights into using *C. mas* in the treatment of rUTI (Table 1). The administration of cornelian cherry tablet (500 mg) for six months in women affected by rUTI decreased dysuria and frequent urination. However, positive urine culture in the group that received *C. mas* (19%) compared to the group that received placebo (33.4%) was not statistically significant [78]. More research must be conducted to clarify the action mechanism of cornelian cherry extract in rUTI cure. Still, polyphenols, anthocyanidins, iridoids, and vitamin C with antioxidant effects seem to be the active molecules involved. At the same time, clinical trials are necessary to use a combination of plant extract and a combination of plant extract and pre- and probiotics to prove their effectiveness [79].

Table 1. Anti-uropathogenic activities of extracts from pomegranate (*P. granatum* L.), black chockeberry (*A. melanocarpa* Michx.), and cornelian cherry (*C. mas* L.).

Plant Name	Extract/Part Used	Bioactive Compounds	Name of Microorganisms Tested (*E. coli* or Other Uropathogens)	Main Conclusions	Reference
Pomegranate (*P. granatum* L.)	aril extracts from six pomegranate varieties grown in the Mediterranean region of Turkey	polyphenols	*E. coli* DM	- The most acidic cultivar with the second highest phenolic content from all varities tested had the greatest inhibitory effect on *E. coli*.	[40]
	leathery exocarp extract		*E. coli* Migula	- The higher concentration of the aqueous extract (10 mg/mL) showed a significant bactericidal effect against *E. coli*, equal to kanamycin, whereas the lower concentration of the extract (5 mg/mL) showed a higher bactericidal activity against *E. coli* than tetracycline.	[39]
	peel extract	ellagic acid	*E. coli*	- Ellagic acid was able to inhibit biofilm formation by *E. coli*.	[41]
	fruit pericarp extract tested in combination with ciprofloxacint	polyphenols	extended-spectrum β-lactamase (ESBL)-producing E. coli, which were screened for their resistance profile against fluoroquinolone antibiotics	- This study demonstrated a synergy of a ciprofloxacin–methanolic extract combination against ESBL Gram-negative bacilli.	[38]
	extracts from different parts (pericarp, leaves, flowers, and seeds)		*E. coli* type (1), which affected calves and had shown antibiotic resistance	- The alcoholic extracts revealed different antibacterial activities against *E. coli* type (1). - Pericarp etanol extract had the best antibacterial activity. - The water and ether petroleum extracts had no antibacterial effectiveness.	[37]
	peel extract	anthocyanins and ellangitannins	*E. coli* collected from urinary cultures	- The inhibitory activity was found to be dose- and pH-dependent, with a minimum inhibitory concentration (MIC) value of 0.6 mg/mL at the pH of the aqueous extract (3.5). - The assay of adhesion carried out at the MIC showed a reduction of up to 80% of the adhesion index accompanied with reductions in motility and polyamide production.	[36]
	seed extract	tannins, steroids, terpenes, coumarins, flavonoids, andglycosides	uropathogens *E. coli*, *Enterococcus faecalis*, *S. aureus*, and *K. pneumonia*	- This study showed urobactericidal activity against different strains that were clinically isolated from patients suffering from urinary tract infections.	[34]

Table 1. *Cont.*

Plant Name	Extract/Part Used	Bioactive Compounds	Name of Microorganisms Tested (*E. coli* or Other Uropathogens)	Main Conclusions	Reference
Black chockeberry (*A. melanocarpa* Michx.)	ripe fruit extracts	- anthocyanins (cyanidin 3-galactoside, cyanidin 3-glucoside, cyanidin 3-arabinoside, and cyanidin 3- xyloside) - Epicatechin and its dimers and trimers	- *E. coli* ATCC 13202, *P. aeruginosa* ATCC 27853, *E.faecalis* ATCC 29212, and *S. aureus* ATCC 29212 - 18 clinical isolates from patients with urinary tract infections (*E. coli* 2041, *E. coli* 1851, E. coli 1992, *P. aeruginosa* 1908, *P. aeruginosa* 1128, *K. pneumoniae* 2110, *K. pneumoniae* 1074, *K. pneumoniae* 831, *Morganella morganii* 2520, *Acinetobacter baumanii* 1908, *A. baumanii* 2329, *Enterobacter cloacae* 2951, *E. faecalis* 2823, *E. faecium* 2862, *E. faecium* 2980, *E. faecium* 2027, *S. aureus* 14, and *S. aureus* 17	- This extract was the most active against one strain of *E. coli* and one of *M. morganii* at MIC (5 mg/mL). - It inhibited the development of biofilm in the case of an *E. coli* and a *M. morganii*. - Antibiotic disks (amikacin, tetracycline, nitrofurantoin, imipenem, and norfloxacin) supplemented with the extract slowly increased the growth inhibition zone in the cases of *P. aeruginosa*, *E. coli*, and *M. morganii*.	[50]
	plant extract	anthocyanins	*E. coli* LWT	- This extract compromised the integrity of the bacterial cell wall and membrane, bound directly to the bacterial DNA, and interfered with protein homeostasis.	[44]
	crude extracts, subfractions, and compounds from aronia	epicathechin	*E. coli* K12 JM109, and uropathogenic *E. coli* CFT073 (ATCC 700928)	- These extracts inhibited bacterial growth of *E. coli in vitro*. - They possessed an anti-biofilm formation effect.	[45]
	aqueous and ethanolic extracts from fresh, dried, and frozen fruits.		*E. coli* MSCL 332	- No activity against *E. coli* was reported.	[48]
	leaves, berries, and stem extracts		*E. coli*	- *E. coli* was one of the most resistant strains.	[49]
	leaf extract		*E. coli,*	- *E. coli* showed the greatest resistance and was reduced by 23% with 20 µL of aqueous ethanol extract.	[47]
	leaf extract	flavonoids: quercetin derivatives, isorhamnetin derivatives, kaempferol-3-O-rutinoside, and hydroxytyrosol		- No clear relationship was found between the influence of the plant extracts and whether the tested bacteria belonged to the Gram-positive or Gram-negative groups. - The extract had the lowest antibacterial activity on *E. coli*.	[46]
Cornelian cherry (*C. mas* L.)	leaf extract	iridoids, ellagic acid, and ellagi-tannins,	*E. coli*	- The extracts acted as bacteriostatic agents. - *C. mas* showed higher antimicrobial activity than *A. melanocarpa* extract. - *C. mas* leaf extract reduced the growth of all Gram-negative bacteria tested at the lowest concentration (1%). - The authors concluded that some interactions may occur between different bioactive compounds, explaining the stronger effect on Gram-negative bacteria in comparison to Gram-positive bacteria.	[46]
	bark extract		*E. coli*	- The extract did not inhibit the growth of *E. coli*.	[54]

7. Conclusions

For prevention and long-term treatment of rUTI, plant-based alternatives to antibiotics are appealing choices because they are cost-effective, readily available, safe, with fewer side effects, reduce antimicrobial resistance hazards, and help decrease adverse antibiotic effects and different symptoms. The antimicrobial resistance problem that occurs due to conventional antibiotic treatments has not been reported yet for natural plant remedies that contain a wide range of active phytochemical biomolecules responsible for their beneficial effects. Plant polyphenols, such as anthocyanidins and proanthocyanidins, flavonoids, ellagitannins, and monoterpenoids, such as iridoids, are the main phytoconstituents accountable for the treatment of rUTI. Still, current research only enlisted them, and information about their mechanisms of action is scarce.

Therefore, screening medicinal plants with therapeutic potential for rUTI cure should be conducted, as well as further studies at the molecular level to reveal the chemical composition of all potential compounds responsible for pharmaceutical activities and their mechanisms. Currently, commercial products with cranberry are considered effective plant extracts to prevent rUTI, but new studies proved that other plants with less side effects than cranberry might be considered. Pomegranate (*P. granatum* L.), black chokeberry (*A. melanocarpa* Michx.), and cornelian cherry (*C. mas* L.), illustrated in detail above, have great potential to be used for prevention or in a synergistic antibiotic therapy to cure UTI. The results sustain the opportunity for further studies with these plant extracts that could improve the current treatment, but more scientific evidence and large-scale well-designed clinical trials are required.

The innovation in prophylactic plant-based treatment strategies should target the bottleneck points from UTI pathogenesis and search for molecules that reduce rUTI symptoms without antibiotics and those that increase the host's immune response. Further progress in developing plant-based products to cure rUTI will be supported by the advances in UTI pathogenesis and human-based models to understand their pharmacological activities better.

Author Contributions: Writing, A.M.T.; writing—review and editing; L.D.D. and E.V. All authors have read and agreed to the published version of the manuscript.

Funding: This research received no external funding.

Institutional Review Board Statement: Not applicable.

Informed Consent Statement: Not applicable.

Data Availability Statement: Not applicable.

Acknowledgments: The graphical abstracts in Figures 1 and 2 were created with BioRender.com, last accessed on 28 February 2022. Figure 3 was upscaled, artifacts were removed, and the quality was increased with the Deep Image AI photo enhancer.

Conflicts of Interest: The authors declare no conflict of interest.

References

1. Sihra, N.; Goodman, A.; Zakri, R.; Sahai, A.; Malde, S. Nonantibiotic prevention, and management of recurrent urinary tract infection. *Nat. Rev. Urol.* **2018**, *15*, 750–776. [CrossRef]
2. Murray, B.O.; Flores, C.; Williams, C.; Flusberg, D.A.; Marr, E.E.; Kwiatkowska, K.M.; Charest, J.L.; Isenberg, B.C.; Rohn, J.L. Recurrent Urinary Tract Infection: A Mystery in Search of Better Model Systems. *Front. Cell. Infect. Microbiol.* **2021**, *11*, 691210. [CrossRef] [PubMed]
3. Silverman, J.A.; Schreiber, H.L.; Hooton, T.M.; Hultgren, S.J. From Physiology to Pharmacy: Developments in the Pathogenesis and Treatment of Recurrent Urinary Tract Infections. *Curr. Urol. Rep.* **2013**, *14*, 448–456. [CrossRef] [PubMed]
4. Lewis, A.J.; Richards, A.C.; Mulvey, M.A. Invasion of Host Cells and Tissues by Uropathogenic Bacteria. *Microbiol. Spectr.* **2016**, *4*, 359–381. [CrossRef] [PubMed]
5. Bazzaz, B.S.F.; Fork, S.D.; Ahmadi, R.; Khameneh, B. Deep insights into urinary tract infections and effective natural remedies. *Afr. J. Urol.* **2021**, *27*, 6. [CrossRef]

6. Gatea Kaabi, S.A.; Abdulrazaq, R.A.; Rasool, K.H.; Khassaf, S.A. Western herbal remedies for Urinary Tract infections. *Arch. Urol. Res.* **2020**, *4*, 49–60. [CrossRef]

7. Jacobsen, S.M.; Stickler, D.J.; Mobley, H.L.T.; Shirtliff, M.E. Complicated catheter-associated urinary tract infections due to Escherichia coli and Proteus mirabilis. *Clin. Microbiol. Rev.* **2008**, *21*, 26–59. [CrossRef]

8. Cortese, Y.J.; Wagner, V.E.; Tierney, M.; Devine, D.; Fogarty, A. Review of Catheter-Associated Urinary tract infections and In Vitro Urinary Tract Models. *J. Healthc. Eng.* **2018**, *2018*, 2986742. [CrossRef]

9. Tamadonfar, K.O.; Omattage, S.N.; Spauldine, C.N.; Hultgren, S.J.O.N. Reaching the End of the Line: Urinary Tract Infections. In C. R. Pascale Cossart, Bacteria and Intracellularity. *Microbiol. Spectr.* **2019**, *7*, 83–99. [CrossRef]

10. Klein, R.D.; Hultgren, S.J. Urinary tract infections: Microbial pathogenesis, host-pathogen interactions and new treatment strategies. *Rev. Nat. Rev. Microbiol.* **2020**, *18*, 211–226. [CrossRef]

11. Storme, O.; Tirán Saucedo, J.; Garcia-Mora, A.; Dehesa-Dávila, M.; Naber, K.G. Risk factors and predisposing conditions for urinary tract infection. *Therap. Adv. Urol.* **2019**, *11*, 1756287218814382. [CrossRef] [PubMed]

12. Zaffanello, M.; Malerba, G.; Cataldi, L.; Antoniazzi, F.; Franchini, M.; Monti, E.; Fanos, V. Genetic risk for recurrent urinary tract infections in humans: A systematic review. *J. Biomed. Biotechnol.* **2010**, *2010*, 321082. [CrossRef] [PubMed]

13. Godaly, G.; Ambite, I.; Svanborg, C. Innate immunity and genetic determinants of urinary tract infection susceptibility. *Curr. Opin. Infect. Dis.* **2015**, *28*, 88–96. [CrossRef] [PubMed]

14. Ragnarsdóttir, B.; Lutay, N.; Grönberg-Hernandez, J.; Köves, B.; Svanborg, C. Genetics of innate immunity and UTI susceptibility. *Nat. Rev. Urol.* **2011**, *8*, 449–468. [CrossRef]

15. Kawalec, A.; Zwolińska, D. Emerging Role of Microbiome in the Prevention of Urinary Tract Infections in Children. *Int. J. Mol. Sci.* **2022**, *23*, 870. [CrossRef]

16. Yang, S.B.F. Pathophysiology of UTIs. In *Female Urinary Tract Infections in Clinical Practice*; Yang, S.F.B., Ed.; Springer International Publishing: Berlin/Heidelberg, Germany, 2020; pp. 1–10. Available online: https://link.springer.com/book/10.1007/978-3-030-27909-7 (accessed on 24 February 2022).

17. Davenport, M.; Mach, E.K.; Shortliffe, L.M.D.; Banaei, N.; Wang, T.H.; Liao, J.C. New and developing diagnostic technologies for urinary tract infections. *Nat. Rev. Urol* **2017**, *14*, 296–310. [CrossRef]

18. Bergeron, M.G.; Ouellette, M. Preventing antibiotic resistance through rapid genotypic identification of bacteria and of their antibiotic resistance genes in the clinical microbiology laboratory. *J. Clin. Microbiol.* **1998**, *36*, 2169–2172. [CrossRef]

19. Mitsakakis, K.; Kaman, W.E.; Elshout, G.; Specht, M.; Hays, J.P. Challenges in identifying antibiotic resistance targets for point-of-care diagnostics in general practice. *Fut. Microbiol.* **2018**, *13*, 1157–1164. [CrossRef]

20. Stork, C.; Kovács, B.; Rózsai, B.; Putze, J.; Kiel, M.; Dorn, A.; Kovács, J.; Melegh, S.; Leimbach, A.; Kovács, T.; et al. Characterization of Asymptomatic Bacteriuria Escherichia coli Isolates in Search of Alternative Strains for Efficient Bacterial Interference against Uropathogens. *Front. Microbiol.* **2018**, *9*, 214. [CrossRef]

21. Köves, B.; Salvador, E.; Grönberg-Hernández, J.; Zdziarski, J.; Wullt, B.; Svanborg, C.; Dobrindt, U. Rare Emergence of Symptoms during Long-Term Asymptomatic Escherichia coli 83972 Carriage without an Altered Virulence Factor Repertoire. *J. Urol.* **2014**, *19*, 191. [CrossRef]

22. Ceprnja, M.; Oros, D.; Melvan, E.; Svetlicic, E.; Skrlin, J.; Barisic, K.; Starcevic, L.; Zucko, J.; Starcevic, A. Modeling of Urinary Microbiota Associated With Cystitis. *Front. Cell. Infect. Microbiol.* **2021**, *11*, 140. [CrossRef]

23. Georgopoulos, N.T.; Kirkwood, L.A.; Varley, C.L.; MacLaine, N.J.; Aziz, N.; Southgate, J. Immortalisation of normal human urothelial cells compromises differentiation capacity. *Eur. Urol.* **2011**, *60*, 141–149. [CrossRef]

24. Baker, S.C.; Shabir, S.; Southgate, J. Biomimetic urothelial tissue models for the in vitro evaluation of barrier physiology and bladder drug efficacy. *Mol. Pharm.* **2014**, *11*, 1964–1970. [CrossRef] [PubMed]

25. Hatina, J.; Schulz, W.A. Stem cells in the biology of normal urothelium and urothelial carcinoma. *Neoplasma* **2012**, *59*, 728–736. [CrossRef] [PubMed]

26. DyStar Colours Distribution GmbH. Mixtures of Disperse Dyes. U.S. Patent US8906116B2, 9 December 2014.

27. Vamanu, E.; Dinu, L.D.; Luntraru, C.M.; Suciu, A. In Vitro Coliform Resistance to Bioactive Compounds in Urinary Infection, Assessed in a Lab Catheterization Model. *Appl. Sci.* **2021**, *11*, 4315. [CrossRef]

28. Chua, R.Y.R.; Lim, K.; Leong, S.S.J.; Tambyah, P.A.; Ho, B. An in-vitro urinary catheterization model that approximates clinical conditions for evaluation of innovations to prevent catheter-associated urinary tract infections. *J. Hosp. Infect.* **2017**, *97*, 66–73. [CrossRef] [PubMed]

29. Flaherty, R.A.; Lee, S.W. Implementation of a Permeable Membrane Insert-based Infection System to Study the Effects of Secreted Bacterial Toxins on Mammalian Host Cells. *J. Vis. Exp.* **2016**, *114*, e54406. [CrossRef]

30. Dadgostar, P. Antimicrobial Resistance: Implications and Costs. *Infect. Drug Resist.* **2019**, *12*, 3903–3910. [CrossRef]

31. Machowska, A.; Stålsby Lundborg, C. Drivers of Irrational Use of Antibiotics in Europe. *Int. J. Environ. Res. Public Health* **2018**, *16*, 27. [CrossRef]

32. Foxman, B.; Buxton, M. Alternative approaches to conventional treatment of acute uncomplicated urinary tract infection in women. *Curr. Infect. Dis. Rep.* **2013**, *15*, 124–129. [CrossRef]

33. Sofowora, A.; Ogunbodede, E.; Onayade, A. The role and place of medicinal plants in the strategies for disease prevention. *Afr. J. Trad. Complement. Alter. Med.* **2013**, *10*, 210–229. [CrossRef] [PubMed]

34. Terlizzi, M.E.; Gribaudo, G.; Maffei, M.E. UroPathogenic *Escherichia coli* (UPEC) Infections: Virulence Factors, Bladder Responses, Antibiotic, and Non-antibiotic Antimicrobial Strategies. *Front. Microbiol.* **2017**, *8*, 1566. [CrossRef] [PubMed]
35. Belkaid, Y.; Hand, T.W. Role of the microbiota in immunity and inflammation. *Cell* **2014**, *157*, 121–141. [CrossRef] [PubMed]
36. Gadisa, E.; Tadesse, E. Antimicrobial activity of medicinal plants used for urinary tract infections in pastoralist community in Ethiopia. *BMC Complement. Med. Ther.* **2021**, *21*, 74. [CrossRef] [PubMed]
37. Shaheen, G.; Akram, M.; Jabeen, F.; Ali Shah, S.M.; Munir, N.; Daniyal, M.; Riaz, M.; Tahir, I.M.; Ghauri, A.O.; Sultana, S.; et al. Therapeutic potential of medicinal plants for the management of urinary tract infection: A systematic review. *Clin. Exp. Pharmacol. Physiol.* **2019**, *46*, 613–624. [CrossRef]
38. Das, S. Natural therapeutics for urinary tract infections-a review. *Future Pharm. J. Sci.* **2020**, *6*, 64. [CrossRef]
39. Peng, M.M.; Fang, Y.; Hu, W.; Huang, Q. The pharmacological activities of compound salviplebeia granules on treating urinary tract infection. *J. Ethnopharmacol.* **2010**, *129*, 59–63. [CrossRef]
40. Cela-López, J.M.; Camacho Roldán, C.J.; Gómez-Lizarraga, G.; Martínez, V. A Natural Alternative Treatment for Urinary Tract Infections: Itxasol©, the Importance of the Formulation. *Molecules* **2021**, *26*, 4564. [CrossRef]
41. Sarecka-Hujar, B.; Szulc-Musioł, B. Herbal Medicines—Are They Effective and Safe during Pregnancy? *Pharmaceutics* **2022**, *14*, 171. [CrossRef]
42. Brown, P.N.; Turi, C.E.; Shipley, P.R.; Murch, S.J. Comparisons of large (Vaccinium macrocarpon Ait.) and small (Vaccinium oxycoccos L., Vaccinium vitis-idaea L.) cranberry in British Columbia by phytochemical determination, antioxidant potential, and metabolomic profiling with chemometric analysis. *Planta Med.* **2012**, *78*, 630–640. [CrossRef]
43. Howell, A.B.; Foxman, B. Cranberry juice and adhesion of antibiotic-resistant uropathogens. *JAMA* **2002**, *287*, 3082–3083. [CrossRef] [PubMed]
44. McMurdo, M.E.; Argo, I.; Phillips, G.; Daly, F.; Davey, P. Cranberry or trimethoprim for the prevention of recurrent urinary tract infections? A randomized controlled trial in older women. *J. Antimicrob. Chemother.* **2009**, *63*, 389–395. [CrossRef] [PubMed]
45. Montefusco, A.; Durante, M.; Migoni, D.; De Caroli, M.; Ilahy, R.; Pék, Z.; Helyes, L.; Fanizzi, F.P.; Mita, G.; Piro, G.; et al. Analysis of the Phytochemical Composition of Pomegranate Fruit Juices, Peels and Kernels: A Comparative Study on Four Cultivars Grown in Southern Italy. *Plants* **2021**, *10*, 2521. [CrossRef] [PubMed]
46. Howell, A.B.; D'Souza, D.H. The pomegranate: Effects on bacteria and viruses that influence human health. *Evid.-Based Complementary Altern. Med. eCAM* **2013**, *2013*, 606212. [CrossRef]
47. Fahmy, H.; Hegazi, N.; El-Shamy, S.; Farag, M.A. Pomegranate juice as a functional food: A comprehensive review of its polyphenols, therapeutic merits, and recent patents. *Food Funct.* **2020**, *11*, 5768–5781. Available online: https://pubs.rsc.org/en/content/articlelanding/2020/fo/d0fo01251c (accessed on 24 February 2022). [CrossRef]
48. Pirzadeh, M.; Caporaso, N.; Rauf, A.; Shariati, M.A.; Yessimbekov, Z.; Khan, M.U.; Imran, M.; Mubarak, M.S. Pomegranate as a source of bioactive constituents: A review on their characterization, properties and applications. *Crit. Rev. Food Sci. Nutr.* **2021**, *61*, 982–999. [CrossRef]
49. Das, S.; Panigrahi, S.; Panda, P. Antiurobacterial activity of *Punica granatum* L. seed extract. *Eur. J. Med. Plants* **2018**, *22*, 1–12. [CrossRef]
50. Abdel-Salam, F.F.; El_deen Moharram, Y.G.; El-Zalaki, E.M. Characterization of Wastes from Pomegranate (*Punica granatum* L.) Juice and Its Use as a Functional Drink. *Egypt J. Food Sci.* **2018**, *46*, 91–100. Available online: https://ejfs.journals.ekb.eg/article_46941.html (accessed on 24 February 2022).
51. Zam, W.; Khaddour, A. Anti-virulence effects of aqueous pomegranate peel extract on *E. coli* urinary tract infection. *Progr. Nutr.* **2017**, *19*, 98–104. [CrossRef]
52. AlFadel, F.; Allaham, S.A.; Alkhatib, R. The Anti-Bacterial Activity of Various Parts of Punica granatum on Antibiotics Resistance Escherichia coli. *Int. J. Pharmacog. Phytochem. Res.* **2014**, *6*, 79–85. Available online: https://brief.land/jjm/articles/56335.html (accessed on 24 February 2022).
53. Dey, D.; Debnath, S.; Hazra, S.; Ghosh, S.; Ray, R.; Hazra, B. Pomegranate pericarp extract enhances the antibacterial activity of ciprofloxacin against extended-spectrum β-lactamase (ESBL) and metallo-β-lactamase (MBL) producing Gram-negative bacilli. *Food Chem. Toxicol.* **2012**, *50*, 4302–4309. [CrossRef] [PubMed]
54. Elshafie, H.S.; Caputo, L.; De Martino, L.; Sakr, S.H.; De Feo, V.; Camele, I. Study of Bio-Pharmaceutical and Antimicrobial Properties of Pomegranate (*Punica granatum* L.) Leathery Exocarp Extract. *Plants* **2021**, *10*, 153. [CrossRef] [PubMed]
55. Duman, A.D.; Ozgen, M.; Dayisoylu, K.S.; Erbil, N.; Durgac, C. Antimicrobial Activity of Six Pomegranate (Punica granatum L.) Varieties and Their Relation to Some of Their Pomological and Phytonutrient Characteristics. *Molecules* **2009**, *14*, 1808–1817. [CrossRef]
56. Bakkiyaraj, D.; Nandhini, J.R.; Malathy, B.; Pandian, S.K. The anti-biofilm potential of pomegranate (*Punica granatum* L.) extract against human bacterial and fungal pathogens. *Biofouling* **2013**, *29*, 929–937. [CrossRef] [PubMed]
57. Vlachojannis, C.; Zimmermann, B.F.; Chrubasik-Hausmann, S. Efficacy and safety of pomegranate medicinal products for cancer. *eCAM* **2015**, *2015*, 258598. [CrossRef] [PubMed]
58. Kokotkiewicz, A.; Jaremicz, Z.; Luczkiewicz, M. Aronia plants: A review of traditional use, biological activities, and perspectives for modern medicine. *J. Med. Food.* **2010**, *13*, 255–269. [CrossRef]
59. Valcheva-Kuzmanova, S.V.; Belcheva, A. Current knowledge of *Aronia melanocarpa* as a medicinal plant. *Folia Med.* **2006**, *48*, 11–17.

60. Smeriglio, A.; Barreca, D.; Bellocco, E.; Trombetta, D. Proanthocyanidins and hydrolysable tannins: Occurrence, dietary intake and pharmacological effects. *Br. J. Pharmacol.* **2017**, *174*, 1244–1262. [CrossRef]

61. Available online: https://handle.nal.usda.gov/10113/43336 (accessed on 17 February 2022).

62. Deng, H.; Zhu, J.; Tong, Y.; Kong, Y.; Tan, C.; Wang, M.; Wan, M.; Meng, X. Antibacterial characteristics and mechanisms of action of *Aronia melanocarpa* anthocyanins against *Escherichia coli*. *LWT* **2021**, *150*, 112018. [CrossRef]

63. Bräunlich, M.; Økstad, O.A.; Slimestad, R.; Wangensteen, H.; Malterud, K.E.; Barsett, H. Effects of *Aronia melanocarpa* constituents on biofilm formation of *Escherichia coli* and *Bacillus Cereus*. *Molecules* **2013**, *18*, 14989–14999. [CrossRef]

64. Efenberger-Szmechtyk, M.; Nowak, A.; Czyżowska, A.; Kucharska, A.Z.; Fecka, I. Composition and Antibacterial Activity of *Aronia melanocarpa* (Michx.) Elliot, *Cornus mas* L. and *Chaenomeles superba* Lindl. Leaf Extracts. *Molecules* **2020**, *25*, 2011. [CrossRef]

65. Tian, Y.; Puganen, A.; Alakomi, H.L.; Uusitupa, A.; Saarela, M.; Yang, B. Antioxidative and antibacterial activities of aqueous ethanol extracts of berries, leaves, and branches of berry plants. *Food Res. Int.* **2018**, *106*, 291–303. [CrossRef]

66. Liepiņa, I.; Nicolajeva, V. Antimicrobial activity of extracts from fruits of *Aronia melanocarpa* and *Sorbus aucuparia*. *Environ. Exp. Biol.* **2013**, *11*, 195–199. Available online: https://americanaronia.org/antimicrobial-activity-extracts-fruits-aronia-melanocarpa-sorbus-aucuparia/ (accessed on 24 February 2022).

67. Cvetanovi'c, A.; Zengin, G.; Zekovi'c, Z.; Švarc-Gaji'c, J.; Raži'c, S.; Damjanovi'c, A.; Maškovi'c, P.; Miti'c, M. Comparative in vitro studies of the biological potential and chemical composition of stems, leaves and berries *Aronia melanocarpa*'s extracts obtained by subcritical water extraction. *Food Chem. Toxicol.* **2018**, *121*, 458–466. [CrossRef]

68. Dorneanu, R.; Cioanc, O.; Chifiriuc, O.; Albu, E.; Tuchilu, C.G.; Mircea, C.; Salamon, I.; Hancianu, M. Synergic benefits of *Aronia melanocarpa* anthocyanin–rich extracts and antibiotics used for urinary tract infections. *Farmacia* **2017**, *65*, 778–783. Available online: https://farmaciajournal.com/issue-articles/synergic-benefits-of-aronia-melanocarpa-anthocyanin-rich-extracts-and-antibiotics-used-for-urinary-tract-infections/ (accessed on 24 February 2022).

69. Bräunlich, M.; Slimestad, R.; Wangensteen, H.; Brede, C.; Malterud, K.E.; Barsett, H. Extracts, anthocyanins and procyanidins from Aronia melanocarpa as radical scavengers and enzyme inhibitors. *Nutrients* **2013**, *5*, 663–678. [CrossRef] [PubMed]

70. Handeland, M.; Grude, N.; Torp, T.; Slimestad, R. Black chokeberry juice (*Aronia melanocarpa*) reduces incidences of urinary tract infection among nursing home residents in the long term–a pilot study. *Nutr. Res.* **2014**, *34*, 518–525. [CrossRef] [PubMed]

71. Kazimierski, M.; Regula, J.; Molska, M. Cornelian cherry (*Cornus mas* L.)–characteristics, nutritional and pro-health properties. *Acta Sci. Pol. Technol. Aliment.* **2019**, *18*, 5–12. [CrossRef]

72. Bayram, H.M.; Ozturkcan, S.A. Bioactive components and biological properties of cornelian cherry (Cornus mas L.): A comprehensive review. *J. Funct. Foods* **2020**, *75*, 104252. [CrossRef]

73. Dinda, B.; Dinda, M.; Dinda, S.; Kyriakopoulos, M.A.; Markopoulos, C.; Thomaidis, S.N.; Velegraki, A.; Zoumpourlis, V. Cornus mas L. (cornelian cherry), an important European and Asian traditional food and medicine: Ethnomedicine, phytochemistry and pharmacology for its commercial utilization in drug industry. *J. Ethnopharmacol.* **2016**, *193*, 670–690. [CrossRef]

74. Dulger, B.; Gonuz, A. Antimicrobial activity of some Turkish medicinal plants. *Pak. J. Biol. Sci.* **2004**, *7*, 1559–1562. [CrossRef]

75. Danielewski, M.; Matuszewska, A.; Nowak, B.; Kucharska, A.Z.; Sozański, T. The Effects of Natural Iridoids and Anthocyanins on Selected Parameters of Liver and Cardiovascular System Functions. *Oxid. Cell Longev.* **2020**, *2020*, 2735790. [CrossRef] [PubMed]

76. Sozański, T.; Kucharska, A.Z.; Rapak, A.; Szumny, D.; Trocha, M.; Merwid-Ląd, A.; Dzimira, S.; Piasecki, T.; Piórecki, N.; Magdalan, J.; et al. Iridoid-loganic acid versus anthocyanins from the *Cornus mas* fruits (cornelian cherry): Common and different effects on diet-induced atherosclerosis, PPARs expression and inflammation. *Atherosclerosis* **2016**, *254*, 151–160. [CrossRef] [PubMed]

77. Szczepaniak, O.M.; Kobus-Cisowska, J.; Kusek, W.; Przeor, M. Functional properties of *Cornelian cherry* (*Cornus mas* L.): A comprehensive review. *Eur. Food Res. Technol.* **2019**, *245*, 2071–2087. [CrossRef]

78. Dadkhah, N.; Shirani, M.; Etemadifar, S.; Mirtalebi, M. The effect of *Cornus mas* in preventing recurrent urinary tract infections in women'. *Adv. Herb. Med.* **2017**, *3*, 67–76. Available online: https://naldc.nal.usda.gov/catalog/43336 (accessed on 24 February 2022).

79. Pugliese, D.; Acampora, A.; Porreca, A.; Schips, L.; Cindolo, L. Effectiveness of a novel oral combination of D-Mannose, pomegranate extract, prebiotics and probiotics in the treatment of acute cystitis in women. *Arch. Ital. Urol. Androl.* **2020**, *92*, 34–38. [CrossRef]

Review

Effects of Heat Treatment Processes: Health Benefits and Risks to the Consumer

Miguel Cascais [1,2], Pedro Monteiro [1], Diana Pacheco [1], João Cotas [1], Leonel Pereira [1], João Carlos Marques [1] and Ana M. M. Gonçalves [1,3,*]

[1] University of Coimbra, MARE—Marine and Environmental Sciences Centre, Department of Life Sciences, Calçada Martim de Freitas, 3000-456 Coimbra, Portugal; 9474@eshte.pt (M.C.); pmdmonteiro1992@gmail.com (P.M.); diana.pacheco@uc.pt (D.P.); jcotas@uc.pt (J.C.); leonel.pereira@uc.pt (L.P.); jcmimar@ci.uc.pt (J.C.M.)

[2] Higher Institute for Tourism and Hotel Management of Estoril, Av. Condes de Barcelona, n.808, 2769-510 Estoril, Portugal

[3] Department of Biology and CESAM, University of Aveiro, 3810-193 Aveiro, Portugal

* Correspondence: amgoncalves@uc.pt; Tel.: +351-239-240-700 (ext. 262-286)

Abstract: Macroalgae are a biological group that has mainly been used in Asian countries; however, the interest shown by Western society is recent, its application in the industrial sector having increased in the last few decades. Seaweeds are filled with properties which are beneficial to our health. To use them as food and enhance these properties, heat has been used on them. This process alters the bioactive compounds. If we study the levels of moisture, they can vary according to the drying methods used. High values of moisture can lead to a short shelf life due to oxidation, microbial or enzyme activity, so controlling these values is highly recommended. Heat causes enzymatic activity as well as oxidation, which leads to degradation of phenolic compounds in comparison with freeze-drying, which causes fewer losses of these components. Due to the same occurrences, lipid content can also vary, modifying the bioactive compounds and their benefits. Pigments are some of the components most affected by heat, since, through this process, seaweeds or seaweed products can suffer a change in color. Iodine in macroalgae can decrease drastically; on the other hand, protein yield can be greatly enhanced. Some studies showed that the amount of arsenic in raw seaweeds was higher than when they were heat processed, and that arsenic values varied when different heat treatments were applied. Additionally, another study showed that heat can alter protein yield in specific species and have a different effect on other species.

Keywords: heat treatment; seaweeds; bioactive compounds; food safety; consumer health

Citation: Cascais, M.; Monteiro, P.; Pacheco, D.; Cotas, J.; Pereira, L.; Marques, J.C.; Gonçalves, A.M.M. Effects of Heat Treatment Processes: Health Benefits and Risks to the Consumer. *Appl. Sci.* **2021**, *11*, 8740. https://doi.org/10.3390/app11188740

Academic Editor: Emanuel Vamanu

Received: 28 August 2021
Accepted: 14 September 2021
Published: 19 September 2021

Publisher's Note: MDPI stays neutral with regard to jurisdictional claims in published maps and institutional affiliations.

1. Introduction

It is estimated that the world population will grow to 10 billion in the next three decades. This will require a 70% increase in food production [1]. Moreover, the production of meat will double in the same time span. Because of this, the search for new food sources is essential [2]. Intensive farming and agriculture have contributed to the saturation of arable lands and reduced access to fresh water [3,4]. These actions are some of the causes of climate change.

Seaweeds are one of the most promising foods for sustainability. Their ability to capture CO_2 from the ocean and the atmosphere represents an excellent method to battle climate change. Furthermore, they have the potential for fast growth in the ocean, which facilitates the method of producing it while reducing the cost of production [3]. Even though seaweeds have been consumed for centuries in Eastern countries such as China, Japan and Korea, in Western societies, their introduction as food source only occurred a few decades ago. This happened after World War II when, due to the exponential population growth, it was noticed there was insufficient protein consumption. To fight that lack of nutrients, seaweeds played a major role, being rich in several macronutrients and micronutrients, including vitamins, minerals and proteins [5,6]. Because of their polymers, such as carrageenan, agar and alginate, and their properties for gelling, emulsifying and thickening, they have been used as a novel food [7,8].

Seaweeds are an excellent source of bioactive compounds with health-promoting benefits, such as antioxidants, dietary fiber, essential fatty acids, vitamins and minerals, consumption of which has been associated with a lower occurrence of some chronic diseases, such as diabetes, obesity, heart disease and cancer [9,10]. Due to these beneficial properties, it is necessary to assess how the culinary processes influence and modify the valuable compounds that are sought after in their consumption. With the growing interest in consumption of these organisms, it is crucial to assess how culinary treatment, specifically thermal treatment, alters the bioactive content of edible algae, as the particular methodology has an impact on the biochemical composition [9].

Utilization of seaweeds normally requires post-harvest dehydration to reduce the water content, which can be up to 85–90% of the biomass [9], preventing decomposition and microbial contamination, increasing shelf life and consumer safety while also aiding the extraction of important chemical constituents [11]. Several techniques have been described for the drying of seaweeds. Sun-drying and oven-drying are the most used, due to their accessibility and relatively low operation cost [11,12]. However, solar drying is dependent on the climate and exposure to airborne contamination, which can compromise the hygiene of the product [11,12]. Oven-drying requires a large space, has high energy consumption and leads to component degradation, especially heat-labile compounds [11,12]. Other methods, such as freeze- or vacuum-drying, are not widely used for commercial purposes, as they are expensive techniques used for the extraction of specific components for specific industries. This review will focus on the available literature regarding drying processes and phytochemical alterations derived from these processes.

2. Moisture

Usually, different drying techniques will result in different moisture levels that remain after treatment. This residual moisture affects the stability of bioactive compounds in dried materials, possibly due to oxidation, enzymes or microbial activities that successfully degrade the remaining bioactive compounds, which are an important factor when considering transportation and storage [13–15].

Regarding *H. banksia* (Phaeophyceae), there is a significant difference in the impacts of physical properties under different drying treatments, with lower moisture being obtained using freeze-drying methods rather than oven-drying, with the highest residual moisture found in sun-dried samples (5.7, 7.7 and 16.2%, respectively) [16]. The extraction yield was, in comparison, higher for freeze-drying (29%), followed by oven drying (20%) and, lastly, sun drying (18%). For *A. taxiformis* (Rhodophyta), the results are the opposite, with higher moisture content obtained in freeze-dried in comparison with oven-dried samples. In a study with *Kappapychus alvarezii* (Rhodophyta) and *Sargassum polycystum* (Phaeophyceae), the moisture content of freeze-dried samples was lower than for oven-dried samples. Overall, there is a general agreement that freeze-drying results in lower moisture percentages in comparison with oven-drying techniques [9,13,16–22].

3. Phenolic Content

The phenolic compounds present in seaweed mainly include chlorogenic acid, phloroglucinol and phlorotannins, caffeic acid, kaempferol, benzoic acids, coumaric acids, cirsimaritin, ferulic acid, gallic acid and syringic acid [11,13,17,20]. The extractability of bioactive components from oven-dried *Gracillaria* sp. (Phaeophyceae), *Ulva rigida* (Chlorophyta) and *Fucus vesiculosus* (Phaeophyceae) was studied at different temperatures (25 °C, 40 °C and 60 °C) [18]. However, there was a demonstrated reduction in total phenolic and flavonoid content after drying. In *Hormosira banksia* (Phaeophyceae), total phenolic content (TPC) and total flavonoid content (TFC) were highly affected by the drying technique, with greater preservation of the phytochemical content through freeze-drying, de-humidification and vacuum drying, compared with oven- or sun drying. For *Saccharina latissima* (Phaeophyceae), the drying temperature and humidity had a significant effect on TPC, with a reported drop by 10-fold compared with the fresh algae [13], with the highest amount of phenolic activity in freeze-dried samples and the lowest in samples dried at 70 °C. This variation is possibly due to less oxidation at lower temperatures and the absence of oxygen while vacuum-drying [20]. In accordance with these results, the drying kinetics of *Himanthalia elongata* (Phaeophyceae) when exposed to a range of temperatures was described [16]. After a 24-h drying period, a reduction of 29–51% in total phenolic compounds was observed, with the maximum reduction of 51% observed for the 25 °C drying, while a reduction of 29% was observed for drying at 40 °C. Total flavonoid content followed a similar tendency to the phenolic content in these algae [10,16].

In Badmus et al., the effects of different drying processes (oven-drying at 40–60 °C, freeze-drying and microwave drying at 385, 540 and 700 W) on antioxidant potential, protein, lipid, amino acid and fatty acid content were tested in five brown seaweeds (*Fucus spiralis, Laminaria digitata, Fucus serratus, Halidrys siliquosa* and *Pelvetia canaliculata*) [11]. The results demonstrated an overall loss of compounds, with high values of certain compounds linked to specific drying methods.

Overall, heat causes cellular damage, inducing enzymatic activity and oxidative stress, leading to thermal degradation of the phenolic compounds [20]. These heat-labile compounds are usually bound within carbohydrates, protein and the fatty acid matrix in the food structure [20]. A substantial number of phenolic compounds can be degraded due to light or oxygen exposure, which are more likely to occur in an oven-dryer system in comparison with freeze-drying. This phenolic content is associated with several of the beneficial properties present in seaweeds. Assessing the best methods is crucial for optimization, in order to minimize phenolic content loss [17].

4. Lipid Content

Seaweeds contain fats in the form of saturated fatty acids (SFA) such as myristic acid, palmitic acid and stearic acid; monounsaturated fatty acids (MUFAs) such as oleic acid, palmitoleic acid and eicosenoic acid; and polyunsaturated fatty acids (PUFAs), such as linoleic acid, stearidonic acid, arachidonic acid and eicosapentanoic acid, representing approximately 40% of the total fat/lipid content [11,20]. In regard to PUFAs, some are considered essential fatty acids because they cannot be biosynthesized by the human body and are solely obtained through the diet. The beneficial impacts and properties of these lipidic compounds is well studied, and it is important to understand the variation in these fatty acids when exposed to heat treatment. For *S. latissima* (Phaeophyceae), there was no significant difference in the lipidic content in [20] when dried under the temperature and humidity conditions tested in the study (30 °C, 50 °C and 70 °C temperature and 25–50% humidity). This could mean that under heat treatment, there is no loss of lipids through dripping or oxidation [20]. However, for *Pyropia orbicularis* (Rhodophyta), fatty acid concentration was observed to suffer modifications according to the drying method applied [13]. While there was a yield loss of SFA for sun-drying and oven-drying compared with freeze or vacuum-drying, there was no significant difference in total MUFA content with different drying techniques [13]. Interestingly, for these algae, a higher decrease in PUFA content was reported for samples treated with freeze-drying and vacuum-drying, compared with sun- or convective drying. For *F. spiralis, F. serratus, H. siliquosa* and *P. canaliculata*, the highest lipid content values tended to be obtained through freeze-drying, while for *L. digitata*, they were obtained through oven-drying at 40 °C, and the lowest value of lipid content was identified for both *Fucus* species in oven-dried batches. C16:0, C18:1 and C20:4n6 were the predominant SFA, MUFA and PUFA for all seaweeds, respectively. For SFAs, higher levels were found in *L. digitata* and *H. siliquosa* after freeze-drying and oven-drying and 40 °C, while other species showed little differences in SFA content. MUFA content was the highest in *F. spiralis* and *L. digitata* after oven-drying at 40 °C, followed by freeze-drying and microwave treatment at 385 W. Regarding PUFA levels, *L. digitata* and *H. siliquosa* exhibited higher concentrations after microwave treatment, while both *Fucus* species had higher levels after freeze-drying [11]. The lipid content in *Aspargopsis taxiformis* [9] when exposed to oven-drying (60 °C) or freeze-drying also demonstrated important changes. Overall, total lipid yield for *A. taxiformis* was low, with slightly higher lipid preservation in freeze-dried samples. The fatty acid profile (as a percentage of total fatty acids) was characterized by a high concentration of saturated FAs (76% of total FA), with low levels of monounsaturated fatty acids (4–11% MUFAs) and even lower concentration of polyunsaturated fatty acids (0–11% PUFAs) after drying. In *A. taxiformis*, a relative higher concentration of SFA and MUFAs was obtained through oven-drying at 60 °C, with the exception of PUFAs, which demonstrated a higher concentration in freeze-dried samples [9].

Overall, there have been several results regarding total fatty acids concentrations, a variation that may have resulted from the method applied or due to the intrinsic variation between algae species. During the drying processes, oxidation and degradation of lipidic content can occur via several mechanisms, such as autoxidation, photosensitized or enzymatic reactions. In the case of PUFAs, autoxidation is specially critical, due to the intrinsic low disassociation energy [11,20].

In order to retain bioactive compounds and functional properties of dried seaweed products, it is necessary to undertake a detailed and simultaneous investigation of the seasonal variation and effects of different drying methods and conditions (humidity and temperature) on the physiological properties, phenolic and antioxidant activities of the target seaweed [11,20]. Within the same species of seaweeds, morphological and structural differences in tissues, age, size, environment and seasonality influence the phytochemical components of seaweeds [20].

Freeze-drying works on the principle of sublimation under a vacuum and removal of frozen water, while conventional oven-drying is subjected to humidity, temperature and air velocity inside the drying chamber [20]. Even though it has demonstrated better yield and quality products, in terms of maintaining the integrity or nutrient and phytochemical profile, freeze-drying is not a commercially deployed technique, as its high-energy requirements and costs make it unprofitable for large-scale operations, targeting the extraction of finer components for specific industries [11,20].

5. Rehydration and Cooking

As with most consumable dried products, there is the need for rehydration of the products prior to consumption, in order to restore the fresh properties of the dried product, usually through contact with a liquid phase. In general, this process can be divided into three phases: (1) water absorption by the dried material, (2) swelling of the dried product and (3) loss or diffusion of the soluble components [10].

Rehydration and blanching are performed to increase palatability but can cause a significant loss of phenolic content, which leaches to the boiling water. Cox et al., when studying the rehydration kinetics of *H. elongata*, observed that the rehydration time decreased as temperatures increased the magnitude of absorbed water [10]. Additionally, it was also observed that for *H. elongata*, rehydration of biomass resulted in a decrease in total phenolic compounds in the dried seaweed, from 1.21 ± 0.02 g GAE/100 g db to 0.2 ± 0.009 g GAE /100 g db when the rehydration was finished [10]. Phenolic compounds are heat-labile, and when exposed to boiling water through blanching or cooking, there can be a significant loss of such compounds. It is generally accepted that rehydration is dependent on the degree of cellular and structural disruption. When irreversible damage has been caused, large structures within the cells can collapse and reduce the hydrophilic properties, hence reducing the rehydration properties [10].

The hydration of dehydrated seaweed food products is usually applied before consumption [10]. This rehydration is a complex method used to restore, in the best possible way, the quality and properties of the original fresh product using a liquid phase, normally an aqueous solution. This method is based on three steps: absorption of water by the dried biomass; swelling of the rehydrated product, which becomes similar to the original fresh product; and loss or diffusion of the soluble compounds (co-extraction of the compounds by the liquid phase) [23,24]. In dried raw seaweed material, the rehydration time is longer at lower temperatures. On the other hand, when the temperature is high (above 80 °C), rehydration is quick and the dry biomass regains the original moisture level more quickly; however, the compounds have greater losses in the yield and quality, mainly the bioactive compounds, such as phenolic and flavonoid compounds, which are oxidated and destroyed [10]. Cox et al. [9] demonstrated that the heat treatment at 100 °C for 40 min caused losses of about 83% and 93% in the total phenolic and flavonoid content, respectively, in *H. elongata*. Furthermore, rehydration velocity is not a constant, because in the initial stages, the rehydration is very rapid and there is a decline until equilibrium in the system is reached. In this process, as demonstrated above, the water temperature is a key factor of the rehydration process, where hot water can rehydrate the seaweed rapidly but causing a major change in the tissue structure (mainly cell wall destruction and denaturation of thermolabile compounds) and composition of the seaweed (the blanching technique used in the kitchen) [10]. Blanching is used mainly to make sea and terrestrial vegetables more palatable; however, this technique also leads to co-extraction (leaching) and denaturation of heat-sensitive compounds [25].

Every thermal process is detrimental to the integrity of plant tissues, particularly cellular membranes. Temperatures above the optimal point can lead to damage which will consequently lead to variations in response to the process applied [10].

6. Heat Treatment Influence on Chemical Structure

The heat treatment of seaweed-based ingredients and foods regularly induces a major chemical modification on their compounds, such as nutrients and bioactive molecules [26]. One of the major changes is in the pigment compounds which give color to the seaweed and, after the heat treatment, change the seaweed's color, mostly in brown seaweeds [12].

Fresh seaweeds are very perishable and start to deteriorate very rapidly after harvest, thus drying is an essential path to working with seaweeds as whole food or at food industry [16]. The most commonly used drying method is sun-drying, which reduces bulk handling and microbial growth, maintaining the characteristics if the seaweed is not directly exposed to sunlight (UV rays interfere greatly with the biochemical structure of photo-sensible compounds, such as phenolic compounds and pigments) [27]. However, when this drying method and other methods are used, the high temperatures can cause a rapid dehydration during the drying process, which can result in undesirable changes in the whole biomass (shrinkage) and chemical compound changes caused by enzymatic and non-enzymatic reactions (such as oxidation) [28]. Additionally, the water flux changes the composition, with partial destruction of the seaweed tissue, releasing various compounds, nutrients and minerals into the solution, decreasing the compounds' yield and nutritional value [29,30]. Only when considering the mineral part is the heat process important for diminishing the mineral intake to the recommended daily dosage [31].

It is well known that the bioavailability of vegetable compounds is conditioned by the species and abiotic factors (light, oxygen and temperature), and thus, their stability is essential for maintaining the bioavailability of the compounds and also changing the seaweed's nutritional profile. Heat treatment at low temperatures can also reduce the compounds availability in seaweed, as demonstrated by Gupta et al. [16], where *H. elongata* lost nearly 30% of the compounds compared with seaweeds subject to 25 °C during rehydration. Mainly, the bioactive compounds are lost in direct response to the rise in the heat, due to their chemical structures' heat sensitivity. On the other hand, the antioxidant power also demonstrated a direct response, where the heat reduced the bioactivity, mostly to denaturation of the bioactive molecules and reduction of the compounds to low-molecular-weight compounds by hydrolysis [10,32]. Thus, some heat processing methods have a different impact on antioxidant activity, as demonstrated by Rossi et al.: using blanching increased free radical scavenging in fruits [33].

The pigments also suffer chemical deterioration, and the sugars and polysaccharides are mainly solubilized, reducing the carbohydrates in the processed seaweed; the same happens to vitamins, amino acids (proteins) and minerals [34]. Although it is known that different heat treatments and temperatures significantly change the preservation of seaweed compounds, there is a general lack of knowledge about this complex topic using chemical characterization techniques, mainly spectroscopy and chromatography [10,12]. Even when similar seaweeds were analyzed, their reaction to the same heat treatment method was very different due to the variation in the seaweeds' compounds and biochemical profile. This was demonstrated by Amorim et al. [35], using *Laminaria* sp. and *Undaria pinnatifida*, where the same bioactive compounds and antioxidants were used. For example, the polyphenol compounds of *Laminaria* sp. were more stable throughout the heat treatment compared with the polyphenols of *U. pinnatifida* [35].

In the case of proteins, one of the most important nutrients of seaweeds, the bioavailability and content depend on the species analyzed. Maehre et al. demonstrated that heat treatment does not interfere in brown seaweeds; however, in red seaweeds, the protein level was more bioavailable after heat treatment [36].The iodine content is one of main risk factors of the seaweed overdosage [37], but the blanching technique (80 °C for 120 s) reduced the iodine content in *S. latissima* from 4605 to 293 mg iodine kg^{-1} dw^{-1} [38].

On the other hand, this technique enhanced the protein yield. Using blanching technique of 60 °C for 300 s enhanced the yield of PUFAs, demonstrating that heat treatment can have various interesting results when multiple variations of time and temperature are used. In the literature, the temperature varies from 30 °C to 110 °C, and the time varies from 60 s up to 3 h. Consequently, there are several dataset with different results, indicating various hypotheses regarding seaweed's reactions to heat treatment; however, the data demonstrate that there is no linear correlation among the techniques, temperatures, durations and biochemical relations of seaweed, because the methods and techniques between the studies are not similar [16,38].

7. The Effect of Heat on Seaweed-Based Products

Seafood, including seaweeds, fishes, mollusks and crustaceans are the biggest contributors of arsenic in our diets. This chemical is highly toxic for humans. Sartal et al. [39] studied the use of heat on four different species of seaweeds to see its influence on the quantity of arsenic. This study presented the following graphic, showing the analysis made of seaweeds, before and after heat treatment.

As seen in Figure 1, in all species, the amount of arsenic reduced when the seaweeds were subjected to heat treatment. The authors also analyzed the cooking water and, as shown, these waters contained large amounts of arsenic [39].

Another study was made to evaluate the proteins in seaweeds. The seaweeds used were *Palmaria palmata* (Rhodophyta) and *Alaria esculenta* (Phaeophyceae), and they were subjected to a heat process, namely boiling. In this study, the samples were also submitted to a gastrointestinal tract simulator to see if, during digestion, the measured proteins changed. In the first species, the results showed that there was an increase from 86% to 109% in the amino acid content. This means that by boiling this seaweed for 15–30 min, the amount of protein available was much greater then when it was raw. It has also been shown that, compared with meat protein, it supplies an adequate intake of protein for our body. If we examine the data on *A. esculenta*, the proteins in this species did not change, which means that in this seaweed, the process of heating did not influence its content [36].

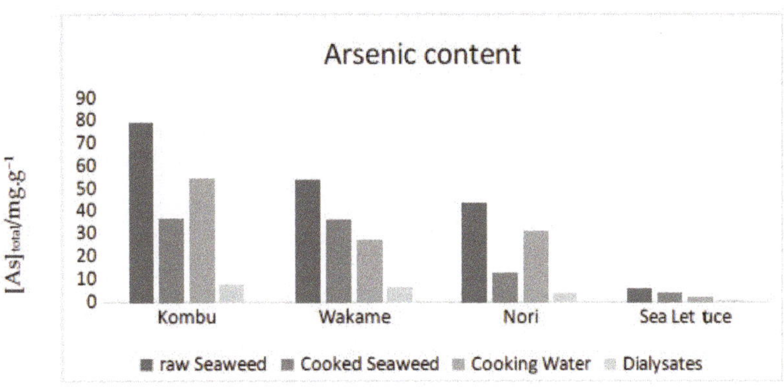

Figure 1. Total arsenic concentrations in kombu, wakame, nori and sea lettuce seaweed samples before and after being cooked, and in their cooking water and dialysates obtained after subjecting these cooked seaweeds to an in vitro gastrointestinal digestion [39].

In another study, Indonesian brown seaweeds were analyzed to see if heat treatment would affect four different components, these being fucoxanthin, antioxidant activity, total phenolic content and color stability; however, in this review, we will highlight the first and the second compounds [12]. These seaweeds were subjected to four different heat processes: boiling, blanching, steaming and sterilizing (Figure 2). Fucoxanthin is abundant in brown seaweeds and it has several health benefits. By putting the seaweeds through the four heat treatments, we obtained different results. In Figure 2, we can see these differences accordingly [12]. When raw/fresh, the amount of fucoxanthin in the seaweed was very low; however, when heated, these values increase. As we can see, the heat process that is most beneficial is blanching, followed by boiling, then steaming and sterilizing [12].

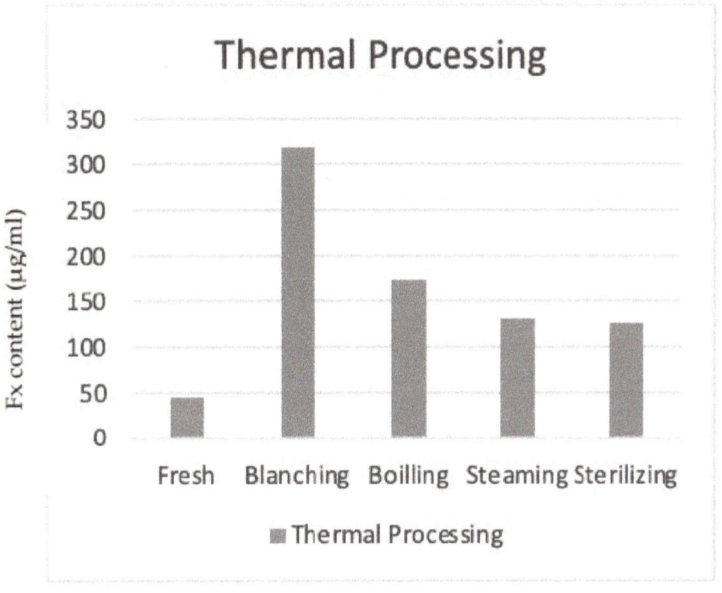

Figure 2. Fucoxanthin content in fresh and thermally processed in *S. ilicifolium*. Values are express in µg/mL [12].

Brown seaweeds contain phloroglucinol phenolics and phlorotannin, which have highly beneficial biological activities, such as anticancer, antioxidant and antidiabetic activities, and others. Therefore, finding ways to enhance these benefits is very important. In another study, the results showed that the total phenolic content decreased; however, they explained it by the high temperature that the algae were submitted to, showing that heat treatment can have benefits when applied correctly, but also have the contrary effect when applied incorrectly [12].

8. Problems of Heat Treatment Analyses and Approaches to Food Safety: A Future Road

The overall effects on the chemical structure of the compounds (mainly molecular weight, bioactivity and toxicity) when seaweed is heat-processed are not well known, due to general analysis of the compounds in their natural state and not in the hydrolysate stage (caused by heat) with low molecular weight. Moreover, this information about the kinetics of heat processed seaweed vs. stability may be essential to obtain valuable insights into the effects of the processes, providing higher food safety levels and maintaining a similar product with stable compounds which can be bioavailable for humans [3,31]. Susanto et al. demonstrated that three different ways of cooking brown seaweeds promoted different results. The blanching method promoted the stability of fucoxanthin, the boiling method produced higher antioxidant activity and steaming minimized the color change [12]. This supports the concept that chemical modification occurs during the heat process, which was demonstrated by the fucoxanthin content and the color after different types of heat process. Thus, it is important to have scientific data about the heat treatment process along with chemical characterization of the resulting product to obtain the best method of maintaining the important properties of seaweed; it is also important to find new methods of heat treatment that prevent the destruction of seaweeds' nutritional value. Furthermore, the performance of the heat treatment can make it hard to obtain purified compounds to be analyzed and characterized by a chemical method in a very rigorous manner [31]. As it is a complex compound with two distinct molecule fractions that are identified and characterized by different chemical methods, the possible combination of compounds or hydrolyzed molecules producing novel compounds can be achieved by the Maillard's reaction combining amino acids and reducing sugars, giving particular characteristics to food, such as color, texture and flavor [40].

Moreover, the presence harmful compounds, such toxins or heavy metals, in seaweeds needs to be carefully checked [41]. In this case, it was found that drying the seaweed and other types of processing (such as washing or cooking) reduced toxins and the concentration other volatile compounds (such as iodine or arsenic) in seaweeds, mainly solar drying, boiling and seaweed dehydration [42–44]. Despite these reports, there is a long road to understanding how thermal treatment affects the toxic concentration in seaweeds and, more importantly the toxic bioavailability and bioaccessibility [43]. At this moment, there is a need for more research to fully understand the effect of heat treatment on toxins and harmful compounds in seaweeds.

Above all, there is a question that remains before the scale-up of seaweed consumption, namely how the heat process affects the objectives, because the techniques have advantages and disadvantages scientifically. However, to scale up and define the best method of preparing seaweed using heat, it is also necessary to see if it is practicable at a large scale because the system/process costs cannot surpass the economic potential of the commercial method.

9. Conclusions

Even though seaweed has been consumed for hundreds of years, there is still a long way to go to reach the optimization of protocols which can maximize seaweeds' potential. Each species intended for processing has unique characteristics that vary according to environmental and/or internal factors, such as seasons, reproductive/lifecycle stage, temperature, nutrients and environmental competition, among others. On top of this, preprocessing and processing of seaweeds also show some degree of specificity, as different species can show different behaviors under different drying and rehydration processes. Overall, for specific compounds used for niche industries that target specific and highly purified phytochemicals, such as for the pharmaceutical or cosmetic industries, freeze-drying or vacuum-drying are good solutions, demonstrating relatively good yields in comparison with fresh samples, and a good degree of compound preservation. However, the application of this technology to food or feed production is not economically sustainable; thus, the most common type of drying method is sun-drying or oven-drying/convective air-drying, though the latter is also economically challenging. While some solutions are being developed, e.g., microwave drying, these also show limitations, such as high potency and cellular damage in this specific case. Additionally, understanding how toxic metabolites can be mobilized by heat treatment is paramount. Seaweeds are known bio-accumulators that can act as concentration hubs of toxic elements, of which the most important are iodine and arsenic species.

Author Contributions: Conception and design of the idea: M.C. and A.M.M.G.; organization of the team: A.M.M.G.; writing and bibliographic research: M.C., P.M., D.P. and J.C.; supervision and manuscript revision, L.P., J.C.M. and A.M.M.G. All authors have read and agreed to the published version of the manuscript.

Funding: This work was financed by national funds through the FCT—Foundation for Science and Technology, I.P., within the scope of the projects UIDB/04292/2020 granted to MARE—Marine and Environmental Sciences Centre and UIDP/50017/2020 + UIDB/50017/2020 (by FCT/MTCES) granted to CESAM—Centre for Environmental and Marine Studies. This research was co-financed by the project MENU—Marine Macroalgae: Alternative recipes for a daily nutritional diet (FA_05_2017_011), funded by the Blue Fund under Public Notice No. 5—Blue Biotechnology. João Cotas thanks the European Regional Development Fund through the Interreg Atlantic Area Program, under the project NASPA (EAPA_451/2016). Diana Pacheco thanks PTDC/BIA-CBI/31144/2017—POCI-01 project-0145-FEDER-031144—MARINE INVADERS, co-financed by the ERDF through POCI (Operational Program Competitiveness and Internationalization) and by the Foundation for Science and Technology (FCT, IP). Ana M. M. Gonçalves acknowledges the University of Coimbra for the contract IT057-18-7253.

Institutional Review Board Statement: Not applicable.

Informed Consent Statement: Not applicable.

Data Availability Statement: Not applicable.

Conflicts of Interest: The authors declare no conflict of interest.

References

1. Geada, P.; Moreira, C.; Silva, M.; Nunes, R.; Madureira, L.; Rocha, C.M.R.; Pereira, R.N.; Vicente, A.A.; Teixeira, J.A. Algal Proteins: Production Strategies and Nutritional and Functional Properties. *Bioresour. Technol.* **2021**, *332*, 125125. [CrossRef]
2. Dopelt, K.; Radon, P.; Davidovitch, N. Environmental Effects of the Livestock Industry: The Relationship between Knowledge, Attitudes, and Behavior among Students in Israel. *Int. J. Environ. Res. Public Health* **2019**, *16*, 1359. [CrossRef]
3. Leandro, A.; Pacheco, D.; Cotas, J.; Marques, J.C.; Pereira, L.; Gonçalves, A.M.M. Seaweed's Bioactive Candidate Compounds to Food Industry and Global Food Security. *Life* **2020**, *10*, 140. [CrossRef] [PubMed]
4. Parodi, A.; Leip, A.; Slegers, P.M.; Ziegler, F.; Herrero, M.; Tuomisto, H.; Valin, H.; Commission, E.; Scientific, C.; Lucia, S.; et al. Future Foods: Towards a Sustainable and Healthy Diet for a Growing Population. *Nat. Sustain.* **2018**, *1*, 782–789. [CrossRef]
5. Klnc, B.; Cirik, S.; Turan, G.; Tekogul, H.; Koru, E. Seaweeds for Food and Industrial Applications. In *Food Industry*; InTech: London, UK,, 2013.
6. Shannon, E.; Abu-Ghannam, N. Seaweeds as Nutraceuticals for Health and Nutrition. *Phycologia* **2019**, *58*, 563–577. [CrossRef]
7. Ferrara, L. Seaweeds: A Food for Our Future. *J. Food Chem. Nanotechnol.* **2020**, *6*, 56–64. [CrossRef]
8. Alba, K.; Kontogiorgos, V. Seaweed Polysaccharides (Agar, Alginate Carrageenan). *Encycl. Food Chem.* **2018**, 240–250. [CrossRef]
9. Regal, A.L.; Alves, V.; Gomes, R.; Matos, J.; Bandarra, N.M.; Afonso, C.; Cardoso, C. Drying Process, Storage Conditions, and Time Alter the Biochemical Composition and Bioactivity of the Anti-Greenhouse Seaweed Asparagopsis taxiformis. *Eur. Food Res. Technol.* **2020**, *246*, 781–793. [CrossRef]

10. Cox, S.; Gupta, S.; Abu-Ghannam, N. Effect of Different Rehydration Temperatures on the Moisture, Content of Phenolic Compounds, Antioxidant Capacity and Textural Properties of Edible Irish Brown Seaweed. *LWT* **2012**, *47*, 300–307, ISBN 3531402757.

11. Badmus, U.O.; Taggart, M.A.; Boyd, K.G. The Effect of Different Drying Methods on Certain Nutritionally Important Chemical Constituents in Edible Brown Seaweeds. *J. Appl. Phycol.* **2019**, *31*, 3883–3897. [CrossRef]

12. Susanto, E.; Fahmi, A.S.; Agustini, T.W.; Rosyadi, S.; Wardani, A.D. Effects of Different Heat Processing on Fucoxanthin, Antioxidant Activity and Colour of Indonesian Brown Seaweeds. *IOP Conf. Ser. Earth Environ. Sci.* **2017**, *55*, 012063. [CrossRef]

13. Uribe, E.; Vega-Gálvez, A.; García, V.; Pastén, A.; Rodríguez, K.; López, J.; Scala, K. Di Evaluation of Physicochemical Composition and Bioactivity of a Red Seaweed (Pyropia orbicularis) as Affected by Different Drying Technologies. *Dry. Technol.* **2020**, *38*, 1218–1230. [CrossRef]

14. Kane, S.N.; Mishra, A.; Dutta, A.K. Preface: International Conference on Recent Trends in Physics (ICRTP 2016). *J. Phys. Conf. Ser.* **2016**, *755*. [CrossRef]

15. Sánchez-García, F.; Mirzayeva, A.; Roldán, A.; Castro, R.; Palacios, V.; G Barroso, C.; Durán-Guerrero, E. Effect of Different Cooking Methods on Sea Lettuce (Ulva rigida) Volatile Compounds and Sensory Properties. *J. Sci. Food Agric.* **2021**, *101*, 970–980. [CrossRef]

16. Gupta, S.; Cox, S.; Abu-Ghannam, N. Effect of Different Drying Temperatures on the Moisture and Phytochemical Constituents of Edible Irish Brown Seaweed. *LWT - Food Sci. Technol.* **2011**, *44*, 1266–1272. [CrossRef]

17. Dang, T.T.; Van Vuong, Q.; Schreider, M.J.; Bowyer, M.C.; Altena, I.A.V.; Scarlett, C.J. The Effects of Drying on Physico-Chemical Properties and Antioxidant Capacity of the Brown Alga (Hormosira banksii (Turner) Decaisne). *J. Food Process. Preserv.* **2017**, *41*, 1–11. [CrossRef]

18. Silva, A.F.R.; Abreu, H.; Silva, A.M.S.; Cardoso, S.M. Effect of Oven-Drying on the Recovery of Valuable Compounds from Ulva rigida, Gracilaria sp. and Fucus vesiculosus. *Mar. Drugs* **2019**, *17*, 90. [CrossRef]

19. Lage-Yusty, M.A.; Alvarado, G.; Ferraces-Casais, P.; López-Hernández, J. Modification of Bioactive Compounds in Dried Edible Seaweeds. *Int. J. Food Sci. Technol.* **2014**, *49*, 298–304. [CrossRef]

20. Sappati, P.K.; Nayak, B.; VanWalsum, G.P.; Mulrey, O.T. Combined Effects of Seasonal Variation and Drying Methods on the Physicochemical Properties and Antioxidant Activity of Sugar Kelp (Saccharina latissima). *J. Appl. Phycol.* **2019**, *31*, 1311–1332. [CrossRef]

21. Gallagher, J.A.; Turner, L.B.; Adams, J.M.M.; Barrento, S.; Dyer, P.W.; Theodorou, M.K. Species Variation in the Effects of Dewatering Treatment on Macroalgae. *J. Appl. Phycol.* **2018**, *30*, 2305–2316. [CrossRef] [PubMed]

22. Nurshahida, M.S.F.; Aini, M.A.N.; Faizal, W.I.W.M.; Hamimi, I.A.; Nazikussabah, Z. Effect of Drying Methods on Nutrient Composition and Physicochemical Properties of Malaysian Seaweeds. *AIP Conf. Proc.* **2018**, *2030*, 1–5. [CrossRef]

23. Marín B, E.; Lemus M, R.; Flores M, V.; Vega G, A. La Rehidratación De Alimentos Deshidratados. *Rev. Chil. Nutr.* **2006**, *33*. [CrossRef]

24. Lee, K.T.; Farid, M.; Nguang, S.K. The Mathematical Modelling of the Rehydration Characteristics of Fruits. *J. Food Eng.* **2006**, *72*, 16–23. [CrossRef]

25. Amin, I.; Norazaidah, Y.; Hainida, K.I.E. Antioxidant Activity and Phenolic Content of Raw and Blanched Amaranthus Species. *Food Chem.* **2006**, *94*, 47–52. [CrossRef]

26. Shahidi, F. Nutraceuticals and Functional Foods: Whole versus Processed Foods. *Trends Food Sci. Technol.* **2009**, *20*, 376–387. [CrossRef]

27. Mota, C.L.; Luciano, C.; Dias, A.; Barroca, M.J.; Guiné, R.P.F. Convective Drying of Onion: Kinetics and Nutritional Evaluation. *Food Bioprod. Process.* **2010**, *88*, 115–123. [CrossRef]

28. Maskan, M. Drying, Shrinkage and Rehydration Characteristics of Kiwifruits during Hot Air and Microwave Drying. *J. Food Eng.* **2001**, *48*, 177–182. [CrossRef]

29. Krokida, M.K.; Karathanos, V.T.; Maroulis, Z.B. Compression Analysis Of Dehydrated Agricultural Products. *Dry. Technol.* **2000**, *18*, 395–408. [CrossRef]

30. Lewicki, P.P.; Jakubczyk, E. Effect of Hot Air Temperature on Mechanical Properties of Dried Apples. *J. Food Eng.* **2004**, *64*, 307–314. [CrossRef]

31. García-Sartal, C.; del CarmenBarciela-Alonso, M.; Moreda-Piñeiro, A.; Bermejo-Barrera, P. Study of Cooking on the Bioavailability of As, Co, Cr, Cu, Fe, Ni, Se and Zn from Edible Seaweed. *Microchem. J.* **2013**, *108*, 92–99. [CrossRef]

32. Olivera, D.F.; Viña, S.Z.; Marani, C.M.; Ferreyra, R.M.; Mugridge, A.; Chaves, A.R.; Mascheroni, R.H. Effect of Blanching on the Quality of Brussels Sprouts (Brassica oleracea L. Gemmifera DC) after Frozen Storage. *J. Food Eng.* **2008**, *84*, 148–155. [CrossRef]

33. Rossi, M.; Giussani, E.; Morelli, R.; Lo Scalzo, R.; Nani, R.C.; Torreggiani, D. Effect of Fruit Blanching on Phenolics and Radical Scavenging Activity of Highbush Blueberry Juice. *Food Res. Int.* **2003**, *36*, 999–1005. [CrossRef]

34. García-Pascual, P.; Sanjuán, N.; Melis, R.; Mulet, A. Morchella esculenta (Morel) Rehydration Process Modelling. *J. Food Eng.* **2006**, *72*, 346–353. [CrossRef]

35. Amorim, K.; Lage-Yusty, M.-A.; López-Hernández, J. Changes in Bioactive Compounds Content and Antioxidant Activity of Seaweed after Cooking Processing. *CyTA-J. Food* **2012**, *10*, 321–324. [CrossRef]

36. Maehre, H.K.; Edvinsen, G.K.; Eilertsen, K.-E.; Elvevoll, E.O. Heat Treatment Increases the Protein Bioaccessibility in the Red Seaweed Dulse (Palmaria palmata), but Not in the Brown Seaweed Winged Kelp (Alaria esculenta). *J. Appl. Phycol.* **2016**, *28*, 581–590. [CrossRef]
37. Milinovic, J.; Rodrigues, C.; Diniz, M.; Noronha, J.P. Determination of Total Iodine Content in Edible Seaweeds: Application of Inductively Coupled Plasma-Atomic Emission Spectroscopy. *Algal Res.* **2021**, *53*, 102149. [CrossRef]
38. Nielsen, C.W.; Holdt, S.L.; Sloth, J.J.; Marinho, G.S.; Sæther, M.; Funderud, J.; Rustad, T. Reducing the High Iodine Content of Saccharina latissima and Improving the Profile of Other Valuable Compounds by Water Blanching. *Foods* **2020**, *9*, 569. [CrossRef] [PubMed]
39. García Sartal, C.; del CarmenBarciela-Alonso, M.; Bermejo-Barrera, P. Effect of the Cooking Procedure on the Arsenic Speciation in the Bioavailable (Dialyzable) Fraction from Seaweed. *Microchem. J.* **2012**, *105*, 65–71. [CrossRef]
40. Rajauria, G.; Jaiswal, A.K.; Abu-Ghannam, N.; Gupta, S. Effect of Hydrothermal Processing on Colour, Antioxidant and Free Radical Scavenging Capacities of Edible Irish Brown Seaweeds. *Food Sci. Technol.* **2010**, *45*, 2485–2493. [CrossRef]
41. Moreira Leite, B.S. Novas Alternativas Para o Uso de Macroalgas Da Costa Portuguesa Em Alimentação. Master's Thesis, Universidade Nova de Lisboa, Lisboa, Portugal, 2017.
42. Vucko, M.J.; Magnusson, M.; Kinley, R.D.; Villart, C.; de Nys, R. The Effects of Processing on the in Vitro Antimethanogenic Capacity and Concentration of Secondary Metabolites of Asparagopsis taxiformis. *J. Appl. Phycol.* **2017**, *29*, 1577–1586. [CrossRef]
43. Banach, J.L.; Hoek-van den Hil, E.F.; Fels-Klerx, H.J. Food Safety Hazards in the European Seaweed Chain. *Compr. Rev. Food Sci. Food Saf.* **2020**, *19*, 332–364. [CrossRef]
44. Nitschke, U.; Stengel, D.B. Quantification of Iodine Loss in Edible Irish Seaweeds during Processing. *J. Appl. Phycol.* **2016**, *28*, 3527–3533. [CrossRef]

applied sciences

Article

Extraction of Bioactive Compounds from *Ulva lactuca*

Sofia Pappou [1,2], Maria Myrto Dardavila [1,*], Maria G. Savvidou [1], Vasiliki Louli [1], Kostis Magoulas [1] and Epaminondas Voutsas [1]

[1] Laboratory of Thermodynamics and Transport Phenomena, Zografou Campus, School of Chemical Engineering, National Technical University of Athens, 9 Iroon Polytechniou Str., 15780 Athens, Greece; mard18005@marine.aegean.gr (S.P.); msavvid@central.ntua.gr (M.G.S.); svlouli@chemeng.ntua.gr (V.L.); mag@chemeng.ntua.gr (K.M.); evoutsas@chemeng.ntua.gr (E.V.)

[2] Department of Marine Sciences, University of Aegean, University Hill, Lesvos Island, 81100 Mytilene, Greece

* Correspondence: mirtodar@mail.ntua.gr

Abstract: Macroalgae *Ulva lactuca*, has been employed as a natural source for the production of extracts with potent bioactivity. The biochemical characterization showed that the macroalgae biomass contains a remarkable amount of the polysaccharide Ulvan (49.9 wt%) which is a valuable chemical compound well known for its benefits in human health. Four nontoxic solvents, water, ethyl acetate, ethanol, and an ethanol/water mixture (70:30 *v/v*) were examined for their recovery efficiency of total carotenoid and phenolic contents. Experimental results showed that the aqueous mixture of ethanol was the most efficient solvent in the recovery of bioactive compounds with extraction yield of 10–15% dw. The effect of extraction parameters, namely time, temperature, and the ratio of biomass to solvent, on the carotenoid and phenolic compounds' content, antioxidant activity, and extraction yield, was investigated, using the ethanol/water mixture as a solvent. The extract obtained under 60 °C, 3 h of extraction time and 1:10 biomass to solvent mass ratio showed the highest antioxidant activity. This extract maintained its antioxidant capacity almost stable for five days of storage under cool and dark conditions. Finally, specific phenolic and carotenoid compounds in the *U. lactuca* extracts were identified using the High-Performance Liquid Chromatography (HPLC) technique.

Keywords: *U. lactuca*; extraction; non-toxic solvents; carotenoids; phenolics; antioxidant activity

Citation: Pappou, S.; Dardavila, M.M.; Savvidou, M.G.; Louli, V.; Magoulas, K.; Voutsas, E. Extraction of Bioactive Compounds from *Ulva lactuca*. *Appl. Sci.* **2022**, *12*, 2117. https://doi.org/10.3390/app12042117

Academic Editors: Emanuel Vamanu and Alessandra Durazzo

Received: 17 November 2021
Accepted: 10 February 2022
Published: 17 February 2022

Publisher's Note: MDPI stays neutral with regard to jurisdictional claims in published maps and institutional affiliations.

1. Introduction

Macroalgae (seaweeds), are macroscopic algae with high growth rate that can be cultivated with minimal use of freshwater at infrastructures installed on non-arable land. However, their cultivation cost is considerably lower compared to microalgae [1]. Macroalgae have great potential for several industrial applications [2] and therefore the interest in algal biotechnology has been increasing in the past few decades, particularly, on the exploitation of various macroalgae species for the production of a vast variety of products in the food, cosmetic and pharmaceutical industry. Consequently, during the last few decades, world macroalgae production increased at a rate of 8.7% per year [2].

Ulva lactuca is a widespread macroalgae growing along the Mediterranean coast which belongs to the phylum *Chlorophyta*, commonly known as "Sea lettuce" [3]. *U. lactuca*, contains commercially valuable chemical compounds that can be exploited in cosmetic, pharmaceutical, chemical, food, and energy applications. It has been reported that up to 60% carbohydrates, 10–47% proteins, 1–3% lipids, and 7–38% mineral ash are contained in *U. lactuca* [4]. Bioactive compounds of major industrial importance found in *U. lactuca* are phenolics, pigments (chlorophylls and carotenoids), and polysaccharides [5].

Phenolic compounds of macroalgae origin have been proven to confer antiallergic, skin antiaging, and whitening properties to cosmetic products [6]. Moreover, the consumption of an adequate level of polyphenols, such as myricetin, morin, and quercetin found in *U. lactuca* [5], could result in the prevention of diseases such as obesity, metabolic syndrome, Alzheimer's, or cancer [7].

Carotenoids present in *U. lactuca*, such as astaxanthin, β-carotene, fucoxanthin, and lutein, have been proven to have anti-inflammatory, antiaging, antioxidant, and other activities. They are frequently added in vitamin supplements and cosmetics, while they are used as natural food dyes and fish and poultry feed additives [6,8]. Carotenoids possess anti-inflammatory properties due to their ability to neutralize free radicals, creating a chemical protection against the proliferation of cancer cells [7].

Ulva spp. contain a significant amount of polysaccharides, varying from 15 to 65% of the total dry mass. These polysaccharides include ulvans, sulphated polysaccharides with rhamnose, uronic acids, and xylose as major components, as well as glucans including starch [4]. Ulvan polysaccharide and its oligosaccharides have anti-viral, antioxidant, anti-tumour, anti-coagulant, antihyperlipidemic, hepato protective, immuno-stimulating, anti-depressant, and anti-anxiolytic activities [4]. Consequently, Ulvan has industrial applications in the chemical, biomedical, and agricultural sectors [4].

Regarding the outcome of the bioactive compounds' extraction process, different variables are important, such as solvent type, solvent to solid ratio, temperature, and extraction time. These operational conditions require individual or combined study in order to maximize yields, the extraction rate, and the bioactive potency of the extracts [9,10].

The extraction yield of the targeted bioactive compounds is solvent dependent and different solvents can be used according to the polarity and location of those compounds. An aqueous solvent is suitable for the separation of polysaccharides, while phenolics and carotenoids are usually extracted with organic solvents [10,11]. In most cases, the extracts obtained with organic solvents appear more bioactive, while it has been confirmed that extracts derived from polar solvents have higher antibacterial activity [11].

When the bioactive extracts are targeted for cosmetics, pharmaceutical, and food industries, the range of the most convenient solvents is limited. The employed solvents, apart from being efficient in separating different compounds from their natural sources, must also be non-toxic [12]. Ethanol, water, and their mixtures are ideal solvents for the production of extracts with high antioxidant capacity, while they are generally recognized as safe (GRAS) and have a lower environmental impact in comparison to other solvents [13].

Regarding *U. lactuca*, experimental studies have been focused on the extraction and characterization of polysaccharides (ulvan) and proteins [3,14–21], while some others have been focused on the study of *U. lactuca* extracts' content in antioxidant chemical compounds (phenolics, carotenoids, tocopherols, etc.) and of their antioxidant capacity [8,22,23].

In this study, a complete biomass characterization of *U. lactuca* (total lipids, proteins, polysaccharides, and mineral ash contents) was performed. Moreover, the effectiveness of four nontoxic solvents (water, ethanol, ethyl acetate, and ethanol/water mixture 70:30 *v/v*) on the total carotenoids and total phenolics content, was studied. The ethanol/water mixture was determined as the most efficient among the four for extracting the targeted bioactive compounds from *U. lactuca*, thus it was employed for further investigation. The aim of the present study is the acquisition of *U. lactuca* extracts rich in carotenoid and phenolic compounds that exhibit potent antioxidant activity, appropriate for use in a variety of cosmetics, pharmaceutical, and food products. To that end, using the ethanolic solvent, the effects of extraction parameters, namely time, temperature, and the ratio of biomass to solvent on the carotenoid and phenolic compounds content, antioxidant activity, and extraction yield, were investigated in order to identify the optimum extraction conditions. To date, there has not been documented a study for *U. lactuca* targeting the investigation of the individual and combined effects of important extraction parameters regarding the carotenoid and phenolic content, as well as the antioxidant activity.

2. Materials and Methods

2.1. Chemicals and Reagents

The chemical reagents used in the present study were all of analytical grade and are presented in the Appendix A.

2.2. Macroalgal Strain

U. lactuca seaweed was purchased in July 2020, from Lusalgae Ltd., Figueira da Foz, Portugal. According to the supplier, the cultivated biomass was washed several times with both filtered and tap water, and after the excess water removal, it was dried on a ventilated drying oven at 60 °C for 48 h. It was then milled on a commercial mill and vacuum packed. The biomass was stored in our laboratory facilities under dry and dark conditions, avoiding direct contact with sunlight.

2.3. Methods

2.3.1. Biochemical Characterization

Lipid Content

Lipid extraction was conducted based on the Folch method [24]. Particularly, 0.5 g of seaweed were extracted with 15 mL of chloroform/methanol (1:2) mixture, in an ultrasonic bath (Elma TRANSONIC DIGITAL) for 4 min and left overnight at room temperature in the dark. Then, the sample was centrifuged (HERMLE Z 206 A) at 3000 rpm for 10 min. The supernatant was transferred into a separating funnel where the chloroform phase was collected in a pre-weighed flask. The chloroform phase was then concentrated in a rotary evaporator (Hei-VAP Advantage HL G3, Heidolph, Germany) at 45 °C and 150 mbar to recover the lipids. Total lipids were gravimetrically determined as wt% of the biomass.

Protein Content

For the determination of the protein content, the extraction procedure described by Kazir et al., 2019 [25] was followed. *U. lactuca* biomass was stirred overnight at 4 °C with distilled water 5% *w/v*. The supernatant was separated using a centrifuge at 5000 rpm for 15 min. Mercaptethanol (0.5% *v/v*) was added to the pellet, pH was adjusted to 12.0 with the addition of 1 M NaOH solution and left for 2 h under stirring at room temperature. The supernatant was separated again via centrifugation. The supernatants from both centrifugation cycles were mixed and the pH was adjusted to 7.0 using 1 M HCl solution. Solid ammonium sulfate was added to the supernatant up to 85% saturation. The solution was kept for 1 h at room temperature and then centrifuged in order to separate the precipitate. The sediment was then washed several times with deionized water, oven-dried (at 100 °C overnight) and measured in order to calculate the protein content gravimetrically as wt% of the biomass.

Polysaccharide Content

The polysaccharide content was determined based on the method of Mao et al., 2006 [26]. According to this method, 100 mL of distilled water was added to five grams of previously defatted *U. lactuca* biomass and the extraction was carried out at 80–90 °C under continuous stirring for 2 h. The supernatant was collected after centrifugation at 3000 rpm for 10 min in a pre-weighed flask. The supernatant was then concentrated using a rotary evaporator at 50 °C and 74 mbar and precipitated with 4 volumes of ethanol. Centrifugation and concentration were repeated. The extracted polysaccharide was washed with distilled water, freeze-dried, and weighed.

U. lactuca's polysaccharide content was gravimetrically determined as wt% of the biomass.

Determination of Mineral Ash

Ash content was determined gravimetrically. Specifically, 2.5 g of *U. lactuca* biomass were weighed in a crucible and combusted for 3 h at 550 °C using a muffle furnace (Thermolyne 47900). Water and other volatile materials are vaporized, and organic substances are burned in the presence of oxygen to CO_2, H_2O, and N_2. After the combustion completion, the crucible was placed in a desiccator and weighed soon after cooling.

The mineral ash content of *U. lactuca* macroalgae was gravimetrically determined as wt% of the biomass.

2.3.2. Solvent Extractions

Pure ethanol, water, and ethyl acetate, as well as a mixture of ethanol/water 70:30 (*v/v*) were investigated for their efficiency in the extraction of carotenoids and phenolics from *U. lactuca* biomass.

For the solvent screening, preliminary extractions were conducted. Specifically, 100 mL of each solvent was added to 10 g biomass and the extractions were carried out under continuous magnetic stirring in double-walled glass vessels for 12 h. A thermostat was employed to control the extraction temperature which was set at 25 °C.

Using the same experimental setup, the effect of some of the most influential extraction parameters, namely time, temperature, and the ratio of biomass to solvent, was investigated using the ethanol/water mixture as a solvent. The extraction conditions are presented in detail in Table 1. After the completion of each extraction, the extract was collected via centrifugation at 4000 rpm for 15 min. Subsequently, total phenolics, total carotenoids, antioxidant activity of the extracts, and the extraction yield were measured.

Table 1. Extraction conditions using ethanol/water (70:30 *v/v*) as solvent.

Extract Abbreviation	Extraction Duration (Hours)	Temperature (°C)	Biomass to Solvent Ratio (*w/v*)
E1	6	25	1:10
E2	10	25	1:10
E3	12	25	1:10
E4	12	25	1:20
E5	12	25	1:40
E6	16	25	1:10
E7	16	25	1:20
E8	16	25	1:40
E9	24	25	1:10
E10	12	40	1:10
E11	12	40	1:40
E12	3	60	1:10
E13	6	60	1:10
E14	12	60	1:10

Determination of Total Phenolics Content (TPC)

Total phenolics content in *U. lactuca* extracts was estimated spectrophotometrically using Folin-Ciocalteu reagent as described by Singleton et al., 1965 [27]. Specifically, 7.9 mL of distilled water and 0.1 mL of extract were transferred in glass vials and the mixture was homogenized by vortexing. Afterwards, the addition of 0.5 mL of Folin-Ciocalteu reagent and homogenization were realized, followed by the addition of 1.5 mL (20% *w/v*) Na_2CO_3 solution. The final mixture was incubated for 30 min in a water bath at 40 °C and its absorbance was subsequently measured at 765 nm using a SHIMADZU UV-1900, UV-VIS spectrophotometer and compared to a gallic acid calibration curve. The measurement was conducted in triplicate.

Determination of Total Carotenoids Content (TCC)

The estimation of total carotenoids content was carried out spectrophotometrically according to Association of Official Analytical Collaboration (AOAC) methods [28]. After each extraction, the absorbance of 3 mL of extract was measured at 450 nm and total carotenoids content was calculated according to Equation (1) acquired by the standard *β*-carotene calibration curve:

$$TCC = (6.9691\,Abs\,450\,nm - 0.1286) \tag{1}$$

Determination of Extraction Yield

After each extraction, the collected extract was employed in order to determine the extraction yield achieved. Particularly, the extracts were weighed, placed in a round bottom flask and the ethanol/water 70:30 *v/v* solvent was evaporated using a rotary evaporator at 45 °C and 100 mbar. Consequently, the remaining residue was dried at 100 °C for 1 h and the obtained dry extract was weighed. The extraction yield was gravimetrically determined as wt% of the biomass.

Determination of Antioxidant Activity (IC50)

Antioxidant activity was assessed using the 2,2-Diphenyl-1-Picrylhydrazyl (DPPH) assay [29], which is a simple, well-established method, and among the most frequently used in the literature. Briefly, 100 µL of *U. lactuca* extracts were added to a 3 mL methanolic solution of DPPH (0.03% *w/v*). Following on, the absorbance of the mixture was recorded at 515 nm after incubation for 20 min at room temperature. The IC50 values reported in the present study denote the concentration of sample that is required to scavenge 50% of DPPH free radicals (Sharma and Bhat, 2009). All the measurements were performed in triplicate.

HPLC Analysis

Carotenoid and phenolic compounds were identified using an HPLC system composed of a Jasco LG 1580-04 gradient unit equipped with a Jasco PU 1580 Intelligent HPLC pump and a Rheodyne 20 µL loop injection valve, connected in series with an SPD M20A diode array detector. The detection of the carotenoid compounds was realized using a YMC C30 reversed-phase column (5 µm, 250 mm × 4.6 mm i.d.), while a ThermoFisher C18 reversed-phase column (5 µm, 250 mm × 4.6 mm i.d.) was employed for the detection of the phenolic compounds. The columns' temperature was maintained at 30 °C and the flow rate of the eluents was adjusted at 1 mL min^{-1}. Prior to injection, the extracts were filtered using a syringe filter PTFE/L 25 mm, 0.45 µm (Membrane Solutions).

For the carotenoids profile study, the mobile phase consisted of methanol (solvent A) and MTBE (solvent B) mixture. An A:B linear gradient was applied starting from 95:5, changing to 70:30 in 30 min, then to 50:50 in 20 min [30]. The detection of carotenoids was performed at a wavelength of 480 nm. The identification of the carotenoids peaks was realized using external standards of all-trans neoxanthin, all-trans violaxanthin, all-trans astaxanthin, all-trans lutein, and *9-cis* astaxanthin. Comparison with HPLC literature data was performed when no external standards were available.

In the case of the phenolics profile study, water containing 0.1% acetic acid (solvent A) and acetonitrile (solvent B) were used as the mobile phase [31]. For the first 5 min, an isocratic elution of 90% of A and 10% of B was applied, followed by an A:B linear gradient changing to 84:16 in 18 min, then to 82:18 in 26 min, then to 72:28 in 31 min, then to 60:40 in 32 min. Subsequently, an isocratic elution of 60% A and 40% B until 40 min and a linear gradient of 90:10 in 43 min were applied [31]. External standards of gallic acid, caffeic acid, and catechin were used for the identification of the chromatograph's phenolics peaks, and their detection was recorded at 320 nm.

3. Results and Discussion

3.1. Biochemical Composition of U. lactuca

The biochemical composition of macroalgae *U. lactuca* tested in this work is shown in Table 2. Polysaccharides were the most abundant components of the biomass (49.9 wt%) and they were obtained in the form of a dark yellow colored gel. Also, *U. lactuca* exhibited a high ash content (27.7 wt%). Total lipids and total extracted proteins were calculated as 3.5% and 8.4% of the biomass, respectively.

Table 2. Biochemical composition of macroalgae *U. lactuca*.

Chemical Component	Content $g\,g^{-1}$ Biomass
Total lipids	0.035
Polysaccharides	0.499
Mineral ash	0.277
Total proteins	0.084
Other *	0.105

* Fibers and other carbohydrates.

The estimated contents of polysaccharides, lipids, and mineral ash are found close to the ones reported in the literature for *U. lactuca*. Particularly, values up to 60%, 3%, and 38% have been reported for these compounds respectively [4]. The protein content of the *U. lactuca* employed in the present work (8.4%), is considered among the lowest when compared to the corresponding values found in the literature (10–47%) which refer to several macroalgae [4]. Nevertheless, it should be mentioned that the biochemical composition of marine macroalgae varies significantly between species, growth conditions, and growth phase. Therefore, even when the biomass of the same species is examined, differences may occur because of the different environmental conditions of growth such as salinity, water temperature, depth, and pollution [32].

3.2. Extraction of Phenolics and Carotenoids—Selection of Solvent

Studies have confirmed that alcoholic solutions and/or hydrophilic solvent mixtures provide extracts with better antioxidant activity [32,33]. This can be attributed to the selective extraction of polar chemical substances, such as phenolic compounds, that exhibit significant antioxidant capacity [11]. In this study, four different solvents (water, ethanol, ethyl acetate, and a mixture of ethanol/water 70:30 *v/v*) were tested in order to evaluate their ability to recover the bioactive compounds of interest.

According to Figure 1, solvent selectivity was encountered both for carotenoids and phenolics. Water was the best solvent among the four for phenolic compounds and the ethanolic mixture for carotenoids extraction. Pure ethanol resulted in a lower solvation of phenolics, while carotenoids exhibited an extraction ability in between that of water and the ethanol/water mixture. Ethyl acetate exhibited the worst results in all bioactive compounds' extraction.

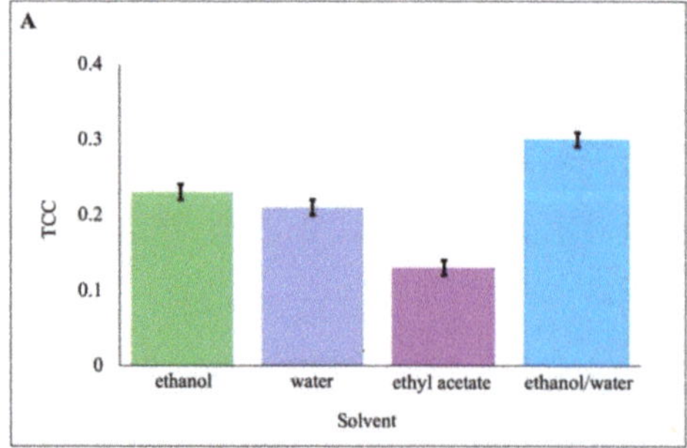

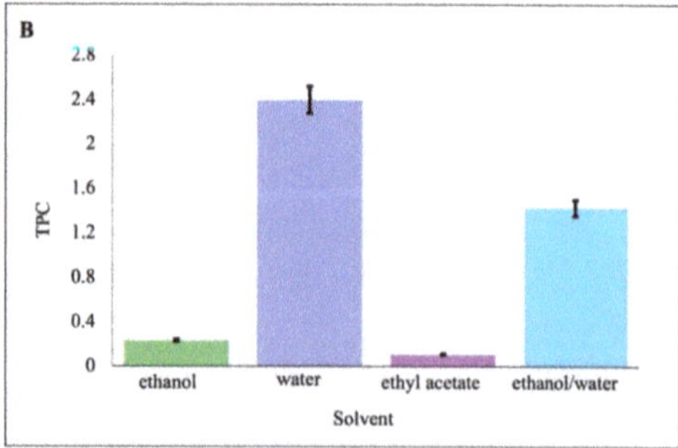

Figure 1. Total carotenoids (**A**) and phenolics (**B**) content of the *U. lactuca* extracts (mg g^{-1}) using different solvents (ethanol, water, ethyl acetate, and ethanol/water 70:30 *v/v*). Presented data are mean values of three replications ± standard deviation ($n = 3 \pm$ s.d).

According to the above results, it can be concluded that among the four tested solvents the most suitable ones, for the purpose of the present work, are water and the ethanol/water mixture. Nevertheless, the required energy for the recovery of the solvent used, and the acquisition of the dry extract is of great importance, and it should be accounted for the design of a feasible extraction process. Among the two most effective solvents, the ethanol/water 70:30 *v/v* mixture exhibits a lower enthalpy of vaporization (ΔH_v) [34]. Therefore, considering the most efficient recovery of all targeted bioactive compounds, environmental impact, GRAS characteristics of solvents, and extraction process feasibility, the ethanol: water 70:30 *v/v* mixture, would be the recommended solvent for the study of the extraction parameters.

3.3. Effect of Extraction Parameters on Carotenoid and Phenolic Content, Antioxidant Capacity and Extraction Yield

The effect of the operational conditions, i.e., time, temperature, and biomass to solvent ratio, using ethanol/water as solvent was examined. The obtained extracts are evaluated according to the total carotenoids content, total phenolic content, antioxidant activity, and extraction yield as shown below (Table 3).

Table 3. Experimental results from the extraction of *U. lactuca* using ethanol/water (70:30 *v/v*) as solvent.

Extract Abbreviation	TCC [a] $(\text{mg g}_{\text{biomass}}^{-1})$	TPC [b] $(\text{mg GAE g}_{\text{biomass}}^{-1})$	Extraction Yield [c] (%)	IC50 [d] $(\text{g}_{\text{biomass}}\,\text{mL}_{\text{solvent}}^{-1})$
E1	0.047	0.979	11.8	0.720
E2	0.044	0.954	12.9	0.390
E3	0.074	1.079	11.5	0.164
E4	0.058	0.979	12.8	0.350
E5	0.062	1.205	15.6	0.480
E6	0.061	1.004	11.7	0.500
E7	0.072	1.104	15.4	0.900
E8	0.082	1.330	15.0	1.200
E9	0.063	0.979	10.9	0.840
E10	0.122	1.230	10.8	0.308
E11	0.137	1.807	15.1	0.227
E12	0.086	1.506	9.8	0.128
E13	0.105	1.757	10.4	0.263
E14	0.118	1.857	12.0	0.291

[a] $\sigma_{\text{TCC}} = \pm 0.004$, [b] $\sigma_{\text{TPC}} = \pm 0.062$, [c] $\sigma_{\text{extraction yield}} = 0.629$, [d] $\sigma_{\text{IC50}} = 0.036$.

3.3.1. Effect of Extraction Time

The effect of extraction time (h) on total carotenoids and phenolics content, antioxidant activity, and extraction yield is presented in Figure 2 and Table 3. For the evaluation of the extraction time effect on each dependent variable (TCC, TPC, IC50, and Yield), two different series of extractions were performed, at 25 °C (6, 10, 12, 16 and 24 h) and 60 °C (3, 6, and 12 h), while solvent to biomass ratio was set at 1:10 *w/v*. The increase in temperature from 25 °C to 60 °C that led to the better recovery of bioactive compounds, resulted in a decreased extraction time of a maximum of 12 h at 60 °C, in order to avoid the use of extra energy. As shown in Figure 2, the most and least affected dependent variable by time is the antioxidant activity (IC50) of the extracts and the extraction yield, respectively. At 25 °C, the best antioxidant activity was measured at the extract obtained after 12 h (E3), while at 60 °C, 3 h were sufficient in order to obtain the best extract as far as antioxidant activity is concerned (E15). The extraction yield exhibits small variations with time at 25 °C, while at 60 °C showed an incremental trend. Total carotenoid and phenolic content at 60 °C exhibited an increase with time, while at 25 °C the variations of the recovered bioactive compounds over time are negligible.

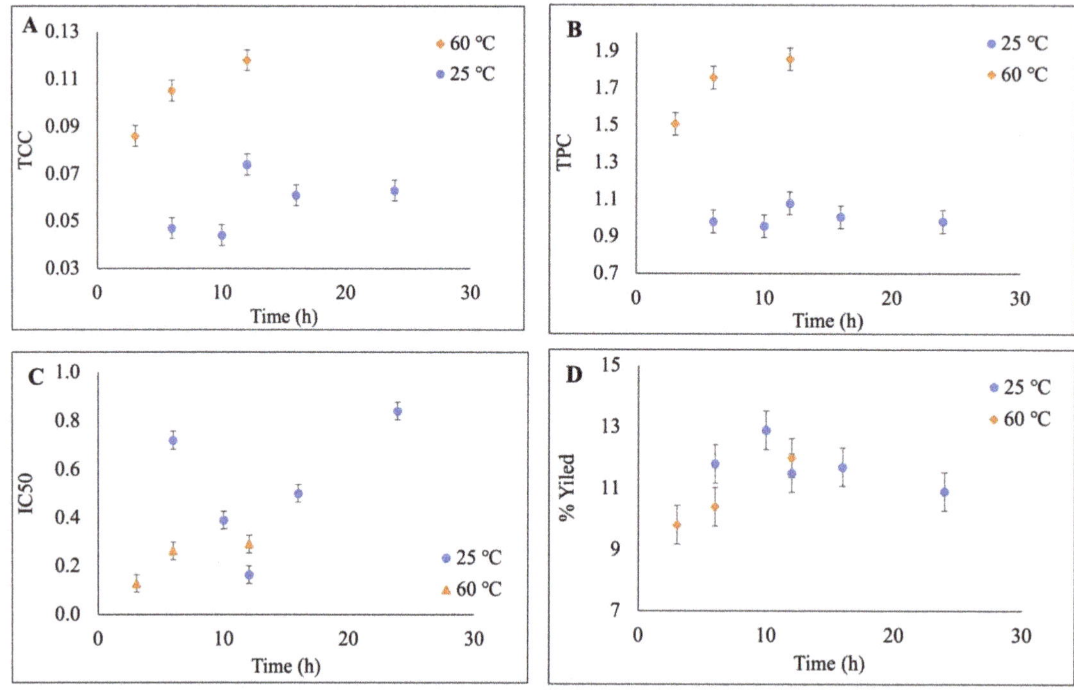

Figure 2. Effect of extraction duration (h) on carotenoids content (**A**), total phenolic content (**B**), antioxidant activity (**C**) and extraction yield (**D**) of *U. lactuca* extracted with ethanol/water 70:30 *v/v* at a biomass to solvent ratio 1:10 *w/v*, under 25 °C and 60 °C. Experimental data are the mean values of three replications.

According to Teramukai et al., 2020 [35], carotenoids' content increased at all temperatures that were examined for extraction during the first 12 h, a finding that agrees with the results of this study. Because of the presence of many conjugated double bonds, carotenoids are easily degraded at higher temperatures and long incubation time, although a higher temperature and longer extraction time can increase the extraction rate of carotenoids [35]. Therefore, an optimal temperature and extraction time should exist in order to achieve the most effective extract as far as antioxidant activity is concerned. Sachindra et al., 2005 [36] reported that the optimized temperature and time for astaxanthin extraction from shrimp waste were 70 °C and 2.5 h, respectively, while the astaxanthin content decreased after 2.5 h incubation at temperatures higher than 70 °C, supporting the findings of this work.

3.3.2. Effect of Biomass to Solvent Ratio

Figure 3 provides the results obtained for total carotenoids and phenolics, antioxidant activity, and extraction yield at different biomass to solvent ratios (r = 1:10, 1:20 and 1:40). Two different series of extractions were performed, for 12 and 16 h, while the temperature was set at 25 °C. Antioxidant activity was significantly affected by the biomass to solvent ratio, while the 1:10 *w/v* ratio appeared to be the optimum one among the three ratios tested for both sets of extraction time. The carotenoid recovery did not seem to be affected significantly by the alteration of the biomass to solvent ratio for the 12 h extraction, while it slightly increased for the 16 h one. Finally, the extracts' phenolic content and the extraction yield increase by increasing the biomass to solvent ratio. Topuz et al., 2016 [37] reported that the most affected dependent variable from the biomass to solvent ratio parameter was the total phenolic compounds content.

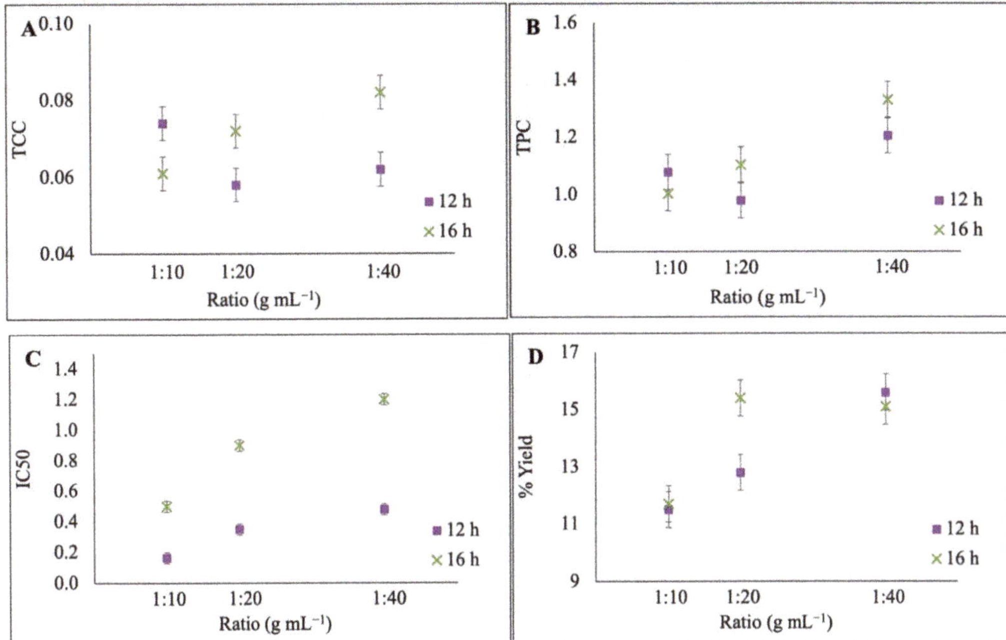

Figure 3. Effect of biomass to solvent ratio on carotenoids content (**A**), total phenolic content (**B**), antioxidant activity (**C**) and total extraction yield (**D**) of *U. lactuca* extracted with ethanol/water 70:30 *v/v* at 25 °C for 12 and 16 h extraction time. Experimental data are the mean values of three replications.

3.3.3. Effect of Temperature

The results obtained regarding total carotenoids and phenolics, antioxidant activity, and extraction yield at different temperatures (25, 40 and 60 °C), for 12 h extraction and biomass to solvent ratio 1:10 *w/v*, are presented in Figure 4. Total carotenoids were increased from 25 to 40 °C and then remained practically constant (Figure 4A), while total phenolic content showed a significant increase with temperature (Figure 4B). Results showed that temperature increase led to a negative effect on the antioxidant activity for the solvent ratio 1:10 *w/v* and for 12 h extraction time (Figure 4C). However, this negative effect of temperature on the antioxidant activity is not observed for other ratios and extraction times, e.g., E1–E13 and E5–E11. Furthermore, the increase in temperature has a negligible effect on the overall extraction yield (Figure 4D). Generally, plant and algal cells are surrounded by a rigid cell wall that acts as a mechanical barrier to extract carotenoids, while the strong interaction between carotenoids and other macromolecules prevents their mass transfer during the extraction [37]. Therefore, in order to break down the cell wall, the extraction occasionally involves treatments, such as heating [37]. According to Strati et al., 2010 [38], the improved extractability of carotenoids as temperature increases, is possibly related to the destruction of the cellular structure. On the other hand, as the extraction temperature increases, several factors can affect the extracted carotenoids, such as degradation and isomerization from all-trans to *cis* isomers [38]. Thus, the extraction conditions of carotenoids should be optimized, since some are more unstable than others, depending on their chemical structure. Mäki-Arvela et al., 2014 [39] suggested carotenoid extraction temperatures lower than 70 °C, in order to minimize their degradation. Moreover, under higher extraction temperatures (40–50 °C), a significant increase in the total phenolic content was observed [37]. In like manner, it can be supported by the results presented in the current study (Figures 3 and 4), that under higher temperature and a higher biomass to solvent ratio, the solubility of the seaweed's phenolic compounds is enhanced. Interestingly, some studies have shown that an increase in temperature favors the extraction process by enhancing both the solubility and the diffusion coefficient of the solutes; nevertheless, the phenolic compounds can be denatured beyond a certain temperature value (>50 °C) [40]. Consequently, the elevated carotenoid and phenolic recovery observed with the increase in the extraction temperature in the present study, does not necessarily result in a respective enhancement of antioxidant activity.

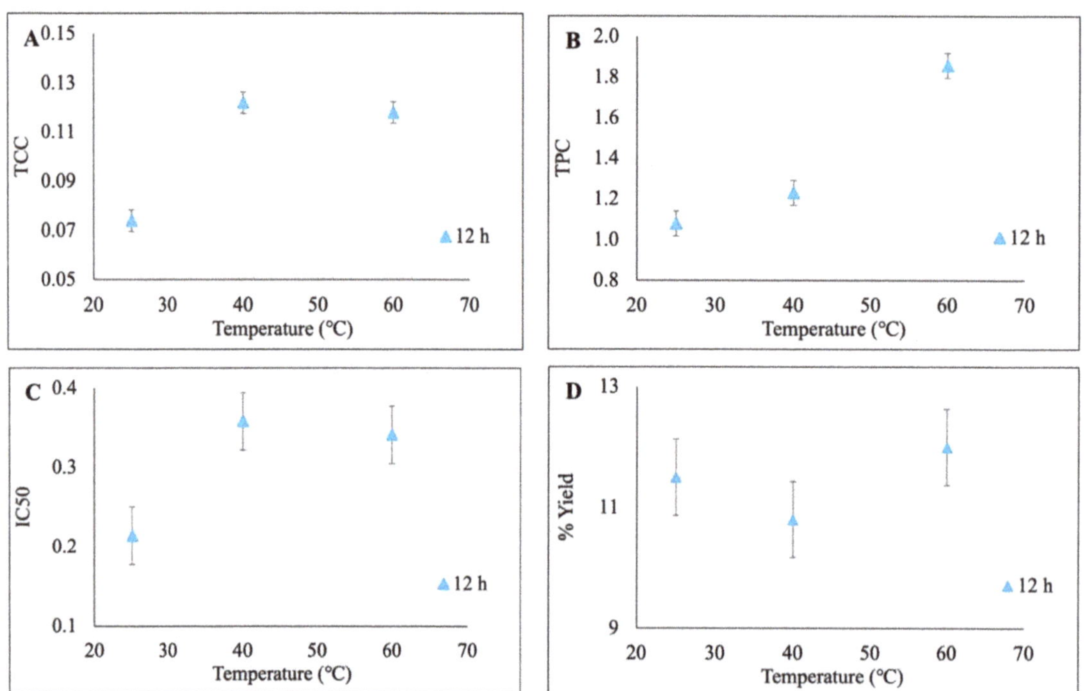

Figure 4. Effect of different extraction temperatures on carotenoids content (**A**), total phenolic content (**B**), antioxidant activity (**C**) and total extraction yield (**D**) of *U. lactuca* extracted with ethanol/water 70:30 *v/v* at a biomass to solvent ratio 1:10 *w/v* and for 12 h extraction time. Experimental data are the mean values of three replications.

3.3.4. Overall Evaluation of the Extraction Parameters Effect

The total carotenoids' content did not fluctuate significantly for the different biomass to solvent ratios applied, while extraction time and temperature showed a slightly higher impact. Particularly, the increase in temperature to 60 °C favored the carotenoids recovery, and the highest carotenoid content was obtained from the 12 h extraction under 60 °C. The total phenolic content increased under increased temperature and biomass to solvent ratio. The effect of extraction time was less significant at 25 °C and more intense at 60 °C, whereas the 12 h extraction was the optimum one for both temperatures. A straight comparison regarding the antioxidant capacity of seaweed extracts between different studies is difficult due to biological variation of the raw material and extraction methods; however, previous studies have shown that phenolic compounds are the main contributors to the antioxidant activity of various seaweeds [37]. In this study, the antioxidant activity was the most influenced factor by all the studied parameters, as depicted in Figures 2–4. Therefore, with regards to the total bioactive potency of the extracts, the best biomass to solvent ratio was the 1:10 *w/v*, while the extraction times of 12 h and 3 h were the two most favorable under

25 °C and 60 °C, respectively. The stability of carotenoids is one of the crucial parameters in the consideration of developing commercial products from algae extract. Some carotenoids are thermally and chemically labile; therefore, the process parameters should be optimized in order to produce high-quality products, stable against exposure to light and heat for a prolonged time [39].

The extraction efficiency expressed as the extraction yield showed a slight increase with the biomass to solvent ratio increase, while no particular effect was found from the parameters of extraction time and extraction temperature. Such findings were also mentioned by the work of Ummat et al., 2020 [41], where no statistical differences were observed within each extraction parameter on the extraction yield (%), while the recovery of the individual bioactive compounds was altered according to the parameter extractions such as duration and temperature. In the same study, the yields obtained from conventional extraction of several species of seaweed were in the range of 10.5–19.3%, while in the current study *U. lactuca* extraction yield ranged between 10% and 15%, depending on the extraction parameters.

With the main objective of this work being the optimum antioxidant activity, the extracts E3 and E12 are proposed as the optimum ones. The antioxidant activity of these two extracts was equal; nevertheless, the extract obtained under 60 °C, at 3 h (E12), exhibited higher recovery of total carotenoid and phenolic compounds, and thus it is evaluated as the best one.

3.3.5. Study of the Extracts' Antioxidant Activity over Time

The evaluation of the antioxidant activity of the best *U. lactuca* extract (E12) over time was studied at two levels. First, the extract's scavenging activity was examined versus time when in contact with free radicals. Second, the period during which the extract preserves its antioxidant activity when kept in storage was determined.

The antioxidant activity of the E12 extract was measured exactly after 20 min, 30 min, 60 min, and 90 min of DPPH incubation to determine how rapidly the antioxidant compounds act. As shown in Figure 5A, the scavenging activity of the extract exhibited a linear decrease throughout the 90 min of study. The highest rate of DPPH decay occurred within the first 20 min of incubation, denoting a rapid reaction of the mixture of antioxidant species present in the extract with the free radicals. According to Savatovic et al., 2012 [42], the expression of the results in terms of a kinetic approach not only considers the activity of the antioxidants but also provides information on how quickly they act, which is probably the result of the different kinetic behavior that the various antioxidants species present in the extract exhibit.

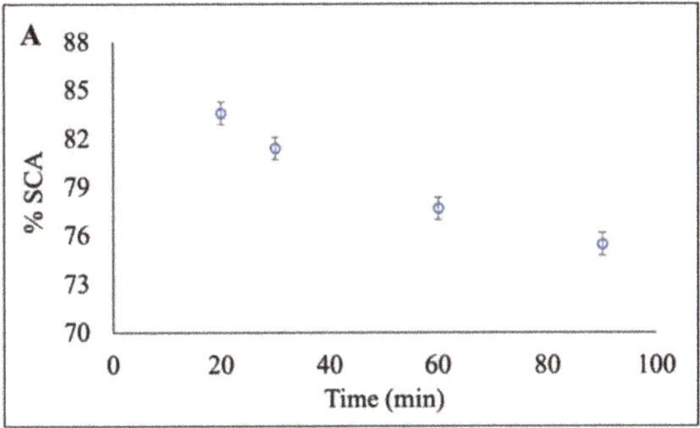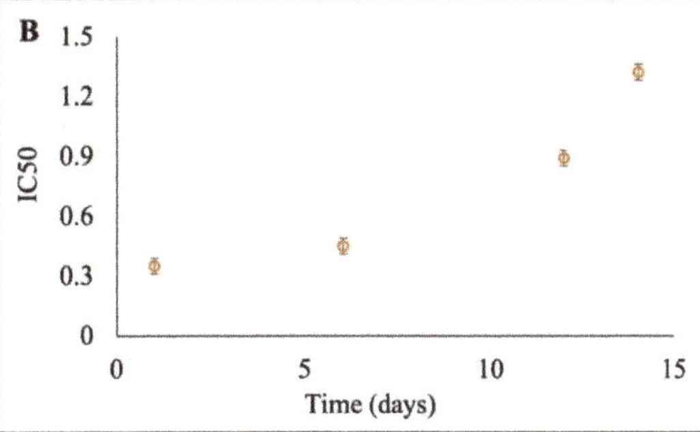

Figure 5. Percent scavenging activity (% SCA) of *U. lactuca* extract E12 versus time (min) (**A**) and antioxidant activity (IC50) of *U. lactuca* extract E12 when stored at 4 °C in the dark for 15 days (**B**). Experimental data are mean values of three replications.

Additionally, in order to test the effect of the storage time on the antioxidant activity, the optimum extract's IC50 was measured in frequent intervals for 15 days. When stored at 4 °C in the dark to minimize the degradation of carotenoids [39], only a slight decrease in the extract's antioxidant capacity was evident for the first 5 days of storage, followed by a drastic decrease thereon, as depicted in Figure 5B.

3.4. Carotenoids Profile

Figure 6 presents the HPLC chromatogram of *U. lactuca* extract E12, in which seven carotenoids were identified. According to the data presented by Fernandes et al., 2020 [30] whose HPLC protocol for carotenoids separation and detection was followed by the present work, peaks 2 and 7 correspond to 9-cis neoxanthin and 9-cis lutein, respectively. All the other peaks were identified by using external standards. According to these standards, peak 1 corresponds to *all-trans*-neoxanthin, peak 3 to *all-trans*-violaxanthin, peak 4 to *all-trans*-astaxanthin, peak 5 to *all-trans*-lutein, and peak 6 to *9-cis*-astaxanthin. In general, the carotenoids reported for *Ulva* spp. in the published literature are lutein, violaxanthin, fucoxanthin, zeaxanthin, neoxanthin, astaxanthin, α- and β-carotene and their isomers, which are in accordance with the carotenoids commonly found in Chlorophyta [21,43].

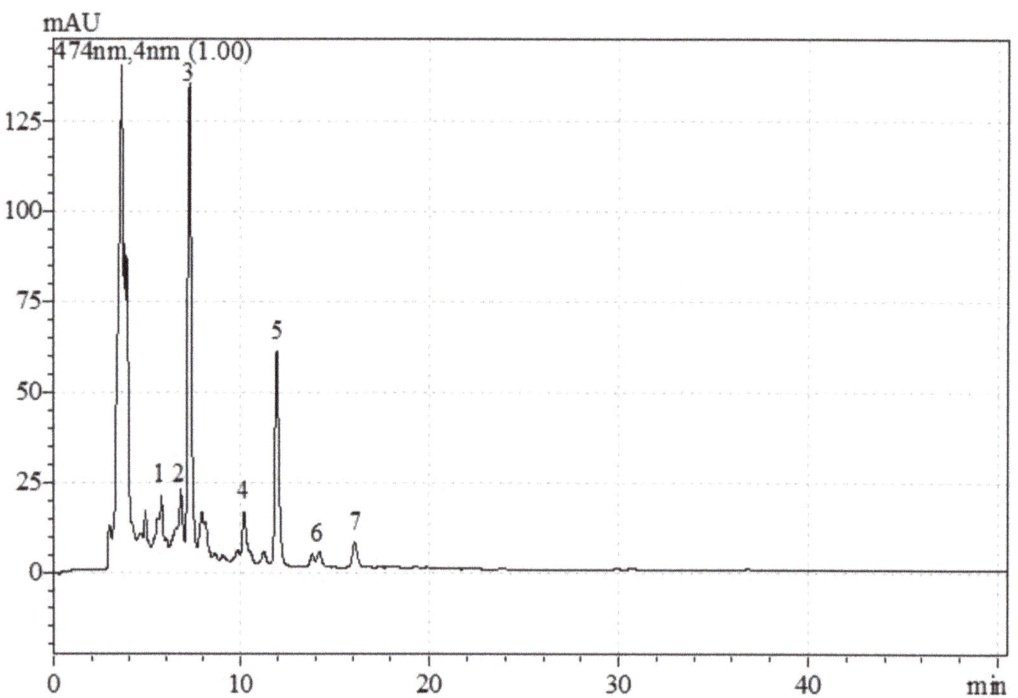

Figure 6. HPLC chromatogram of the E12 *U. lactuca* extract. Peak 1: *all-trans*-neoxanthin; Peak 2: *9-cis*-neoxanthin; Peak 3: *all-trans*-violaxanthin; Peak 4: *all-trans*-astaxanthin; Peak 5: *all-trans*-lutein; Peak 6: *9-cis* astaxanthin; Peak 7: *9-cis*-lutein.

In conclusion, the extract that exhibited the most potent antioxidant activity was found to contain important carotenoid compounds [43]. Particularly, *all-trans*-violaxanthin possesses anti-inflammatory effects and *all-trans*-lutein can help in the prevention of cataracts, coronary syndromes, strokes, and retinitis, as well as in the treatment of macular disintegration [44].

3.5. Phenolics Profile

Several different phenolic compounds are present in most macroalgae species. As shown in Figure 7, the HPLC analysis of the *U. lactuca* E12 extract resulted in the identification of four phenolic compounds: gallic acid (Peak 1), caffeic acid (Peak 2), catechin (Peak 3), and rutin (Peak 4).

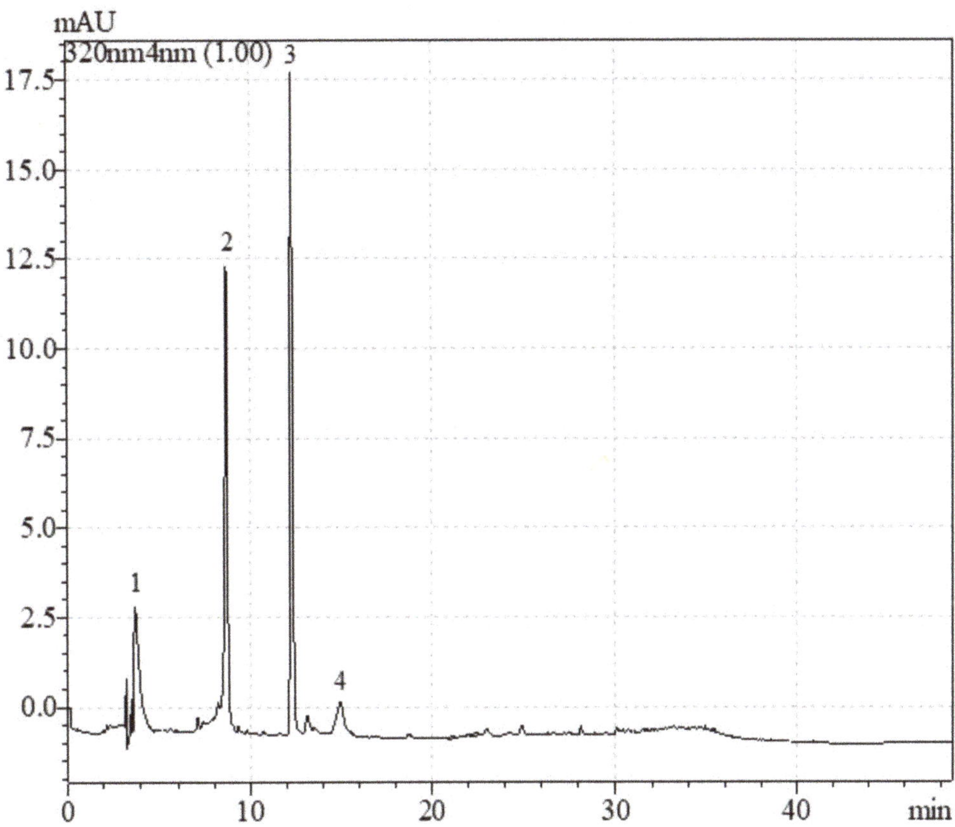

Figure 7. HPLC chromatogram of the E12 *U. lactuca* extract. Peaks: 1 gallic acid, 2 caffeic acid, 3 catechin, 4 rutin.

Gallic acid is one of the most abundant phenolic acids and is frequently used as a standard in the quantification of extracts' total phenolic content, as in the present work. The presence of gallic acid has been previously reported in methanol, ethanol, and aqueous *U. lactuca* extracts [45,46]. Caffeic acid is among the hydroxycinnamic acids that have been found in *U. lactuca* extract obtained using water as a solvent [46]. Catechin is a well-known flavonoid abundant in brown algae [47], and also present in methanolic extracts of macroalgae *U. fasciata* and *U. lactuca* [46,48] according to the literature. Rutin is another flavonoid whose presence in *U. lactuca* extracts obtained from different methanolic solvent systems has been reported [47].

These compounds, being powerful antioxidants, exhibit further diverse bioactivities such as antiviral, antimicrobial, antitumoral, anti-inflammatory, etc., which are important for human well-being [7].

4. Conclusions

The *U. lactuca* biochemical characterization showed that the predominant component of the employed macroalgae was polysaccharide (ulvan) which accounts for 49.9 wt% of the total dry algal biomass. The corresponding values of mineral ash, total proteins, and total lipids were 27.4 wt%, 8.4 wt% and 3.5 wt%, respectively. The screening of four non-toxic solvents namely, water, ethyl acetate, ethanol, and ethanol/water (70:30 *v/v*), showed that the aqueous ethanolic mixture was the most efficient in the recovery of carotenoids and phenolics from *U. lactuca* via the classic solid–liquid extraction method. The study of the effect of the extraction parameters, i.e., temperature, time, and biomass to solvent ratio, on the carotenoid and phenolic compounds recovery, antioxidant activity, and extraction yield, resulted in the determination of the most efficient parameter values. Particularly, the *U. lactuca* extract that was obtained under 60 °C, 3 h extraction time and using a biomass to solvent ratio equal to 1:10, exhibited the highest antioxidant activity and recovery of the targeted bioactive compounds. Carotenoid and phenolic compounds were identified via HPLC analysis. According to the findings of the present work, the antioxidant activity was the most influenced factor by all the studied parameters. The best extract maintained its antioxidant capacity almost stable for five days under storage in cool and dark conditions. Considering the ease and effectiveness of the extraction process, the nontoxic nature of the ethanol/water solvent, the good antioxidant activity of the *U. lactuca* extracts, and their content of high added value compounds, it can be suggested that the obtained extracts could be exploited as possible additives/constituents in cosmetic, pharmaceutical and food products. In that context, the present study serves as a useful steppingstone for further investigations on the optimization and the upscaling of the whole process.

5. Highlights

- *U. lactuca* is a rich source of bioactive compounds and dietary supplements
- High polysaccharides content (49.9 wt%) was found
- Ethanol/water 70:30 *v/v* was the most efficient solvent. The optimum extraction parameters were 60 °C, biomass to solvent ratio 1:10 and 3 h extraction time
- The best extract maintained its high antioxidant capacity for 5 days

Author Contributions: S.P. methodology, formal analysis, investigation, writing—original draft; M.M.D. methodology, data curation, formal analysis, validation, writing—original draft; M.G.S. methodology, data curation, formal analysis, visualization, validation, writing—review and editing. V.L. data curation, writing—review and editing; K.M. supervision; E.V. conceptualization, resources, writing—review and editing, supervision. All authors have read and agreed to the published version of the manuscript.

Funding: This research has been co-financed by the European Union and Greek national funds through the Operational Program Competitiveness, Entrepreneurship, and Innovation, under the call, "Special Actions, Aquaculture-Industrial Materials-Open Innovation in Culture", MIS: 5045790.

Institutional Review Board Statement: Not applicable.

Informed Consent Statement: Not applicable.

Data Availability Statement: Not applicable.

Conflicts of Interest: No potential conflict of interest was reported by the authors.

Ethics Statement: No human or animal rights are applicable to this study.

Appendix A

Table A1. List of chemical reagents.

Chemical Reagents	Supplier	Purity/Concentration
2,2-Diphenyl-1-picrylhydrazyl	Alfa Aesar	95%
Folin-Ciocalteu reagent	Carlo Erba reagents	Special grade
Methanol	Fisher Scientific	≥99.8%
Chloroform	Fisher Scientific	Analytical reagent grade
Ethanol	Fisher Scientific	≥99.8%
Ethyl acetate	Merck	HPLC grade
MTBE	Fisher Scientific	≥99.5%
2-Mercaptoethanol	Sigma-Aldrich	≥98%
Water	Fisher Scientific	HPLC grade
β-carotene	Alfa Aesar	99%
all-trans Astaxanthin	Acros Organics	≥98%
all-trans Lutein	Extrasynthese	≥95%
all-trans Neoxanthin	Supelco	Analytical standard
all-trans Violaxanthin	Sigma-Aldrich	≥90.0%
9-cis Astaxanthin	Alga Technologies	Natural CO_2 extract of Haematococcus Pluvialis microalgae
Gallic acid	Acros Organics	98%
Caffeic acid	Acros Organics	99%+
Catechin	Sigma-Aldrich	≥98%

References

1. Chemodanov, A.; Robin, A.; Jinjikhashvily, G.; Yitzhak, D.; Liberzon, A.; Israel, A.; Golberg, A. Feasibility study of *Ulva* sp. (Chlorophyta) intensive cultivation in a coastal area of the Eastern Mediterranean Sea. *Biofuel. Bioprod. Biorefin.* **2019**, *13*, 864–877. [CrossRef]
2. White, W.L.; Wilson, P. World seaweed utilization: A summary. *J. Aquac. Res. Dev.* **2015**, *6*, 7–25. [CrossRef]
3. Postma, P.R.; Cerezo-Chinarro, O.; Akkerman, R.J.; Olivieri, G.; Wijffels, R.H.; Brandenburg, W.A.; Eppink, M.H.M. Biorefinery of the macroalgae Ulva lactuca: Extraction of proteins and carbohydrates by mild disintegration. *J. Appl. Phycol.* **2018**, *30*, 1281–1293. [CrossRef] [PubMed]
4. Dominguez, H.; Loret, E. Ulva lactuca, a source of troubles and potential riches. *Mar. Drugs* **2019**, *17*, 357. [CrossRef]
5. Caf, F.; Yilmaz, Ö.; Durucan, F.; Özdemir, N.Ş. Biochemical components of three marine macroalgae (Padina pavonica, Ulva lactuca and Taonia atomaria) from the Levantine Sea coast of Antalya, Turkey. *J. Biodivers. Environ. Sci.* **2015**, *6*, 401–411.
6. Morais, T.; Cotas, J.; Pacheco, D.; Pereira, L. Seaweeds compounds: An Eco sustainable source of cosmetic ingredients? *Cosmetics* **2021**, *8*, 8. [CrossRef]
7. Jimenez-Lopez, C.; Pereira, A.G.; Lourenço-Lopes, C.; Garcia-Oliveira, P.; Cassani, L.; Fraga-Corral, M.; Prieto, M.A.; Simal-Gandara, J. Main bioactive phenolic compounds in marine algae and their mechanisms of action supporting potential health benefits. *Food Chem.* **2021**, *341*, 128262. [CrossRef]
8. Spolaore, P.; Joannis-Cassan, C.; Duran, E.; Isambert, A. Commercial applications of microalgae. *J. Biosci. Bioeng.* **2006**, *101*, 87–96. [CrossRef]
9. Domínguez, R.; Zhang, L.; Rocchetti, G.; Lucini, L.; Pateiro, M.; Munekata, P.; Lorenzo, J. Elderberry as potential source of antioxidants. Characterization, optimization of extraction parameters and bioactive properties. *Food Chem.* **2020**, *330*, 127266. [CrossRef]
10. Strati, I.F.; Oreopoulou, V. Process optimization for recovery of carotenoids from tomato waste. *Food Chem.* **2011**, *129*, 747–752. [CrossRef]
11. Suchinina, T.; Shestakova, T.; Petrichenko, V.; Novikova, V. Solvent polarity effect on the composition of biologically active substances, UV spectral characteristics, and antibacterial activity of Euphrasia brevipila herb extracts. *Pharm. Chem. J.* **2011**, *44*, 683–686. [CrossRef]
12. Torres-Valenzuela, L.S.; Ballesteros-Gómez, A.; Rubio, S. Green solvents for the extraction of high added-value compounds from agri-food waste. *Food Eng. Rev.* **2020**, *12*, 83–100. [CrossRef]
13. Monteiro, M.; Santos, R.A.; Iglesias, P.; Couto, A.; Serra, C.R.; Gouvinhas, I.; Barros, A.; Oliva-Teles, A.P.; Enes, P.; Díaz-Rosales, P. Effect of extraction method and solvent system on the phenolic content and antioxidant activity of selected macro-and microalgae extracts. *J. Appl. Psychol.* **2020**, *32*, 349–362. [CrossRef]
14. He, J.; Xu, Y.; Chen, H.; Sun, P. Extraction, structural characterization, and potential antioxidant activity of the polysaccharides from four seaweeds. *Int. J. Mol. Sci.* **2016**, *17*, 1988. [CrossRef] [PubMed]

15. Costa, C.; Alves Paula, A.; Pinto, R.; Sousa, R.A.; Borges da Silva, E.A.; Reis, R.L.; Rodrigues, A.E. Characterization of ulvan extracts to assess the effect of different steps in the extraction procedure. *Carbohydr. Polym.* **2012**, *88*, 537–546. [CrossRef]

16. Ortiz, J.; Romero, N.; Robert, P.; Araya, J.; Lopez-Hernandez, J.; Bozzo, C.; Navarrete, E.; Osorio, A.; Rios, A. Dietary fiber, amino acid, fatty acid and tocopherol contents of the edible seaweeds Ulva lactuca and Durvillaea antarctica. *Food Chem.* **2006**, *99*, 98–104. [CrossRef]

17. Yaich, H.; Garna, H.; Besbes, S.; Paquot, M.; Blecker, C.; Attia, H. Effect of extraction conditions on the yield and purity of ulvan extracted from Ulva lactuca. *Food Hydrocoll.* **2013**, *3*, 375–382. [CrossRef]

18. Yaich, H.; Garna, H.; Besbes, S.; Barthélemy, J.-P.; Paquot, M.; Blecker, C.; Attia, H. Impact of extraction procedures on the chemical, rheological and textural properties of ulvan from Ulva lactuca of Tunisia coast. *Food Hydrocoll.* **2014**, *40*, 53–63. [CrossRef]

19. Yaich, H.; Amir, A.B.; Abbes, F.; Bouaziz, M.; Besbes, S.; Richel, A.; Blecker, C.; Attia, H.; Garna, H. Effect of extraction procedures on structural, thermal and antioxidant properties of ulvan from Ulva lactuca collected in Monastir coast. *Int. J. Biol. Macromol.* **2017**, *105*, 1430–1439. [CrossRef]

20. Tian, H.; Yin, X.; Zeng, Q.; Zhu, L.; Chen, J. Isolation, structure, and surfactant properties of polysaccharides from Ulva lactuca L. from South China Sea. *Int. J. Biol. Macromol.* **2015**, *79*, 577–582. [CrossRef] [PubMed]

21. El-Baky, H.H.A.; El-Baz, F.K.; El-Baroty, G.S. Natural preservative ingredient from marine alga. *Int. J. Food Sci. Tech.* **2009**, *44*, 1688–1695. [CrossRef]

22. Abirami, R.G.; Kowsalya, S. Nutrient and Nutraceutical Potentials of Seaweed Biomass Ulva lactuca and Kappaphycus alvarezii. *J. Agric. Sci. Technol.* **2011**, *5*, 109–115.

23. Wang, T.; Jónsdóttir, R.; Ólafsdóttir, G. Total phenolic compounds, radical scavenging and metal chelation of extracts from Icelandic seaweeds. *Food Chem.* **2009**, *116*, 240–248. [CrossRef]

24. Folch, J.; Lees, M.; Stanley, G. A simple method for the isolation and purification of total lipids from animal tissues. *J. Biol. Chem.* **1957**, *226*, 497–509. [CrossRef]

25. Kazir, M.; Abuhassira, Y.; Robin, A.; Nahor, O.; Luo, J.; Israel, A.; Golberg, A.; Livney, Y. Extraction of proteins from two marine macroalgae, Ulva sp. and Gracilaria sp., for food application, and evaluating digestibility, amino acid composition and antioxidant properties of the protein concentrates. *Food Hydrocoll.* **2019**, *87*, 194–203. [CrossRef]

26. Mao, W.; Zang, X.; Li, Y.; Zhang, H. Sulfated polysaccharides from marine green algae Ulva conglobata and their anticoagulant activity. *J. Appl. Phycol.* **2006**, *18*, 9–14. [CrossRef]

27. Singleton, V.; Rossi, J. Colorimetry of total phenolics with phosphomolybdic-phosphotungstic acid reagents. *Am. J. Enol. Vitic.* **1965**, *16*, 144–1588.

28. Association of Official Analytical Chemists (AOAC). *Official Methods of Analysis of the Association of Official Analytical Chemists International*, 16th ed.; The Association of Official Analytical Chemists: Washington, DC, USA, 1995.

29. Brand-Williams, W.; Cuvelier, M.; Berset, C. Use of a free radical method to evaluate antioxidant activity. *LWT* **1995**, *28*, 25–30. [CrossRef]

30. Fernandes, A.; Petry, F.; Mercadante, A.; Jacob-Lopes, E.; Zepka, L. HPLC-PDA-MS/MS as a strategy to characterize and quantify natural pigments from microalgae. *Curr. Res. Nutr. Food Sci.* **2020**, *3*, 100–112. [CrossRef]

31. Benavente-García, O.; Castillo, J.; Lorente, J.; Ortuño, A.; Del Rio, J. Antioxidant activity of phenolics extracted from Olea europaea L. leaves. *Food Chem.* **2000**, *68*, 457–462. [CrossRef]

32. Bikker, P.; Van Krimpen, M.; Van Wikselaar, P.; Houweling-Tan, B.; Scaccia, N.; Van Hal, J.; Huijgen, W.; Cone, J.; López-Contreras, A. Biorefinery of the green seaweed Ulva lactuca to produce animal feed, chemicals and biofuels. *J. Appl. Phycol.* **2016**, *28*, 3511–3525. [CrossRef] [PubMed]

33. Altemimi, A.; Lakhssassi, N.; Baharlouei, A.; Watson, D.G.; Lightfoot, D.A. Phytochemicals: Extraction, isolation, and identification of bioactive compounds from plant extracts. *Plants* **2017**, *6*, 42. [CrossRef]

34. Newsham, D.M.T.; Mendez-Lecanda, E.J. Isobaric enthalpies of vaporization of water, methanol, ethanol, propan-2-ol and their mixtures. *J. Chem. Thermodyn.* **1982**, *14*, 291–301. [CrossRef]

35. Teramukai, K.; Kakui, S.; Beppu, F.; Hosokawa, M.; Miyashita, K. Effective extraction of carotenoids from brown seaweeds and vegetable leaves with edible oils. *Innov. Food Sci. Emerg. Technol.* **2020**, *60*, 102302. [CrossRef]

36. Sachindra, N.; Mahendrakar, N. Process optimization for extraction of carotenoids from shrimp waste with vegetable oils. *Bioresour. Technol.* **2005**, *96*, 1195–1200. [CrossRef]

37. Topuz, O.K.; Gokoglu, N.; Yerlikaya, P.; Ucak, I.; Gumus, B. Optimization of antioxidant activity and phenolic compound extraction conditions from red seaweed (Laurencia obtuse). *J. Aquat. Food Prod. Technol.* **2016**, *25*, 414–422. [CrossRef]

38. Strati, I.; Oreopoulou, V. Effect of extraction parameters on the carotenoid recovery from tomato waste. *Int. J. Food Sci. Technol.* **2010**, *46*, 23–29. [CrossRef]

39. Mäki-Arvela, P.; Hachemi, I.; Murzin, D. Comparative study of the extraction methods for recovery of carotenoids from algae: Extraction kinetics and effect of different extraction parameters. *J. Chem. Technol. Biotechnol.* **2014**, *89*, 1607–1626. [CrossRef]

40. Spigno, G.; Tramelli, L.; De Faveri, D. Effects of extraction time, temperature and solvent on concentration and antioxidant activity of grape marc phenolics. *J. Food Eng.* **2007**, *81*, 200–208. [CrossRef]

41. Ummat, V.; Tiwari, B.; Jaiswal, A.; Condon, K.; Garcia-Vaquero, M.; O'Doherty, J.; O'Donnell, C.; Rajauria, G. Optimisation of Ultrasound Frequency, Extraction Time and Solvent for the Recovery of Polyphenols, Phlorotannins and Associated Antioxidant Activity from Brown Seaweeds. *Mar. Drugs* **2020**, *18*, 250. [CrossRef]

42. Savatovic, S.; Cetkovic, G.; Canadanovic-Brunet, J.; Djilas, S. Kinetic behavior of DPPH radical scavenging activity of tomato waste extracts. *J. Serbian Chem. Soc.* **2012**, *77*, 1381–1389. [CrossRef]
43. Yalçın, S.; Karakaş, Ö.; Okudan, E.; Başkan, K.; Çekiç, S.; Apak, R. HPLC detection and antioxidant capacity determination of brown, red and green algal pigments in seaweed extracts. *J. Chromatogr. Sci.* **2020**, *59*, 325–337. [CrossRef]
44. Torregrosa-Crespo, J.; Montero, Z.; Fuentes, J.L.; García-Galbis, M.R.; Garbayo, I.; Vílchez, C.; Martínez-Espinosa, R.M. Exploring the Valuable Carotenoids for the Large-Scale Production by Marine Microorganisms. *Mar. Drugs* **2018**, *16*, 203. [CrossRef] [PubMed]
45. Aslan, E.; Aksu, A.; Korkmaz, N.E.; Taskin, O.S.; Caglar, N.B. Monitoring the antioxidant activities by extracting the polyphenolic contents of algae collected from the Bosphorus. *Mar. Pollut. Bull.* **2019**, *141*, 313–317. [CrossRef] [PubMed]
46. Farvin, K.H.S.; Jacobsen, C. Phenolic compounds and antioxidant activities of selected species of seaweeds from Danish coast. *Food Chem.* **2013**, *138*, 1670–1681. [CrossRef]
47. Mekinić, I.G.; Skroza, D.; Šimat, V.; Hamed, I.; Čagalj, M.; Perković, Z.P. Phenolic content of brown algae (Pheophyceae) species: Extraction, identification, and quantification. *Biomolecules* **2019**, *9*, 44–69.
48. Ibrahim, R.; Saber, A.; Hammad, H. The possible role of the seaweed Ulva fasciata on ameliorating hyperthyroidism-associated heart inflammations in a rat model. *Environ. Sci. Pollut. Res.* **2020**, *28*, 6830–6842. [CrossRef]

 *applied sciences*

Article

Tyrosol, at the Concentration Found in Maltese Extra Virgin Olive Oil, Induces HL-60 Differentiation towards the Monocyte lineage

Lucienne Gatt [1,2,*], David G. Saliba [2,3], Pierre Schembri-Wismayer [4] and Marion Zammit-Mangion [1,2]

[1] Department of Physiology and Biochemistry, Faculty of Medicine and Surgery, University of Malta, MSD 2080 Msida, Malta; marion.zammit-mangion@um.edu.mt

[2] Centre for Molecular Medicine and Biobanking, University of Malta, MSD 2080 Msida, Malta; david.saliba@um.edu.mt

[3] Department of Applied Biomedical Science, Faculty of Health Science, University of Malta, MSD 2080 Msida, Malta

[4] Department of Anatomy, Faculty of Medicine and Surgery, University of Malta, MSD 2080 Msida, Malta; pierre.schembri-wismayer@um.edu.mt

* Correspondence: lucienne.gatt@um.edu.mt

Abstract: Tyrosol is a phenolic found in extra virgin olive oil (EVOO). In a Maltese monocultivar EVOO, it was present at a concentration of 9.23 ppm. The HL-60 acute myeloid leukaemia cell line, which can be differentiated to both monocytes and neutrophils, was exposed to tyrosol at this concentration and analysed for evidence of differentiation and effects of cytotoxicity. The polyphenol induced a 1.93-fold increase in cellular oxidative activity (*p*-value 0.044) and enhanced surface expression of CD11b and CD14. This indicates that tyrosol induces monocytic-like differentiation. An RNA-seq analysis confirmed the upregulation of monocyte genes and the loss of neutrophil genes concomitant with the bi-potential promyelocyte precursor moving down the monocytic pathway. A cell cycle analysis showed an accumulation of cells in the Sub G_0/G_1 phase following tyrosol exposure for 5 days, which coincided with an increase in apoptotic and necrotic markers. This indicates differentiation followed by cell death, unlike the positive monocyte differentiation control PMA. This selective cytotoxic effect following differentiation indicates therapeutic potential against leukaemia.

Keywords: HL-60; phenols; differentiation; apoptosis; transcriptome

Citation: Gatt, L.; Saliba, D.G.; Schembri-Wismayer, P.; Zammit-Mangion, M. Tyrosol, at the Concentration Found in Maltese Extra Virgin Olive Oil, Induces HL-60 Differentiation towards the Monocyte lineage. *Appl. Sci.* **2021**, *11*, 10199. https://doi.org/10.3390/app112110199

Academic Editors: Emanuel Vamanu and Alessandra Durazzo

Received: 30 July 2021
Accepted: 24 October 2021
Published: 30 October 2021

Publisher's Note: MDPI stays neutral with regard to jurisdictional claims in published maps and institutional affiliations.

1. Introduction

Leukaemia is a myeloproliferative disorder characterised by uncontrolled proliferation of immature blasts in the bone marrow, eventually penetrating into the bloodstream, preventing normal haematopoiesis [1,2]. Such blasts suffer an obstruction during early development, referred to as a maturation arrest, and fail to reach functional maturity [3,4]. In acute promyelocytic leukaemia (APL), this arrest may be reversed by using all *trans* retinoic acid (ATRA), an approach called differentiation therapy, that has completely altered this leukaemia's prognosis [5,6]. It generally shows less toxic side effects than chemotherapy [7]. Hence, reproducing the success of differentiation therapy on different leukaemia types is a much sought-after goal.

The HL-60 cell line consists of immature cells arrested at the myeloblast-promyelocyte stage and is well-characterised in terms of the induction of terminal differentiation down either the monocytic or granulocytic pathway. Dimethyl sulfoxide (DMSO) and ATRA drive HL-60 to granulocytic differentiation, whilst 1,25-dihydroxy vitamin D3, sodium butyrate and phorbol esters drive them towards monocytes/macrophages [8]. Such maturation is accompanied by a reduction in cell size, the condensation of nuclear material, and changes in cell surface antigens such as CD11b and CD14 surface markers [8–10]. Gene expression analysis in differentiating HL-60 treated cells show an increase in genes associated with the oxidative burst [11,12], cytokine upregulation [13], adhesion, and trans-endothelial migration such ICAM adhesion molecules [12,14]. Following differentiation, an increase in the expression of apoptotic-related genes may follow as these cells naturally die, which is the end point of successful differentiation therapy [14,15].

Polyphenols are a group of highly diverse molecules found extra virgin olive oils (EVOOs) and other plant sources and are associated with chemo-preventive functions and differentiation-inducing properties [16]. They are secondary metabolites that appear to offer protection against solar irradiation as well as anti-microbial and anti-pathogenic activity. They are classified into phenolic acids (hydroxycinnamic and hydroxybenzoic acids), phenolic alcohols, lignans, stilbenes, secoiridoids, coumarins, xanthones, and flavonoids. The major components of the phenolic fraction in EVOOs are tyrosol, hydroxytyrosol, and their derivatives [17]. The former, tyrosol, is known to increase in concentration during olive oil storage [18]. The phenolic profiles of Maltese EVOOs have been characterized by Lia et al., 2019 and Gatt et al., 2021 [19,20].

Differentiation-induction in leukaemia has been attributed to oleuropein, apigenin-7-glucoside, hydroxytyrosol, the dialdehydic compound of elenoic acid linked to hydroxytyrosol, as well as pinoresinol, amongst EVOO-derived polyphenols [21–23]. Moreover, hydroxytyrosol was noted to bring about HL-60 cell proliferation arrest at the G_0/G_1 phase and upregulate the cyclin-dependent protein kinase inhibitors p21 WAF1/CIP1 and p27 KIP1 [24]. Transcriptomic studies in AML models following phenolic administration do not appear to be published.

This study investigated whether tyrosol could induce HL-60 differentiation using biochemical and morphological assays. The effects on the cell cycle, as well as the transcriptomic changes following exposure to tyrosol treatment, were then explored. Changes to the NFκB pathway and the Interferon Regulatory Factors network are identified.

2. Materials and Methods

2.1. Reagents

Dimethyl sulfoxide (DMSO), ethylenediaminetetraacetic acid (EDTA), fetal bovine serum (FBS), hexane, histopaque, hydrochloric acid (HCl), Leishman stain, methanol, nitroblue tetrazolium chloride (NBT), penicillin streptomycin solution (Pen Strep), phorbol-12-myristate 13-acetate (PMA), phosphate buffered saline (PBS), phytohaemagglutinin (PHA), Roswell park memorial institute (RPMI-1640) medium, thiazolyl blue tetrazolium bromide (MTT), trypan blue, and tyrosol were acquired from Sigma-Aldrich® (St Louis, MO, USA). The identification of tyrosol was confirmed by Innovhub Stazioni Sperimentali per l'industria, Milan using HPLC analysis. Absolute ethanol and potassium hydroxide (KOH) were obtained from Scharlau (China). Antibodies and isotypes were obtained from Becton Dickinson, Mountain View, CA, USA.

2.2. Cell Culture

HL-60 cells (Acute myeloid Leukaemia) were purchased from Cell Lines services (CLS) GMBH, Eppelheim, Germany, Cryovial no. 330209. The cells were cultured in Roswell Park Memorial Institute (RPMI-1640) medium supplemented in 10% foetal bovine serum (FBS) and 1% penicillin-streptomycin (pen-strep) at 37 °C, 5% CO_2, and 95% relative humidity.

2.3. Phenolic Extraction and Identification

Phenolics were extracted from a Maltese monocultivar EVOO variety using liquid-liquid extraction (LLE) in a 60:40 (v/v) methanol:water mixture [20]. The phenolic extract was then characterized using HPLC analysis through the use of phenolic standards (Supplementary File S1).

2.4. Induction of Differentiation

The phenolic extracts were tested for differentiating activity at a range of concentrations on HL-60 cells. Exponentially growing cells (50 μL) were seeded at a concentration of

1×10^5 cells/mL into separate wells of a 96-well plate and treated with the crude EVOO phenolic extract at 100 ppm, 10 ppm and 1 ppm. Within the 100ppm crude phenolic extract that was found to be effective at inducing differentiation (Supplementary Figure S1), tyrosol constituted a peak area of 9.23% or a concentration of 9.23 ppm. This concentration of the pure tyrosol purchased from the chemical suppliers was therefore used for all subsequent experiments outlined in this section. Exponentially growing cells (50 µL) were again seeded at a concentration of 1×10^5 cells/mL into separate wells of a 96-well plate and treated with 50 µL test reagent (9.23 ppm tyrosol), positive controls phorbol myristate acetate (PMA) at a concentration of 10 nM for monocytic differentiation, and 1.6% dimethyl sulfoxide (DMSO) for granulocytic differentiation, as well as 0.5% ethanol as the vehicle control. The negative control consisted of untreated cells in RPMI medium. All conditions were prepared in complete RPMI. All tests were carried out in triplicate and repeated as three separate experiments ($n = 9$).

2.5. Evaluation of Differentiation

Cell proliferation was assessed using the MTT assay, which is characterised by the reduction of yellow MTT to purple formazan catalysed by mitochondrial dehydrogenases [25]. After 3 and 5 days, 20 µL of a solution of 5 mg/mL MTT:PBS were added to each well and incubated for 4 h at 37 °C. The formazan crystals were then dissolved by the addition of 120 µL of DMSO, and the absorbance read at 570 nm [26]. The differentiation capacity was tested using the NBT assay. After 3 and 5 days, 100 µL of activated NBT (2 mg/mL NBT:PBS + 1% PMA) was added to each well, and the plate was incubated for 20 min at 37 °C and 5% CO_2. Next, 70 µL of 1M HCl followed by 50 µL of 2M KOH and 150 µL of DMSO were added, and the absorbance read at 630 nm, with 405 nm as the reference wavelength [27]. The index of differentiation was then evaluated by dividing NBT by MTT.

2.6. Assessment of the Specificity of the Anti-Proliferative Action

The effect of tyrosol on lymphocytes was tested to assess whether any anti-proliferative activity exhibited by tyrosol was specific and limited to leukaemia cells. Blood from healthy donors was obtained from the National Blood Transfusion Service (as approved by the University of Malta, Faculty of Medicine and Surgery Research Ethics Committee FRECMDS1718_073) and was processed to isolate mononuclear cells through density-gradient centrifugation using filtered histopaque. The cells were incubated overnight to allow adherence and thus separation of the monocyte fraction. The lymphocytes were collected, stimulated with 2% phytohaemagglutinin for lymphocyte proliferation [28,29], then incubated at a temperature of 37 °C and a 5% CO_2 concentration for 2 days. The MTT assay was repeated to test for proliferative activity.

2.7. Leishman's Stain for Visualization of Morphological Changes

The experimental steps described in Section 2.4 above were repeated. At days 3 and 5, cells were cytocentrifuged at $1000 \times g$ for 5 min using Cytospin 2 (Shandon), fixed by air-drying, and then stained using Leishman's stain and buffer for 15 min, rinsed, and once again dried. They were then viewed via the oil immersion technique using an inverted microscope (Motic, Barcelona, Spain, AE2000).

2.8. Evaluation of the Expression of Differentiation Specific Cell Markers by Flow Cytometry

HL-60 cells were assessed for differentiation by flow cytometry at the designated time points using the FACS Calibur™ flow cytometer. The antibodies selected were the PE-conjugated Mouse Anti-Human CD14 for monocytic differentiation and the Alexa Fluor® 488-conjugated Mouse Anti-Human CD11b for granulocytic and monocytic differentiation. The isotype controls selected were the PE-conjugated Mouse IgG2b, κ and Alexa Fluor® 488-conjugated Mouse IgG1, κ.

After treatment of HL-60 cells, 1×10^6 cells were washed with PBS and incubated at room temperature with blocking agent (2 mM EDTA and 10% FBS in PBS) for 10 min. The cells were then incubated with the antibodies or isotypes respectively for 30 min, on ice and in the dark. Following three consecutive washes with Tween wash buffer (0.1% Tween in PBS) and resuspension in staining buffer (2% FBS, 1% human serum and 0.1% sodium azide in 0.05M Tris Buffered Saline solution at pH 7.4), the samples were read using CellQuest™ Pro (Becton Dickinson, Mountain View, CA, USA). Histogram statistics were used to determine the percentage of HL-60 cells that were CD11b and CD14 positive (% positive gated cells). Statistical tests were carried out using IBM® SPSS® Statistics Version 21 (Armonk, NY, USA) to identify significantly different values.

2.9. Cell Cycle Analysis by Flow Cytometry

HL-60 cells were synchronized in the G_0/G_1 phase for the cell cycle analysis by serum starvation for 24 h. Then, cells (1.5 mL) were seeded in 12-well plates at a concentration of 1×10^6 viable cells/mL along with 1.5 mL treatment and incubated at a temperature of 37 °C and 5% CO_2. On Days 1, 3, and 5 of the experiment, cells were harvested then washed with PBS and fixed in 70% ethanol at 4 °C for 30 min. For DNA staining, samples were washed twice in PBS and centrifuged at $805 \times g$ for 5 min. To the pellet, 200 µL of Ribonuclease A in PBS at a concentration of 100 µg/mL and 200 µL of a 50 µg/mL propidium iodide solution were added. Samples were then incubated at a temperature of 37 °C and a 5% CO_2 concentration for an hour. Samples were read using the FACS Calibur™ and CellQuest™ Pro software using the FL2 channel. For each sample, 1×10^4 events were acquired. For gating, debris was excluded and the percentage of cells in the Sub G_0/G_1, G_0/G_1, S, and G_2/M phases was determined. Analysis was carried out on four samples per condition.

2.10. Investigation of the Molecular Mechanism of Differentiation

Treated and untreated HL-60 cells were seeded at a final concentration of 1×10^6 viable cells/mL in 6-well plates (1.5 mL cells and 1.5 mL treatment) and incubated at 37 °C and 5% CO_2. Following 1, 6, and 12 h, the cells were harvested by centrifugation at $400 \times g$ for 5 min, the supernatant decanted, and the cell pellet resuspended completely in 350 µL QIAzol Lysis Reagent (Qiagen), then treated according to the manufacturer's instructions and stored at −80 °C. The RNA concentrations and A260/280 values were determined using the NanoDrop 2000 UV-Vis Spectrophotometer (Thermo Scientific, Waltham, MA, USA), while the RNA integrity was determined by electrophoresis on an agarose gel. RNA sequencing using an Illumina HiSeq 2000 was performed by the European Molecular Biology Laboratory (EMBL) in Heidelberg, Germany.

2.11. RNA-Seq Data Analysis

Raw FastQ data were mapped to the human GRCH37_hg19 genome assembly using TopHat. The RNA-seq quantitation pipeline in SeqMonk software (www.bioinformatics. babraham.ac.uk/projects/seqmonk, accessed date 28 June 2020; v1.39.0) was used to quantitate the mapped reads at mRNA level. Opposing strand-specific quantitation was performed using mRNA transcript features, corrected for transcript length; log-transformed and cumulative distributions normalized across samples. Differentially expressed (DE) genes were determined using DEseq2 ($p < 0.01$, with multiple testing correction) and Intensity difference filter ($p < 0.05$, with multiple testing correction). DE genes common to both statistical tests generated the list of high confidence gene list used in downstream analyses. Briefly, STRING analysis of positively DE 101 genes (+Tyrosol/RPMI control) was visualised in Cytoscape (v11.0). Markov clustering algorithm [30] was performed to determine DE genes with high degree of adjacency. Reactome (reactome.org – accessed date 28 June 2020) was used to determine if DE genes are found in particular regulatory pathways than is predicted by chance.

2.12. Statistical Analysis

Using the software IBM® SPSS® Statistics Version 21, all data sets were tested for normality using the Shapiro–Wilk test. The Kruskal–Wallis test was selected as the non-parametric test for pairwise comparisons, with significant comparisons having a p-value less than 0.05.

2.13. Summary of the Study

The design of this study is summarised in Figure 1.

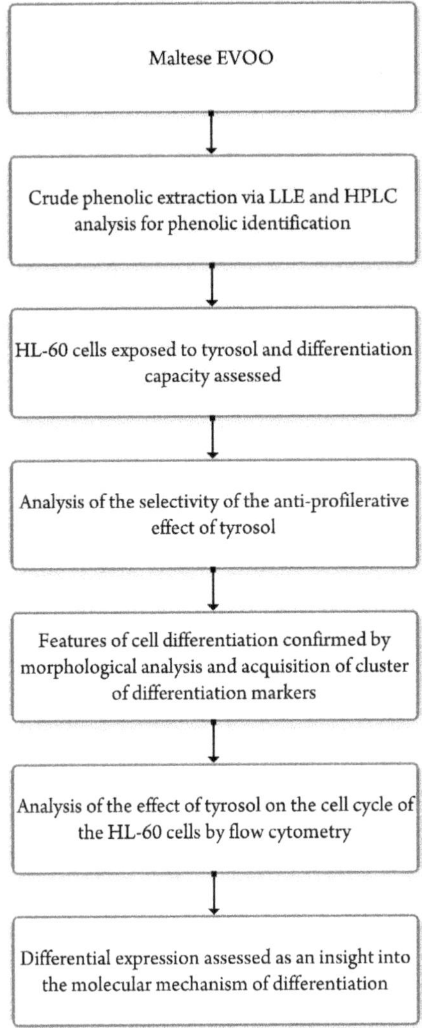

Figure 1. The design of the study.

3. Results

Following the phenolic extraction of a Maltese monocultivar EVOO variety using LLE, its characterization revealed that tyrosol was present at a concentration of 9.23 ppm. As a result, this concentration was used for all subsequent experiments.

3.1. Tyrosol Shows Differentiating Activity in HL-60 Cells

Tyrosol induced a 1.93-fold increase ($p < 0.05$) in cellular oxidative activity compared to the untreated negative control on day 3 (Figure 2), indicating increased myeloid cell maturation. The same concentration of tyrosol was tested on lymphocytes, and no increase in cytotoxicity was noted over a period of 3 days (Figure 3).

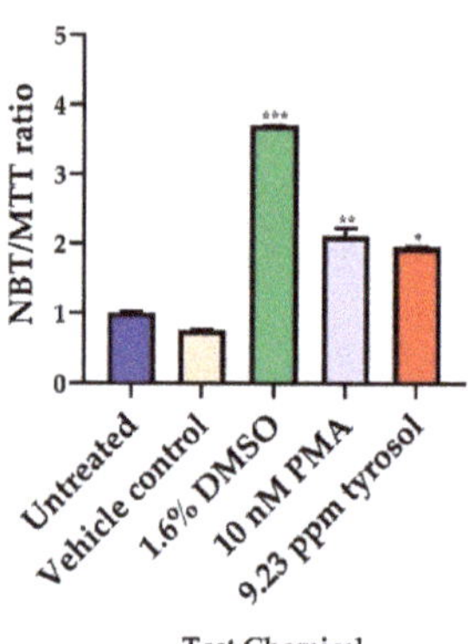

Figure 2. NBT/MTT ratio in HL-60 cells, 3 days following exposure to tyrosol. Untreated = Cells in RPMI medium, Negative control for differentiation; Vehicle control = 0.5% Ethanol; 1.6% DMSO, 10 nM PMA = Positive controls for granulocytic and monocytic differentiation, respectively. Each value is a median value where $n = 9$ ($n = 3$ for 3 separate experiments). Values normalised to medium (RPMI) control. Error bars represent differences between the median and upper and lower quartiles. Statistically significant differences from the negative control are denoted by (*) $p < 0.05$, (**) $p < 0.01$, and (***) $p < 0.001$.

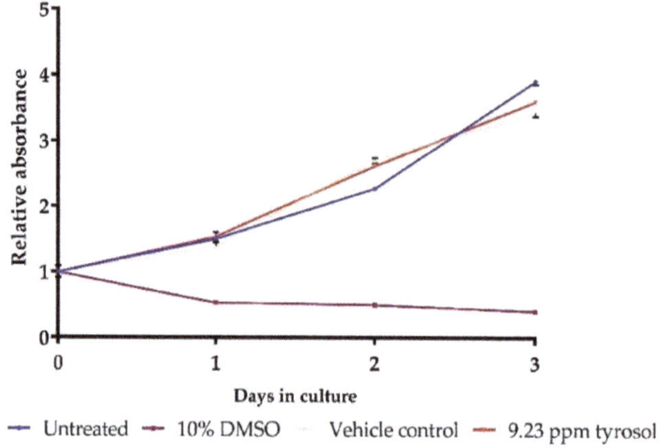

Figure 3. Cytotoxicity curves of activated lymphocytes at 0, 1, 2, and 3 days following tyrosol exposure using MTT assay. Untreated = Negative (medium) control for cytotoxicity; 10% DMSO = Positive control for cytotoxicity; Vehicle control = 0.5% Ethanol. Each value is a median value where $n = 9$ ($n = 3$ for 3 separate experiments). Values normalised to initial cell number. Error bars represent differences between the median and upper and lower quartiles.

3.2. Morphological Analysis of Tyrosol-Exposed HL-60 Cells Show Evidence of Monocytic Differentiation

A morphological analysis of the HL-60 cells post-tyrosol exposure using Leishman's staining (Figure 4) was carried out. This confirmed the presence of morphological changes associated with white cell progenitor (leukaemia blast) differentiation in HL-60 cells treated with the positive controls (Figure 4C–F), as well as with tyrosol (Figure 4G,H). Visible changes included the presence of vacuoles, a decreased nuclear/cytoplasmic ratio, a more irregular shape, a reduction in the number of nucleoli, and reduced cell numbers (due to reduced proliferation). Since, as evident in Figures 2 and 4, no differentiation-inducing effect was seen for the vehicle control, this was not included in subsequent experiments.

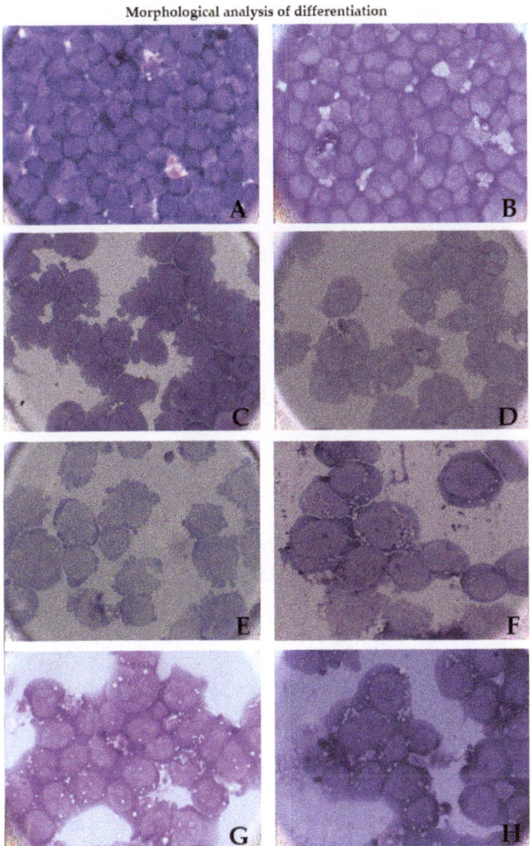

Figure 4. (**A–H**) Leishman's stained HL-60 cells exposed to RPMI—Medium only negative control (**A**), Vehicle control 0.5% Ethanol (**B**), 1.6% DMSO on day 3 (**C**), 1.6% DMSO on day 5 (**D**), 10 nM PMA on day 3 (**E**), 10 nM PMA on day 5 (**F**), 9.23 ppm tyrosol on day 3 (**G**), 9.23 ppm tyrosol on day 5 (**H**). Magnification 1000×.

3.3. Tyrosol Stimulates the Expression of CD11b and CD14 Surface Antigens in Hl-60 Cells

Tyrosol-exposed cells were assessed with flow cytometry for changes in the expression of cell surface markers. Tyrosol treatment was found to enhance the surface expression of CD11b and CD14 with peak expression in both cases happening 3 days after exposure as with the monocytic differentiation positive control PMA (Figure 5a,b, respectively).

CD11b is an integrin marker expressed by leukocytes during the inflammatory response and is associated with leukocyte adhesion and migration. High CD11b populations were recorded at 23.70% (day 3) for tyrosol and were comparable to the 29.31% (day 3) recorded with the positive control (PMA) associated with the induction of monocyte differentiation (Figure 5a). There was no statistically significant difference between CD11b expression following tyrosol treatment and CD11b expression following PMA treatment.

CD14 is an antigen expressed by monocytes and macrophages and is associated with participation in binding of lipopolysaccharide. For CD14, the effect of tyrosol was again similar to that of PMA with CD14 high populations reported at 14.11% (day 3) and 4.38% (day 5) after tyrosol treatment compared to 16.61% (day 3) and 4.03% (day 5) after PMA application (Figure 5b). There was no statistically significant difference between CD14 expression following tyrosol and PMA exposure. Both these results confirm that tyrosol is inducing differentiation along a monocytic pathway.

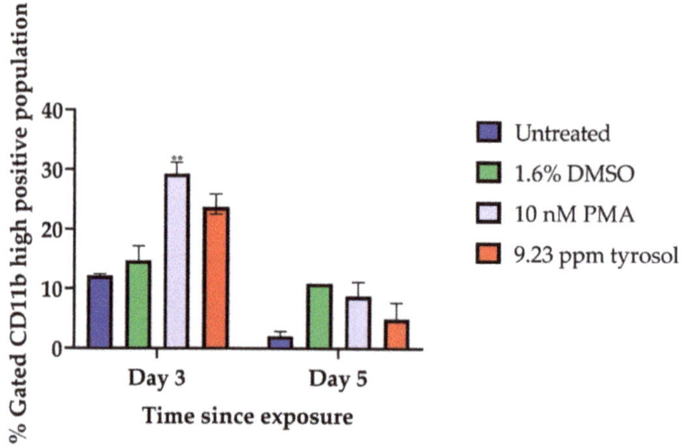

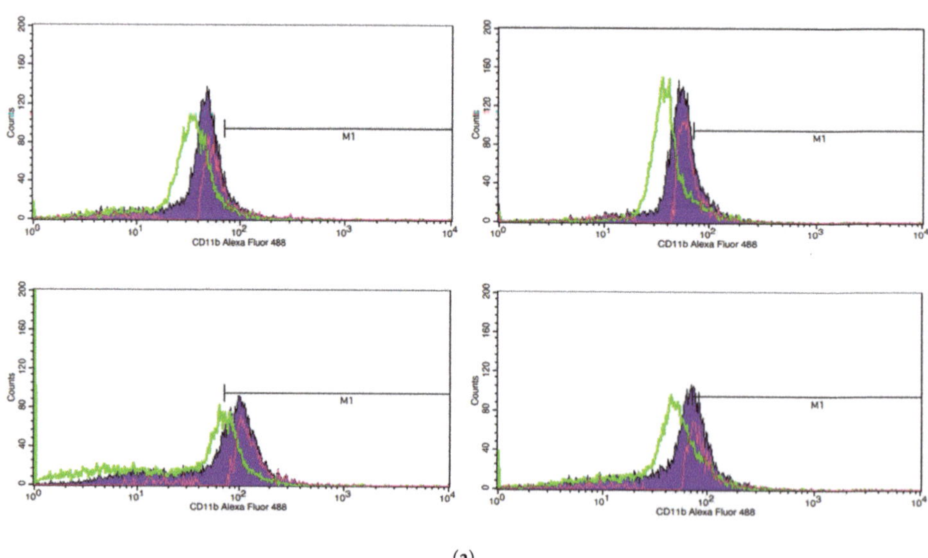

(a)

Figure 5. *Cont.*

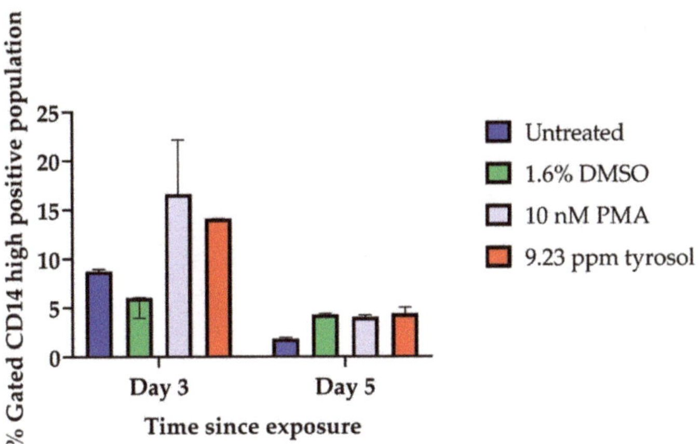

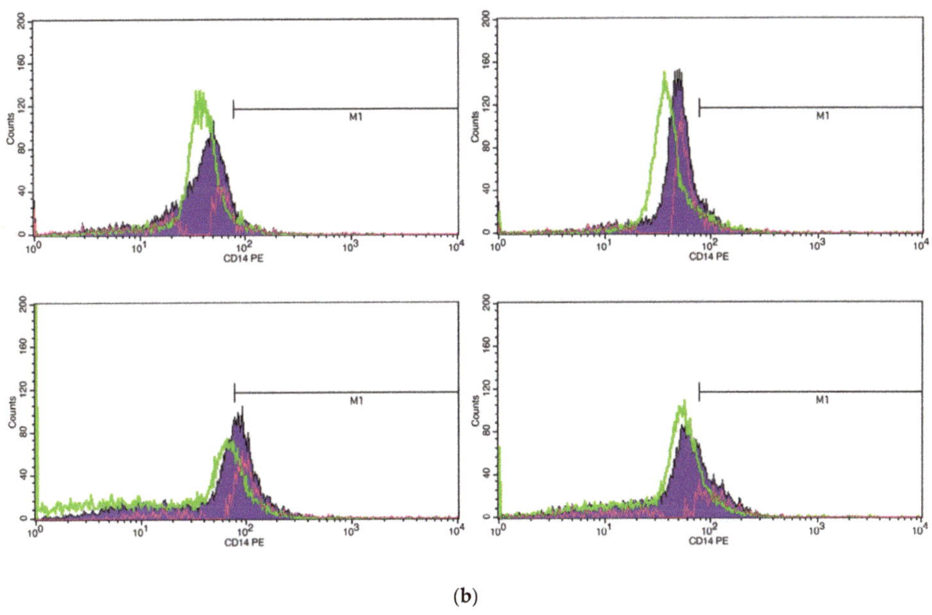

(**b**)

Figure 5. (**a**): A graph representing the % Gated CD11b high population on days 3 and 5 and representative histogram overlays and subtractions for each condition on day 3. Histograms from left to right = Untreated, 1.6% DMSO, 10 nM PMA, and 9.23 ppm tyrosol. Untreated = negative (medium only) control; 1.6% DMSO = granulocytic positive control; 10 nM PMA = monocytic positive control. Each bar represents the median value where *n* = 3 per condition. Error bars represent differences between the median and upper and lower quartiles. Statistically significant differences from the negative control are denoted by (******) *p* < 0.01. (**b**): A graph representing the % Gated CD14 high population on days 3 and 5 and representative histogram overlays and subtractions for each condition on day 3. Histograms from left to right = Untreated, 1.6% DMSO, 10 nM PMA, and 9.23 ppm tyrosol. Untreated = negative (medium only) control; 1.6% DMSO = granulocytic positive control; 10 nM PMA = monocytic positive control. Each bar represents the median value where *n* = 3 per condition. Error bars represent differences between the median and upper and lower quartiles.

3.4. Tyrosol Induces Apoptosis in Hl-60 Associated with a Reduction in Cells in the G1 and S Phases

There are numerous reports of the inhibition of cell proliferation and even cancer chemo-preventive effects by EVOO phenols [24,31,32]. To further explore the observed reduction in cell numbers (MTT assay—Supplementary Figure S2), the cell cycle status of tyrosol-exposed cells was analysed. The exposure of HL-60 cells to tyrosol resulted in a marked increase in cells in the Sub G_0/G_1 section (Figure 6). This suggests a significant difference in mechanism from the positive controls, with tyrosol appearing to induce cell death following induced differentiation. This appears to coincide with a reduction in the G_0/G_1 and S phase, with cells accumulating in the G_2M phase of the cycle. This may suggest that tyrosol-exposed cells cannot progress beyond G_2M, and when attempting to do so, they proceed to cellular demise.

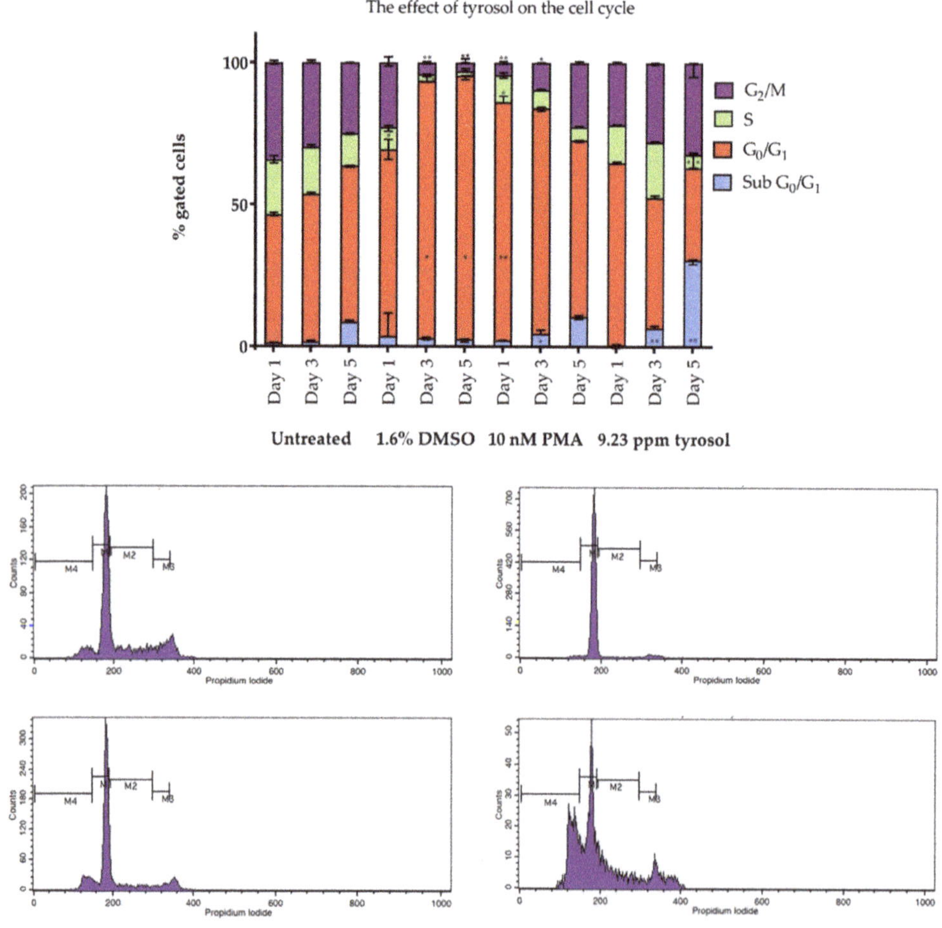

Figure 6. Percentage gated HL-60 cells in the Sub G_0/G_1, G_0/G_1, S, and G_2/M phases of the cell cycle on day 1, day 3, and day 5 and representative histograms for each condition on day 5. Histograms from left to right = Untreated, 1.6% DMSO, 10 nM PMA, and 9.23 ppm tyrosol. Untreated = negative (medium only) control, 1.6% DMSO, 10 nM PMA are positive controls for granulocytic and monocytic differentiation, respectively. Each bar represents the median value where $n = 4$. Error bars represent differences between the median and upper and lower quartiles. Statistically significant differences from the negative control are denoted by (*) $p < 0.05$ and (**) $p < 0.01$.

To assess the mechanism of tyrosol-induced cell death, Annexin V/PI staining was performed and analysed by flow cytometry (Figure 7). Tyrosol exposure causes cells to primarily undergo apoptosis, with cells on day 5 showing late apoptosis (35.52%), as well as necrosis (9.35%). This contrasts with the results obtained for both differentiation controls, which do not appear to induce significant amounts of cell death by day 5, with cells (21.28% and 19.25% for DMSO and PMA, respectively) appearing to be in early apoptosis. These results indicate that the polyphenol tested here arrests the leukaemia cell line, subsequently inducing apoptosis.

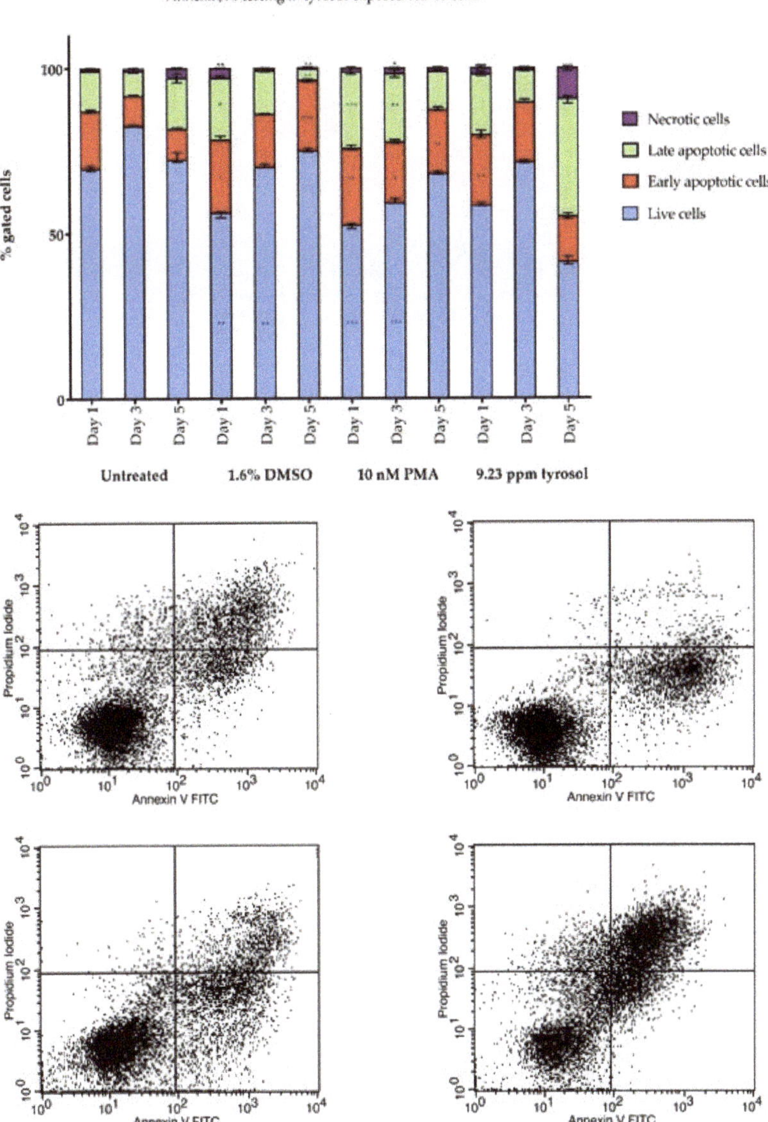

Figure 7. Percentage gated live, early apoptotic, late apoptotic and necrotic HL-60 cells on day 1, day 3, and day 5 and representative quadrants for each condition on day 5. Days 3 and 5 and representative histogram overlays and subtractions

for each condition on day 5. From left to right = Untreated, 10 nM PMA, 1.6% DMSO, and 9.23 ppm tyrosol. Untreated = negative (medium only) control, 1.6% DMSO, 10 nM PMA = positive controls for granulocytic and monocytic differentiation respectively. Each bar represents the median value where $n = 4$. Error bars represent differences between the median and upper and lower quartiles. Statistically significant differences from the negative control are denoted by (*) $p < 0.05$, (**) $p < 0.01$, and (***) $p < 0.001$.

3.5. Tyrosol Downregulates Neutrophil and Cholesterol Biosynthesis Genes and Upregulates Monocytic Differentiation Genes

In order to determine the Tyrosol-mediated global gene expression changes in the ATRA-resistant HL-60 cell line, an RNA-seq analysis was performed after stimulation with Tyrosol for 1, 6, and 12 h. This defined a total of 199 differentially expressed (DE) genes (DE-seq2 $p < 0.01$ and Intensity Difference filter $p < 0.05$, with multiple testing correction), with 142 genes resulting in increased expression, whilst 57 genes were decreased in expression (Supplementary Table S1).

Figure 8A,B show the hierarchical clustering of the negative and positively regulated genes respectively. In order to explore the biological significance of negatively regulated genes we used Reactome pathway analysis (Reactome.org). This revealed Neutrophil Degranulation (FDR: 5.10×10^{-13}) as an important process that is downregulated upon Tyrosol treatment across the timepoints and include the genes CEACAM6, CLEC5A, FPR1, SERPINA1, ELANE, AZU1, and PRG2. Moreover, the tyrosol treatment resulted in decreased expression of Cholesterol biosynthesis (FDR: 1.28×10^{-8}) related genes including LSS, SQLE, ACAT2, DHCR7 HMGCS1, and FDFT1.

We used STRING analysis of the 143 positively regulated genes, which revealed a network of 130 genes (p-value $< 1 \times 10^{-16}$). These are displayed to minimize the energy of the system based on the confidence score for interactions (Figure 8C). Gene Ontology (GO) analysis revealed a Myeloid differentiation process (FDR: 0.0001) with upregulated genes including the genes OSCAR, VEGFA, IFI16, JUNB, GAB2, and RELB (Figure 8C). We observed key monocyte-derived macrophage (Figure 8D) and dendritic cell (Figure 8E) genes, including transcription factors IRF1, IRF7, STAT2, RelB, NFKB2, ATF3, and BCL3 and chemokines CCL3 and CCL4 [33–35].

A Markov clustering algorithm highlighted six modules with five or more nodes displaying a high degree of adjacency [30] (Figure 8F). In addition, the functional annotation of the promoters of the upregulated genes by DAVID (v6.8) revealed NF-κb, AP-1, and ISRE (p-value < 0.001) binding sites highlighting the role of these transcription factors as regulators of the tyrosol induced gene network [36] (Figure 8G).

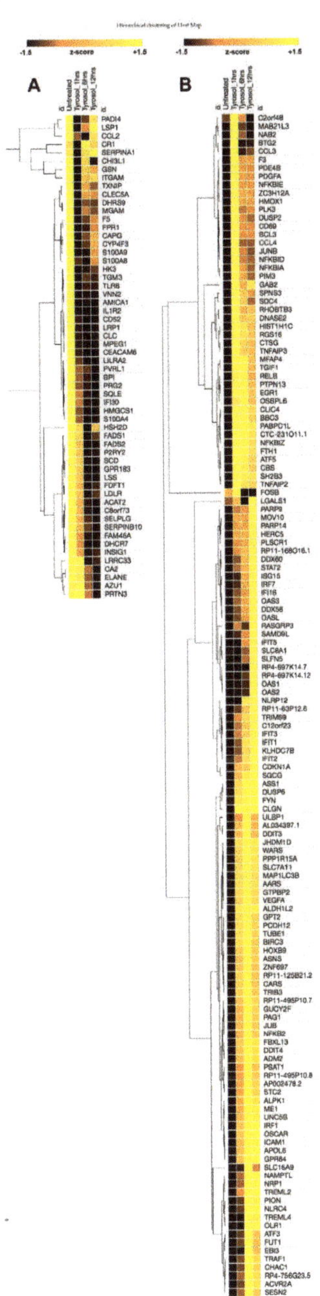

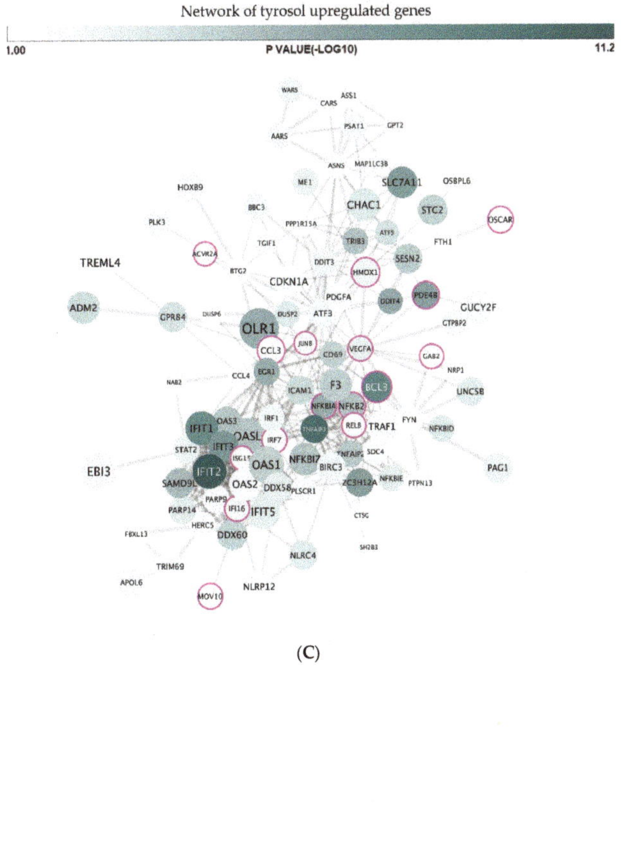

Figure 8. *Cont.*

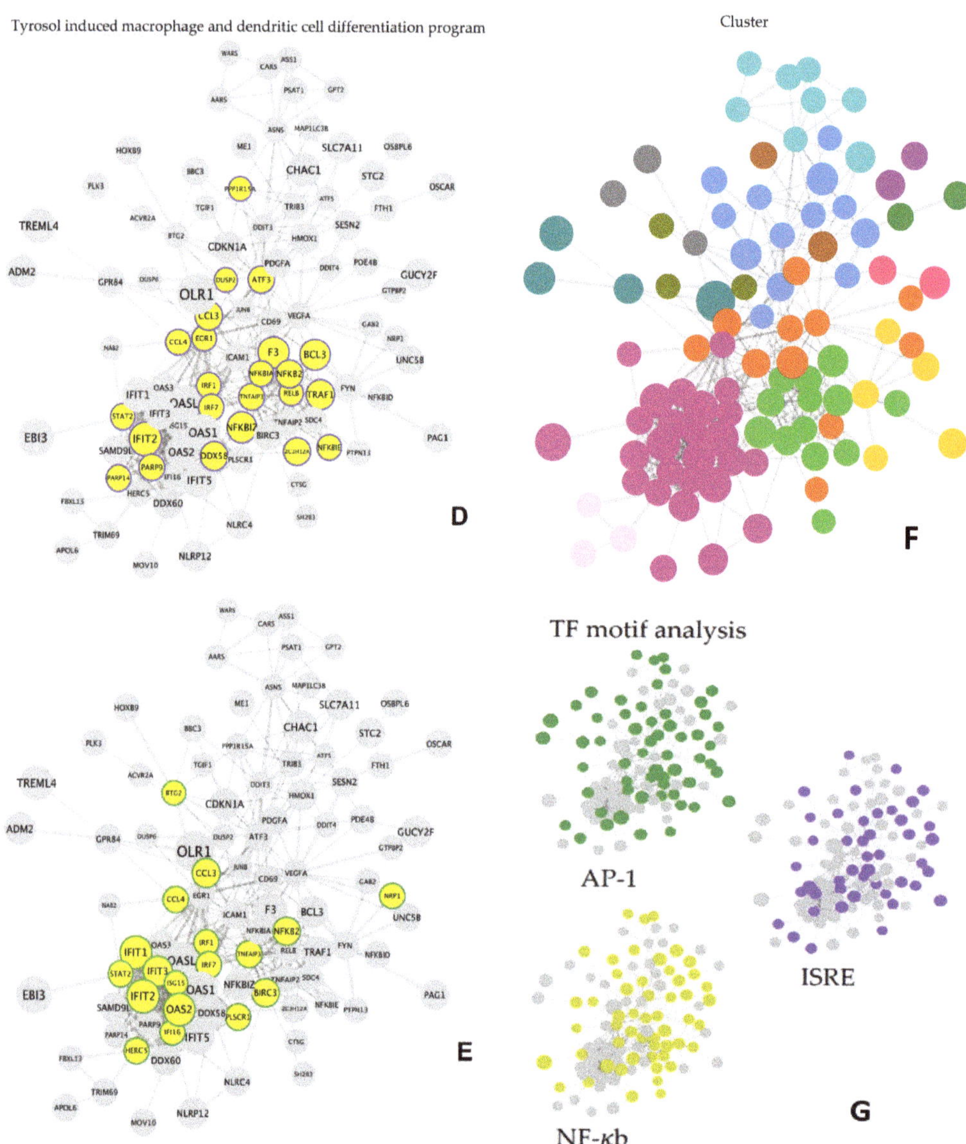

Figure 8. (**A–C**): Tyrosol induces monocyte lineage gene program. Gene expression analysis of tyrosol at 1, 6, and 12 h using RNA-seq as described in Methods. (**A,B**) Hierarchical clustering of Heat Map showing relative expression of negative (**A**) and positively expressed genes (**B**).Scale bar depicts z-score relative expression. (**C**) The network plot is based on known and predicted interactions from the STRING database (v11), with minimal confidence score of 0.4. Node size represents Fold change over untreated cells in two independent experiments. Nodes with magenta border represent genes involved in Myeloid differentiation process. Scale bar depicts −log10 of *p*-value. (**D–G**): (**D**) Network depicted in (**C**) with nodes involved in monocyte-derived macrophage gene program highlighted in yellow and blue border. (**E**) Network depicted in (**C**) with nodes involved in monocyte-derived dendritic gene program highlighted in yellow and green border. (**F**) Modules identified in protein network shown in (**C**) determined using the Markov Cluster algorithm (inflation parameter: 2.5). Each colour represents a separate module (associated adjacency matrix). (**G**) DAVID analysis of network depicted in (**C**) showing enrichment of NF-κb (yellow), AP-1 (green), and ISRE (blue) binding sites in gene promoters of the network.

4. Discussion

In this study, the differentiation-inducing effect of the pure polyphenol tyrosol at the concentration found in the 'Bidni' Maltese EVOO variety is described. Evidence of significant differentiation and a change in cell maturation was shown through biochemical assays, the assessment of cell surface marker expression, the observation of morphological features, and transcription analysis. Critically, we also observed that the anti-proliferative action of tyrosol on HL-60 cells is specific, as no cytotoxicity was observed when the same polyphenol was tested on non-malignant human lymphocytes.

The differentiating activity and inhibition of cell proliferation of HL-60 cells has been reported elsewhere using EVOO crude polyphenol extracts. Fabiani et al., 2006 suggested that this effect is partially a result of dialdehydic forms of elenoic acid linked to hydroxytyrosol (3,4-DHPEA-EDA) and tyrosol (pHPEA-EDA) present in the phenolic extract while Crescimanno et al., 2009 reported that the composition of their crude extract contained 3,4-DHPEA, pHPEA, 3,4-DHPEA-EDA, pHPEA-EDA, and 3,4-HPEA-EDA (oleuropein aglycone) [21,32]. In this study, the pure compound pHPEA was used allowing the confirmation that differentiation was due to a single bioactive agent. In addition, an extended testing time-frame of 3 and 5 days, respectively, was used. Since marked oxidative activity and expression of cell surface markers was seen on day 3, this suggests that tyrosol is driving the HL-60 cells to the monocyte stage as opposed to that of the granulocyte.

The capacity of polyphenols to inhibit cell cycle progression is an important feature shared with many drugs currently used as cancer treatments and was therefore assessed here. Previous work by Fabiani et al., 2008 demonstrated that hydroxytyrosol treatment of HL-60 cells over a period of 25 h resulted in an increase in cells in the G_0/G_1 phase and a decline in cells in the S phase, causing an increase in apoptotic cells [24]. This result is compatible with our findings for tyrosol, where a marked increase in the cells recorded in the sub G_0/G_1 stage was observed, although a direct comparison is not possible as the experimental time-plan differed. We also report an increase in cells undergoing late apoptosis, as well as necrosis on day 5. These results are in contrast with those recorded by Della Ragione et al., 2000 who reported an absence of anti-proliferative activity and apoptosis induction by tyrosol at a concentration of 100 μM for 2 days, however, the timeframe may have been a contributing factor [37].

In a further step, we also analysed the HL-60 transcriptome following polyphenol application in an effort to describe the genetic changes seen during tyrosol-induced monocytic differentiation. The transcriptome analysis, which is reported here for the first time, further confirms the upregulation of myeloid differentiation genes such as OSCAR, RELB, VEGFA, GAB2, JUNB, DNASE2, ICAM, and CCL3. OSCAR is a cell surface receptor expressed on monocytes, overexpression of which is linked to monocyte adhesion and migration [38,39]. The importance of VEGFA in myeloid cell differentiation has also been reported by Huang et al., 2017 while Czepluch et al., 2011 discussed its expression in two monocyte subsets [40,41], showing its role in monocyte chemotaxis. JUNB has been found to regulate the responses of monocytes to different immunostimulatory ligands, and its levels have also been reported to increase during monocyte maturation [42]. Moreover, ICAM that was upregulated across all time points, is a cell surface marker that is expressed by macrophages. The adhesion molecule binds to CD11b, which was observed to increase using fluorescent antibody analysis (Figure 5a). ALDH codes for aldehyde dehydrogenase, an enzyme important for the regulation of HSC differentiation. Chute et al. 2006 reported that ALDH inhibition in NOD/SCID mice resulted in HSC expansion and a delay in HSC differentiation [43].

Additionally, through Markov clustering (Figure 8F), we show that tyrosol treatment results in the upregulation of two tightly clustered networks, one involving Interferon Regulatory Factors (IRFs) and interferon-induced IFIT proteins and the NFκB network. Within the former network, IFIT proteins are expressed in response to a stimulus and gene expression is dependent on both pathogen-associated molecular patterns (PAMPs), as well as the JAK-STAT pathway. This is in line with our results which confirm the upregulation of both IRF7 and STAT2. Matikainen et al., 1997 have also reported the activation of STAT2 in myeloid cell differentiation. They showed that for NB4 and U937 cells treated with ATRA, increased levels of STAT1 and STAT2 were recorded, resulting in these differentiated cells being more responsive to IFNs [44]. Induction of the JAK-STAT pathway following treatment of HL-60 cells by PMA and DMSO has been recorded [45]. Moreover, previous studies have reported the upregulation of IRF1 gene expression by ATRA on the myeloid cell lines NB4 and HL-60, and hence the results presented in this study are compatible with these findings [46,47]. IRF1 is a transcriptional activator of both IFN and IFN inducible genes, and its level is lowest during cell proliferation and highest at cell growth arrest [48]. This upregulation is consistent with the findings outlined in Figure 7 confirming the apoptotic effect of tyrosol.

Within the same network, the upregulation of PARP9 and PARP14 was also observed, with PARP having previously been found to regulate cell differentiation [49]. Our results are again consistent with those obtained by Bhatia, Kirkland, and Meckling-Gill (1995), who observed the upregulation of PARP for both NB4 and HL-60 cells differentiated by PMA and $1,25\text{-}D_3$ to the monocyte/macrophage lineage [50]. Specifically, Iwata et al., 2016 reported the importance of PARP9 and PARP14 in activating macrophages [51].

With regards to the NFκB pathway, two members of the transcription factor family, NFkB2 and RELB, were found to be upregulated following tyrosol treatment. The upregulation of three inhibitors NFκBIA, NFκBIE, NFκBIZ show that tyrosol inhibits the NFκB pathway [52,53]. Inhibition of this pathway has been linked to the treatment of cancer and inflammation. A study by Holmes and Baldwin (2000) has shown that the naturally occurring flavonoid resveratrol exerts its effects by inhibiting the NFκB pathway, thus explaining the benefits of red wine in reducing coronary heart disease and cancer mortality [54]. Its suppressive effects have been attributed to a decline in the activities of both p65, as well as IκB [55]. This compound has also been found to downregulate this pathway in activated macrophages, leading to the inhibition of NO generation [56]. In summary, our results indicate that tyrosol is an effective differentiating agent driving the HL-60 cells to the monocyte stage, followed by apoptosis as cells regain senescence.

5. Conclusions

The phenolic compound tyrosol, tested at the concentration at which it was found to be present in a Maltese EVOO, was found to induce HL-60 differentiation towards the monocytic lineage as well as apoptosis. Transcriptome analysis revealed the upregulation of myeloid differentiation genes such as OSCAR, RELB, VEGFA, GAB2, JUNB, DNASE2, ICAM, and CCL3 following tyrosol treatment. Tyrosol cytotoxicity was found to be selective to HL-60 cells thus making tyrosol a suitable non-cytotoxic candidate for AML differentiation therapy. In this study, we presented the effect of one component found in a Maltese EVOO. Future studies involve an analysis of the biological effect of the rest of the phenolic profile of this EVOO on HL-60 cells.

Supplementary Materials: The following are available online at https://www.mdpi.com/article/10.3390/app112110199/s1, File S1: HPLC method, Figure S1: Extract effect on the index of differentiation, Figure S2: MTT relative absorbance on day 3, Table S1: DE intensity difference.

Author Contributions: Conceptualization, L.G., M.Z.-M. and P.S.-W.; methodology, L.G., M.Z.-M. and P.S.-W.; experimental work and data analysis, L.G.; RNA-seq data analysis, D.G.S.; investigation, L.G.; writing—original draft preparation, L.G. and D.G.S.; writing—review and editing, M.Z.-M. and P.S.-W.; supervision, M.Z.-M. and P.S.-W.; project administration, L.G.; funding acquisition, L.G. All authors have read and agreed to the published version of the manuscript.

Funding: This research was funded through the Malta Government Scholarship Scheme (Postgraduate).

Institutional Review Board Statement: The study was conducted according to the guidelines of the Declaration of Helsinki, and approved by the University of Malta, Faculty of Medicine and Surgery Research Ethics Committee (FRECMDS1718_073).

Informed Consent Statement: Informed consent was obtained from all subjects involved in the study.

Acknowledgments: The authors would like to thank Analisse Cassar, Frederick Lia, and Christian Saliba for their assistance with flow cytometry, HPLC, and RNA extraction, respectively.

Conflicts of Interest: The authors declare no conflict of interest. The funders had no role in the design of the study; in the collection, analyses, or interpretation of data; in the writing of the manuscript, or in the decision to publish the results.

References

1. Virchow, R. Weisses blut. *Froriep's Notizen.* **1847**, *36*, 151–156.
2. Estey, E.H. Acute myeloid leukemia: 2012 update on diagnosis, risk stratification, and management. *Am. J. Hematol.* **2012**, *87*, 89–99. [CrossRef] [PubMed]
3. Pierce, G.B.; Shikes, R.; Fink, L.M. *Cancer: A Problem of Developmental Biology*; Prentice-Hall Inc: Englewood Cliffs, NJ, USA, 1978; p. 242.
4. Nowak, D.; Stewart, D.; Koeffler, H.P. Differentiation therapy of leukemia: 3 decades of development. *Blood* **2009**, *113*, 3655–3665. [CrossRef]
5. Tallman, M.S.; Lefèbvre, P.; Baine, R.M.; Shoji, M.; Cohen, I.; Green, D.; Kwaan, H.C.; Paietta, E.; Rickles, F.R. Effects of all-trans retinoic acid or chemotherapy on the molecular regulation of systemic blood coagulation and fibrinolysis in patients with acute promyelocytic leukemia. *J. Thromb. Haemost.* **2004**, *2*, 1341–1350. [CrossRef] [PubMed]
6. Sell, S. Leukemia: Stem cells, maturation arrest and differentiation therapy. *Stem Cell Rev.* **2005**, *1*, 197–205. [CrossRef]
7. Leszczyniecka, M.; Roberts, T.; Dent, P.; Grant, S.; Fisher, P.B. Differentiation therapy of human cancer: Basic science and clinical applications. *Pharmacol. Ther.* **2001**, *90*, 105–156. [CrossRef]
8. Birnie, G.D. The HL60 cell line: A model system for studying human myeloid cell differentiation. *Br. J. Cancer. Supplement.* **1988**, *9*, 41–45.
9. Salvi, H.Y.; Aalto, B.; Nagy, S.; Knuutila, S.; Pakkala, S. Gene expression analysis of 1,25(OH)$_2$D$_3$-dependent differentiation of HL-60 cells: A cDNA array study. *Br. J. Hematol.* **2002**, *118*, 1065–1070.
10. Chang, H.H.; Oh, P.Y.; Ingber, D.E.; Huang, S. Multistable and multistep dynamics in neutrophil differentiation. *BMC Cell Biol.* **2006**, *7*, 11. [CrossRef] [PubMed]
11. Newburger, P.E.; Subrahmanyam, Y.V.; Weissman, S.M. Global analysis of neutrophil gene expression. *Curr. Opin. Hematol.* **2000**, *7*, 16–20. [CrossRef]
12. Rincón, E.; Rocha-Gregg, B.L.; Collins, S.R. A map of gene expression in neutrophil-like cell lines. *BMC Genom.* **2018**, *19*, 573. [CrossRef] [PubMed]
13. Naegelen, I.; Plançon, S.; Nicot, N.; Kaoma, T.; Muller, A.; Vallar, L.; Tschirhart, E.J.; Bréchard, S. An essential role of syntaxin 3 protein for granule exocytosis and secretion of IL-1α, IL-1β, IL-12b, and CCL4 from differentiated HL-60 cells. *J. Leukoc. Biol.* **2014**, *97*, 557–571. [CrossRef]
14. Mark Welch, D.B.; Jauch, A.; Langowski, J.; Olins, A.L.; Olins, D.E. Transcriptomes reflect the phenotypes of undifferentiated, granulocyte and macrophage forms of HL-60/S4 cells. *Nucleus* **2017**, *8*, 222–237. [CrossRef]
15. Santos-Beneit, A.M.; Mollinedo, F. Expression of genes involved in initiation, regulation, and execution of apoptosis in human neutrophils and during neutrophil differentiation of HL-60 cells. *J. Leukoc Biol.* **2000**, *67*, 712–724. [CrossRef] [PubMed]
16. Dai, J.; Mumper, R.J. Plant phenolics: Extraction, analysis and their antioxidant and anticancer properties. *Molecules* **2010**, *15*, 7313–7352. [CrossRef]
17. Gómez-Romero, M.; García-Villalba, R.; Carrasco-Pancorbo, A.; Fernández-Gutiérrez, A. Metabolism and Bioavailability of Olive Oil Polyphenols (2012). In *Olive Oil—Constituents, Quality, Health Properties and Bioconversions*; Dimitrios, B., Ed.; IntechOpen: London, UK; pp. 333–356.
18. Shendi, E.G.; Ozay, D.S.; Ozkaya, M.T.; Ustunei, N.F. Changes occurring in chemical composition and oxidative stability of virgin olive oil during storage. *Oilseeds Fats Crop. Lipids* **2018**, *25*, 1–8.
19. Lia, F.; Formosa, J.P.; Zammit-Mangion, M.; Farrugia, C. The First Identification of the Uniqueness and Authentication of Maltese Extra Virgin Olive Oil Using 3D-Fluorescence Spectroscopy Coupled with Multi-Way Data Analysis. *Foods* **2020**, *9*, 498. [CrossRef] [PubMed]
20. Gatt, L.; Lia, F.; Zammit-Mangion, M.; Thorpe, S.J.; Schembri-Wismayer, P. First Profile of Phenolic Compounds from Maltese Extra Virgin Olive Oils Using Liquid-Liquid Extraction and Liquid Chromatography-Mass Spectrometry. *J. Oleo Sci.* **2021**, *70*, 145–153. [CrossRef]
21. Fabiani, R.; De Bartolomeo, A.; Rosignoli, P.; Servili, M.; Selvaggini, R.; Montedoro, G.F.; Di Saverio, C.; Morozzi, G. Virgin olive oil phenols inhibit proliferation of human promyelocytic leukemia cells (HL60) by inducing apoptosis and differentiation. *J. Nutr.* **2006**, *136*, 614–619. [CrossRef]
22. Abaza, L.; Talorete, T.P.N.; Yamada, P.; Kurita, Y.; Zarrouk, M.; Isoda, H. Induction of growth inhibition and differentiation of human leukemia HL-60 cells by a Tunisian gerboui olive leaf extract. *Biosci. Biotechnol. Biochem.* **2007**, *71*, 1306–1312. [CrossRef]
23. Sepporta, M.V.; Mazza, T.; Morozzi, G.; Fabiani, R. Pinoresinol inhibits proliferation and induces differentiation on human HL60 leukemia cells. *Nutr. Cancer* **2013**, *65*, 1208–1218. [CrossRef]

24. Fabiani, R.; Rosignoli, P.; De Bartolomeo, A.; Fuccelli, R.; Morozzi, G. Inhibition of cell cycle progression by hydroxytyrosol is associated with upregulation of cyclin-dependent protein kinase inhibitors p21(WAF1/Cip1) and p27(Kip1) and with induction of differentiation in HL60 cells. *J. Nutr.* **2008**, *138*, 42–48. [CrossRef] [PubMed]

25. Abe, K.; Matsuki, N. Measurement of cellular 3-(4,5-dimethylthiazol-2-yl)-2,5-diphenyltetrazolium bromide (MTT) reduction activity and lactate dehydrogenase release using MTT. *Neurosci. Res.* **2000**, *38*, 325–329. [CrossRef]

26. Stoica, S.; Magoulas, G.E.; Antoniou, A.I.; Suleiman, S.; Cassar, A.; Gatt, L.; Papaioannou, D.; Athanassopoulos, C.M.; Schembri-Wismayer, P. Synthesis of minoxidil conjugates and their evaluation as HL-60 differentiation agents. *Bioorganic Med. Chem. Lett.* **2016**, *26*, 1145–1150. [CrossRef] [PubMed]

27. Chen, H.W.; Heiniger, H.J.; Kandutsch, A.A. Relationship between sterol synthesis and DNA synthesis in phytohemagglutinin-stimulated mouse lymphocytes. *Proc. Natl. Acad. Sci. USA* **1975**, *72*, 1950–1954. [CrossRef] [PubMed]

28. Mire-Sluis, A.R.; Wickremasinghe, R.G.; Hoffbrand, A.V.; Timms, A.M.; Francis, G.E. Human T lymphocytes stimulated by phytohaemagglutinin undergo a single round of cell division without a requirement for interleukin-2 or accessory cells. *Immunology* **1987**, *60*, 7–12.

29. Hutchins, D.; Steel, C.M. Phytohaemagglutinin-induced proliferation of human T lymphocytes: Differences between neonate and adults in accessory cell requirements. *Clin. Exp. Immunol.* **1983**, *52*, 355–364.

30. Enright, A.J.; Van Dongen, S.; Ouzounis, C.A.A. An efficient algorithm for large-scale detection of protein families. *Nucleic Acids Res.* **2002**, *30*, 1575–1584. [CrossRef]

31. Kawaii, S.; Lansky, E.P. Differentiation-promoting activity of pomegranate (Punica granatum) fruit extracts in HL-60 human promyelocytic leukemia cells. *J. Med. Food* **2004**, *7*, 13–18. [CrossRef]

32. Crescimanno, M.; Sepporta, M.V.; Tripoli, E.; Flandina, C.; Giammanco, M.; Tumminello, F.T.; Di Majo, D.; Tolomeo, M.; La Guardia, M.; Leto, G. Effects of extra virgin olive oil phenols on HL60 cell lines sensitive and resistant to anthracyclines. *J. Biol. Res.* **2009**, *82*, 34–37. [CrossRef]

33. Baillie, J.K.; Arner, E.; Daub, C.; De Hoon, M.; Itoh, M.; Kawaji, H.; Lassmann, T.; Carninci, P.; Forrest, A.R.; Hayashizaki, Y.; et al. Analysis of the human monocyte-derived macrophage transcriptome and response to lipopolysaccharide provides new insights into genetic aetiology of inflammatory bowel disease. *PLoS Genet.* **2017**, *13*, e1006641. [CrossRef] [PubMed]

34. Saliba, D.G.; Heger, A.; Eames, H.L.; Oikonomopoulos, S.; Teixeira, A.; Blazek, K.; Androulidaki, A.; Wong, D.; Goh, F.G.; Weiss, M.; et al. IRF5:RelA interaction targets inflammatory genes in macrophages. *Cell Rep.* **2014**, *8*, 1308–1317. [CrossRef]

35. Birmachu, W.; Gleason, R.M.; Bulbulian, B.J.; Riter, C.L.; Vasilakos, J.P.; Lipson, K.E.; Nikolsky, Y. Transcriptional networks in plasmacytoid dendritic cells stimulated with synthetic TLR 7 agonists. *BMC Immunol.* **2007**, *8*, 26. [CrossRef] [PubMed]

36. Huang, D.W.; Sherman, B.T.; Lempicki, R.A. Systematic and integrative analysis of large gene lists using DAVID bioinformatics resources. *Nat. Protoc.* **2009**, *4*, 44–57. [CrossRef]

37. Ragione, F.D.; Cucciolla, V.; Borriello, A.; Pietra, V.D.; Pontoni, G.; Racioppi, L.; Manna, C.; Galletti, P.; Zappia, V. Hydroxytyrosol, a natural molecule occurring in olive oil, induces cytochrome c-dependent apoptosis. *Biochem. Biophys. Res. Commun.* **2000**, *278*, 733–739. [CrossRef] [PubMed]

38. Merck, E.; Gaillard, C.; Gorman, D.M.; Montero-Julian, F.; Durand, I.; Zurawski, S.M.; Menetrier-Caux, C.; Carra, G.; Lebecque, S.; Trinchieri, G.; et al. OSCAR is an FcRgamma-associated receptor that is expressed by myeloid cells and is involved in antigen presentation and activation of human dendritic cells. *Blood* **2004**, *104*, 1386–1395. [CrossRef]

39. Goettsch, C.; Kliemt, S.; Sinningen, K.; von Bergen, M.; Hofbauer, L.C.; Kalkhof, S. Quantitative proteomics reveals novel functions of osteoclast-associated receptor in STAT signaling and cell adhesion in human endothelial cells. *J. Mol. Cell. Cardiol.* **2012**, *53*, 829–837. [CrossRef]

40. Huang, Y.; Rajappa, P.; Hu, W.; Hoffman, C.; Cisse, B.; Kim, J.H.; Gorge, E.; Yanowitch, R.; Cope, W.; Vartanian, E.; et al. A proangiogenic signaling axis in myeloid cells promotes malignant progression of glioma. *J. Clin. Investig.* **2017**, *127*, 1826–1838. [CrossRef]

41. Czepluch, F.S.; Olieslagers, S.; van Hulten, R.; Vöö, S.A.; Waltenberger, J. VEGF-A-induced chemotaxis of CD16+ monocytes is decreased secondary to lower VEGFR-1 expression. *Atherosclerosis* **2011**, *215*, 331–338. [CrossRef]

42. Fontana, M.F.; Baccarella, A.; Pancholi, N.; Pufall, M.A.; Herbert, D.R.; Kim, C.C. JUNB is a key transcriptional modulator of macrophage activation. *J. Immunol.* **2015**, *194*, 177–186. [CrossRef]

43. Chute, J.P.; Muramoto, G.G.; Whitesides, J.; Colvin, M.; Safi, R.; Chao, N.J.; McDonnell, D.P. Inhibition of aldehyde dehydrogenase and retinoid signaling induces the expansion of human hematopoietic stem cells. *Proc. Natl. Acad. Sci. USA* **2006**, *103*, 11707–11712. [CrossRef]

44. Matikainen, S.; Ronni, T.; Lehtonen, A.; Sareneva, T.; Melén, K.; Nordling, S.; Levy, D.E.; Julkunen, I. Retinoic acid induces signal transducer and activator of transcription (STAT) 1, STAT2, and p48 expression in myeloid leukemia cells and enhances their responsiveness to interferons. *Cell Growth Differ. Mol. Biol. J. Am. Assoc. Cancer Res.* **1997**, *8*, 687–698.

45. Cohen, S.; Dovrat, S.; Sarid, R.; Huberman, E.; Salzberg, S. JAK-STAT signaling involved in phorbol 12-myristate 13-acetate- and dimethyl sulfoxide-induced 2′-5′ oligoadenylate synthetase expression in human HL-60 leukemia cells. *Leuk. Res.* **2005**, *29*, 923–931. [CrossRef]

46. Matikainen, S.; Ronni, T.; Hurme, M.; Pine, R.; Julkunen, I. Retinoic acid activates interferon regulatory factor-1 gene expression in myeloid cells. *Blood* **1996**, *88*, 114–123. [CrossRef]

47. Gianni, M.; Terao, M.; Fortino, I.; LiCalzi, M.; Viggiano, V.; Barbui, T.; Rambaldi, A.; Garattini, E. Stat1 is induced and activated by all-trans retinoic acid in acute promyelocytic leukemia cells. *Blood* **1997**, *89*, 1001–1012. [CrossRef]
48. Harada, H.; Kitagawa, M.; Tanaka, N.; Yamamoto, H.; Harada, K.; Ishihara, M.; Taniguchi, T. Anti-oncogenic and oncogenic potentials of interferon regulatory factors-1 and -2. *Science* **1993**, *259*, 971–974. [CrossRef] [PubMed]
49. Francis, G.E.; Gray, D.A.; Berney, J.J.; Wing, M.A.; Guimaraes, J.E.; Hoffbrand, A.V. Role of ADP-ribosyl transferase in differentiation of human granulocyte-macrophage progenitors to the macrophage lineage. *Blood* **1983**, *62*, 1055–1062. [CrossRef] [PubMed]
50. Bhatia, M.; Kirkland, J.B.; Meckling-Gill, K.A. Modulation of poly (ADP-ribose) polymerase during neutrophilic and monocytic differentiation of promyelocytic (NB4) and myelocytic (HL-60) leukaemia cells. *Biochem. J.* **1995**, *308*, 131–137. [CrossRef] [PubMed]
51. Iwata, H.; Goettsch, C.; Sharma, A.; Ricchiuto, P.; Goh, W.W.; Halu, A.; Yamada, I.; Yoshida, H.; Hara, T.; Wei, M.; et al. PARP9 and PARP14 cross-regulate macrophage activation via STAT1 ADP-ribosylation. *Nat. Commun.* **2016**, *7*, 12849. [CrossRef] [PubMed]
52. Tieri, P.; Termanini, A.; Bellavista, E.; Salvioli, S.; Capri, M.; Franceschi, C. Charting the NF-κB pathway interactome map. *PLoS ONE* **2012**, *7*, e32678. [CrossRef] [PubMed]
53. Ghosh, S.; May, M.J.; Kopp, E.B. NF-kappa B and Rel proteins: Evolutionarily conserved mediators of immune responses. *Annu. Rev. Immunol.* **1998**, *16*, 225–260. [CrossRef]
54. Holmes-McNary, M.; Baldwin, A.S., Jr. Chemopreventive properties of trans-resveratrol are associated with inhibition of activation of the IkappaB kinase. *Cancer Res.* **2000**, *60*, 3477–3483. [PubMed]
55. Ren, Z.; Wang, L.; Cui, J.; Huoc, Z.; Xue, J.; Cui, H.; Mao, Q.; Yang, R. Resveratrol inhibits NF-kB signaling through suppression of p65 and IkappaB kinase activities. *Die Pharm.-Int. J. Pharm. Sci.* **2013**, *68*, 689–694.
56. Tsai, S.H.; Lin-Shiau, S.Y.; Lin, J.K. Suppression of nitric oxide synthase and the down-regulation of the activation of NFkappaB in macrophages by resveratrol. *Br. J. Pharmacol.* **1999**, *126*, 673–680. [CrossRef] [PubMed]

 *applied sciences*

Article

Anti-Inflammatory Activity of *Cnidoscolus aconitifolius* (Mill.) Ethyl Acetate Extract on Croton Oil-Induced Mouse Ear Edema

Eduardo Padilla-Camberos [1,*], Omar Ricardo Torres-Gonzalez [1], Ivan Moises Sanchez-Hernandez [1], Nestor Emmanuel Diaz-Martinez [1], Oscar Rene Hernandez-Perez [1] and Jose Miguel Flores-Fernandez [2,3,*]

[1] Unit of Medical and Pharmaceutical Biotechnology, Center for Research and Assistance in Technology and Design of the State of Jalisco, A.C. (CIATEJ), Guadalajara 44270, Mexico; polimerasados@gmail.com (O.R.T.-G.); isanchez@ciatej.mx (I.M.S.-H.); ediaz@ciatej.mx (N.E.D.-M.); operez@ciatej.mx (O.R.H.-P.)
[2] Centre for Prions and Protein Folding Diseases, Department of Biochemistry, University of Alberta, 204 Brain and Aging Research Building, Edmonton, AB T6G 2M8, Canada
[3] Department of Research and Innovation, Universidad Tecnológica de Oriental, Oriental 75020, Mexico
* Correspondence: epadilla@ciatej.mx (E.P.-C.); jose.flores@utdeoriental.edu.mx or floresfe@ualberta.ca (J.M.F.-F.); Tel.: +52-(33)-33455200 (ext. 1640) (E.P.-C.); +1-(825)-9931702 (J.M.F.-F.)

Citation: Padilla-Camberos, E.; Torres-Gonzalez, O.R.; Sanchez-Hernandez, I.M.; Diaz-Martinez, N.E.; Hernandez-Perez, O.R.; Flores-Fernandez, J.M. Anti-Inflammatory Activity of *Cnidoscolus aconitifolius* (Mill.) Ethyl Acetate Extract on Croton Oil-Induced Mouse Ear Edema. *Appl. Sci.* **2021**, *11*, 9697. https://doi.org/10.3390/app11209697

Academic Editor: Emanuel Vamanu

Received: 30 August 2021
Accepted: 8 October 2021
Published: 18 October 2021

Publisher's Note: MDPI stays neutral with regard to jurisdictional claims in published maps and institutional affiliations.

Abstract: Nowadays, there is a growing interest in the development of medicinal plant-based therapies to diminish the ravages of the inflammatory process related to diseases and tissue damage. Most therapeutic effects of these traditional medicinal plants are owed to their phenolic and antioxidant properties. *C. aconitifolius* is a traditional medicinal plant in Mexico. Previous characterization reports have stated its high nutritional and antioxidant components. The present study aimed to better understand the biological activity of *C. aconitifolius* in inflammation response. We developed an ethyl acetate extract of this plant to evaluate its anti-inflammatory capacity and its flavonoid content. The topical anti-inflammatory effect of the ethyl acetate extract of *C. aconitifolius* was determined by the croton oil-induced mouse ear edema test, while flavonoid detection and concentration were determined by thin layer chromatography and the aluminum chloride colorimetric assay, respectively. Topical application of the extract showed significant inhibition of the induced-ear edema (23.52 and 49.41% for 25 and 50 mg/kg dose, respectively). The extract also exhibited the presence of flavonoids. The finding of the anti-inflammatory activity exerted by the *C. aconitifolius* and the identification of its active principles may suggest and support its use for inflammation treatment.

Keywords: *Cnidoscolus aconitifolius*; inflammation; croton oil; flavonoids

1. Introduction

Medicinal plants have been widely used for to treat several diseases since ancient times, and their application has spread throughout the world due to social diffusion. Nowadays, there are reports stating that medicinal plant extracts contain a large amount of phytochemical compounds responsible for their medicinal properties, such as terpenoids, essential oils, sterols, alkaloids, polysaccharides, tannins, anthocyanins, and phenolic compounds capable of stimulating the immmune system in disease. Currently, several studies are focused on plant extraction methods for preserving their biological properties and obtaining a good extraction yield [1–4].

In Mexico, there are two plant species known as Chaya: *Cnidoscolus chayamansa* and *Cnidoscolus aconitifolius*. The natural habitat of *C. aconitifolius* is in Mexico and Central America; it belongs to the family Euphorbiaceae and can grow on poor nutrient soils, besides being easily propagated and having a high resistance to pests and diseases [5].

Previously, it has been reported that both possess a high nutritional and antioxidant content, and owing to this, they are considered as a promising alternative to modulate hepatoprotection, lipid storage, insulin levels, pain, and inflammation [6–9].

Inflammation, considered as one of the principal responses of the immune system, involves redness, swelling, pain, heat, and dysfunction [10]. Non-steroidal anti-inflammatory drugs are used to treat inflammation. However, prolonged use may cause side effects. Hence, it is important to explore and develop new alternatives for the treatment of inflammation [11]. Previous studies have reported the physicochemical properties of *C. aconitifolius*. The chemical study carried out by Adeniran and Abimbade (2014) identified a new compound, 2,3-dimethoxy-5-vinylbenzene-1,4-dioic acid, and a known aromatic compound, 1,4-dimethylbenzene-1,4-dicarboxylate, in an ethanolic extract [12]. Godinez et al. (2019) identified for the first time eleven phenols in the species *C. aconitofolius* in aqueous and ethanolic extract [13]. Ajiboye et al. (2019) reported that the aqueous extract of *C. aconitifolius* showed the presence of total phenol and total flavonoids, in addition to determining its antioxidant activity [14]. On the other hand, García et al. (2014) demonstrated that the ethyl acetate extract from the leaf of *Cnidoscolus chayamansa*, a close relative of *C. aconitifolius*, has significant anti-inflammatory (with TPA-induced mouse ear edema) and cardioprotective activity (with ischemia/reperfusion (I/R) rat model), due to the presence of sterols, flavonoids, coumarins, and saponins [15]. The novelty of this work is the evaluation of biological activity in an animal model; therefore, this study aimed to determine the anti-inflamatory effect of an *C. aconitofolius* ethyl acetate extract and its total flavonoid content.

2. Materials and Methods

2.1. Plant Material

Fresh leaves of *C. aconitofolius* were collected from a local shrub located in Tonalá, Jalisco, Mexico, and the identification of the species was carried out by a specialist from the University of Guadalajara, Dr. Arturo Castro.

According to Jiménez-Aguilar and Grusak (2015), *C. aconitofolius* contains several minerals such as calcium, magnesium, potassium, phosphorus, sulfur, iron, sodium, and vitamin C [16]. Some physico-chemical characteristics of vegetable matter are: dark green color, pH of 1% solution 6.15 units.

2.2. Extract Preparation

Leaves were dried in an oven at 60 °C for 72 h, ground into fine powder using a mechanical grinder blender, and passed through an 850 μm sieve. For subsequent extraction, 15 g of powder were used in Soxhlet (Pyrex) at 77 °C using ethyl acetate as the solvent. This solvent is considered semipolar and has been used to extract both polar and non-polar compounds, and several studies have shown its efficacy in extracting polyphenols and flavonoids from plants [17]. The extracts were then concentrated in a rotary vacuum evaporator at 74 rpm and 45 °C. A final concentrate volume of 5 mL was recovered and stored at room temperature without light exposure. These operating conditions are suitable for the type of solvent used [18]. The extraction yield was calculated using the following equation:

$$\text{Yield of extraction}(\%) = \frac{\text{Weight of extract}}{\text{Dry weight of original sample}} \times 100 \qquad (1)$$

where the dry weight of the original sample was 15 g.

2.3. Phytochemical Analysis

2.3.1. Thin-Layer Chromatography

Thin-layer chromatography (TLC) was performed to confirm the presence of bioactive compounds as flavonoids; according to the United States Pharmacopeia [19], Quercetin (0.2 mg/mL) and Rutin (0.6 mg/mL) were used as standard controls. The TLC plate was developed using a mixture of ethyl acetate, water, glacial acetic acid, and formic acid (100:26:11:11), and test samples were diluted in methanol and water (4:1). A volume of 5 μL of each sample was applied to silica gel chromatoplates (Merck KGaA). Samples were run separately, and results were observed under UV light. The Retention factor (Rf) was calculated by dividing the distance traveled by the compound, by the distance traveled by the solvent.

2.3.2. Total Flavonoid Assay

Total flavonoid content was determined by the aluminum chloride colorimetric method as reported by Marinova et al. (2005) [20] and Pekal (2014) [21] with slight modifications. The samples were analyzed in triplicates. One milliliter of the extract was added to 6.4 mL of distilled water, 0.3 mL of 5% $NaNO_2$, 0.3 mL of 10% $AlCl_3$, and 2 mL of 1M NaOH. The absorbance was measured at 510 nm, and the total flavonoid content of *C. aconitifolius* extract was expressed as mg of quercetin equivalents (QE)/mL. A calibration solution containing 0–400 mg/L of quercetin was used in distilled water. Samples were analyzed in triplicates.

2.4. Anti-Inflammatory Study

2.4.1. Animals

Adult male Balb-C mice were housed under controlled temperature an illumination with food and water *ad libitum*. The experiments were conducted according to the guidelines established by the Federal Government of Mexico (NOM-062-ZOO-1999) [22], according to the "Guide for the Care and Use of Laboratory Animals" council for the National Institute of Health. All animal procedures were approved by the internal committee of CIATEJ who reviewed the protocol for the care of laboratory animals (approval number 2018-002-C).

2.4.2. Croton Oil-Induced Assay

Croton oil, indomethacin, and *C. aconitofolius* extract were dissolved in acetone and applied topically in the bundle and underside of the ear. All groups ($n = 6$) had croton oil applied in the right ear (1 mg/20 μL acetone), whereas the left ear received only acetone. Posterior to the croton oil application, indomethacin (10 mg/kg) was applied in the right ears, as well as *C. aconitifolius* extract (25 and 50 mg/kg) doses. After 4 h, mice were sacrificed to obtain ear biopsies (6 mm) to determine the anti-inflammatory activity by calculated weight differences between ears [23]. The control group did not receive any treatment. Data were expressed according to the following equation:

$$\%\text{Inhibition} = \frac{(\text{mean of ear biopsies weight}) \text{ control} - (\text{mean of ear biopsies weight}) \text{ samples}}{(\text{mean of ear biopsies weight}) \text{ control}} \times 100 \qquad (2)$$

2.5. Statistical Analysis

Results are expressed as mean $\pm$ SEM. One-way ANOVA was used to compare differences between mean followed by Dunnett's test as a *post hoc* to compared treated groups versus the control group. Significant differences were considered when $p < 0.05$. Statistical analyses were computed using Prism (GraphPad Version 8) statistical software.

3. Results

3.1. Phytochemical Analysis

The *C. aconitifolius* ethyl acetate extract analysis using TLC revealed a retention factor (Rf) of 0.79 and 0.93; these values were similar values obtained for the standard control Quercetin (Rf: 0.77 and 0.92) and different to values of standard control Rutin (Rf: 0.25 and 0.36) (Table 1).

Table 1. Retention factor values of *C. aconitifolius* extract and standard controls obtained during thin-layer chromatography (duplicate).

Sample	Rf	Presence
C. aconitifolius	0.79	Q
	0.93	Q
Quercetin	0.77	Q
	0.92	Q
Rutin	0.25	R
	0.36	R

Presence: Refer to similar values of Rf between extract and standard controls: Quercetin (Q) and Rutin (R).

In addition, TLC allowed to observe the colors of *C. aconitifolius* ethyl acetate extract and compared them to standard controls under UV light (Table 2 and Figure 1).

Figure 1. TLC results: (**a**) *C. aconitifolius* ethyl acetate extract, (**b**) Quercetin, (**c**) Rutin.

Table 2. Colorimetric determination of flavonoids observed in thin-layer chromatography under UV light.

Sample	Color	Presence
C. aconitifolius	Orange	Q
Quercetin	Orange	Q
Rutin	Yellow	R

Standard controls: Quercetin (Q) and Rutin (R).

Total flavonoids content of *C. aconitifolius* ethyl acetate extract was obtained using a standard calibration curve of quercetin solution as quercetin equivalents (mg QE/mL). *C. aconitifolius* contained 154.23 + 3.35 mg QE/mL as is shown in (Table 3).

Table 3. Total flavonoids determination of *C. aconitifolius* ethyl acetate extract.

Quercetin (mg/mL)	Mean (mg/mL)
154.58	
150.71	154.23 ± 3.35
157.39	

Results are expressed as mean ± standard error of mean (SEM). Values obtain to *C. aconitifolius* extract (10 mg/mL).

3.2. Induced Mouse Ear Edema with Croton Oil Assay

To test the anti-inflammatory effect of *C. aconitifolius* ethyl acetate extract, two doses of the extract (25 and 50 mg/kg) on mouse ear edema induced with croton oil were applied. As is shown in Figure 2, the application of *C. aconitifolius* extracts on-ear induced a significant change in ear weight for both 25 mg/kg (one-way ANOVA; $p < 0.05$) and 50 mg/kg (one-way ANOVA; $p < 0.01$) doses in induced mouse ear edema assay compared to the control group. Indomethacin induced a significant decrease in the change in ear weight compared with the control group (one-way ANOVA; $p < 0.001$). The percent of inhibition of inflammation elicited by the *C. aconitifolius* ethyl acetate extract was 23.52% for a dose of 25 mg/kg and 49.41% for a 50 mg/kg dose. Indomethacin induced 87.05% of inhibition over inflammatory effect elicited by croton oil (Table 4).

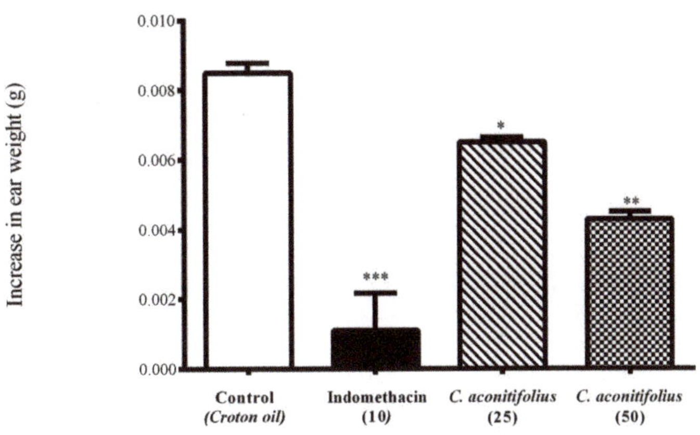

Figure 2. Anti-inflammatory effect by topical application of *C. aconitifolius* ethyl acetate extract on croton oil-induced ear edema. *C. aconitifolius* ethyl acetate extract in both doses decreased the ear edema. Indomethacin was used as positive control. Results are expressed as means ± SEM. N = 6. * $p < 0.05$; ** $p < 0.01$; *** $p < 0.001$. One-way ANOVA was followed by Dunnett's test.

Table 4. Ear weight differences and inhibition percentage in croton-oil-induced ear edema test.

Treatment	Dose (mg/kg)	Change in Ear Weight (g)	Inhibition (%)
Control	-	0.0085 ± 0.00028	0.00
Indomethacin	10	0.0011 ± 0.00106 ***	87.05
C. aconitifolius	25	0.0065 ± 0.00014 *	23.52
C. aconitifolius	50	0.0043 ± 0.00021 **	49.41

Results of change in weight are expressed as mean ± standard error of mean (SEM). N = 6 in each group * $p < 0.05$; ** $p < 0.01$; *** $p < 0.001$. One-way ANOVA. Control treatments refer to application of croton oil to induce acute inflammation.

4. Discussion

Topical application of the ethyl acetate extract of the *C. aconitifolius* leaf showed an anti-inflammatory biological activity in mouse ear edema induced by croton oil by decreasing the ear edema weight (Figure 2). In agreement with this, the percentage inhibition of the extract of *C. aconitifolius* ethyl acetate extract significantly inhibited the ear edema by 23.52% and 49.41% at 25 and 50 mg/kg doses, respectively (Table 4). The approach to determine flavonoid content confirmed the presence of quercetin in the *C. aconitifolius* ethyl acetate extract, which could be responsible for the decrease in inflammation. Flavonoids belong to the group of natural polyphenolic compounds with more than 4000 identified varieties. A large number of epidemiological, in vitro, and in vivo studies have documented the anti-inflammatory properties of a wide variety of flavonoids in different chronic inflammatory conditions, such as autoimmune diseases, cancer, diabetes, cardiovascular disorders, and neurodegenerative diseases [24].

Our results suggest that *C. aconitifolius* extract has anti-inflammatory activity in the induced mouse ear edema by inhibiting the action of croton oil. A previous report describes the edema inhibition by topical application of *Cnidoscolus chayamansa* agents [7], another species of the genus *Cnidoscolus*. Similar results have been reported using an aqueous and ethanolic extract of *C. aconitifolius* with anti-inflammatory activity, decreasing the cytokines TNF-α and IL-6 by 46 and 48.38% in macrophages stimulated by lipopolysaccharide [25]. The in vitro anti-inflammatory activity exherted by *C. aconitifolius* ethyl acetate extract coincides with the effects reported on a carrageenan-induced paw edema of an *C. aconitifolius* ethanolic extract by percolation [1,26].

On other hand, the effect of indomethacin, a non-selective inhibitor of cyclooxygenase (COX) that reduces the production of prostaglandins, promoting pain and inflammation [27,28], as was expected, was the most effective to prevent mouse ear edema. Nevertheless, the outcome of this study will serve as the basis for further investigations in order to isolate bioactive compounds from *C. aconitifolius* ethyl acetate extract and test its anti-inflammatory properties in order to obtain a compound with similar activity to the indomethacin drug. The presence of quercetin flavonoids in *C. aconitifolius* ethyl acetate reported in this study and in the ethanolic extract was determined in the phytochemical characterization using TLC and aluminum chloride colorimetric assay [25]. Among flavonoids, quercetin has been described as the most abundant in vegetables, and their biological effects have been extensively studied [29]. Flavonoids target prostaglandins which are involved in the late phase of acute inflammation [30].

A relation between the presence of quercetin and the anti-inflammatory effect of *C. aconitifolius* is possible based on both in vitro demonstration of the inhibition of inflammation-producing enzymes COX [31] and in vivo amelioration of the inflammation induced by carrageenan [32].

5. Conclusions

The results of this work underline the role of *C. aconitifolius* extract on induced mouse ear edema with croton oil. It is disclosed that the application of 25 and 50 mg/kg of ethyl acetate extract of the *C. aconitifolius* showed anti-inflammatory activity, which can be explained by the quercetin extract that it contains. The obtained results of this study reveal the potential application of *C. aconitifolius* extract in the pharmaceutical field. However, further investigations are required regarding extraction process optimization, isolation and identification of bioactive compounds, and elucidation of cellular and molecular mechanisms in biological activity.

Author Contributions: E.P.-C., O.R.T.-G., N.E.D.-M.: Investigation, Conceptualization, Methodology, Writing— Review and Editing; I.M.S.-H.: Methodology, Software; J.M.F.-F., O.R.H.-P.: Data Curation, Formal Analysis, Resources; I.M.S.-H., O.R.T.-G., N.E.D.-M.: Data Curation and Visualization, Investigation, Formal Analysis; E.P.-C.: Investigation, Writing—Review and Editing, Acquisition of Financial Support for the Project, Resources, Management and Coordination Responsibility for the Research, Activity Planning and Execution. All authors have read and agreed to the published version of the manuscript.

Funding: This research received no external funding.

Institutional Review Board Statement: The study was an experimental trial approved by the internal committee of CIATEJ for animal health, approval number (2018-002-C).

Informed Consent Statement: Not applicable.

Data Availability Statement: The data underlying this article will be shared on reasonable request from the corresponding author.

Conflicts of Interest: The authors declare that there are no conflict of interest.

References

1. Nam, J.S.; Jagga, S.; Sharma, A.R.; Lee, J.H.; Park, J.B.; Jung, J.S.; Lee, S.S. Anti-inflammatory effects of traditional mixed extract of medicinal herbs (MEMH) on monosodium urate crystal-induced gouty arthritis. *Chin. J. Nat. Med.* **2017**, *15*, 561–575. [CrossRef]
2. Bahta, T.; Karim, A.; Periasamy, G.; Gebremedhin, G.; Ur-Rehman, N.; Bitew, H.; Hagazi, K. Analgesic, Anti-inflammatory and In-vitro Hyaluronidase Inhibitory Properties of the Leaf Extract and Solvent Fractions of Otostegia Fruticosa (Forssk.) Schweinf. ex Penzig. *Iran J. Pharm. Res.* **2020**, *19*, 218–230.
3. Alhawarri, M.B.; Dianita, R.; Razak, K.N.A.; Mohamad, S.; Nogawa, T.; Wahab, H.A. Antioxidant, Anti-Inflammatory, and Inhibition of Acetylcholinesterase Potentials of *Cassia timoriensis* DC. Flowers. *Molecules* **2021**, *26*, 2594. [CrossRef]
4. Baldino, L.; Scognamiglio, M.; Reverchon, E. Supercritical fluid technologies applied to the extraction of compounds of industrial interest from Cannabis sativa L. and to their pharmaceutical formulations: A review. *J. Supercrit. Fluids* **2020**, *165*, 104960. [CrossRef]
5. Ross-Ibarra, J. Origen y domesticación de la chaya (*Cnidoscolus aconitifolius* Mill I. M. Johnst): La espinaca Maya. *Mex. Stud.* **2003**, *19*, 287–302. [CrossRef]
6. Loarca-Pina, G.; Mendoza, S.; Ramos-Gomez, M.; Reynoso, R. Antioxidant, antimutagenic, and antidiabetic activities of edible leaves from *Cnidoscolus chayamansa* Mc. Vaugh. *J. Food Sci.* **2010**, *75*, H68–H72. [CrossRef] [PubMed]
7. Perez-Gonzalez, M.Z.; Gutierrez-Rebolledo, G.A.; Yepez-Mulia, L.; Rojas-Tome, I.S.; Luna-Herrera, J.; Jimenez-Arellanes, M.A. Antiprotozoal, antimycobacterial, and anti-inflammatory evaluation of *Cnidoscolus chayamansa* (Mc Vaugh) extract and the isolated compounds. *Biomed. Pharmacother.* **2017**, *89*, 89–97. [CrossRef] [PubMed]
8. Awoyinka, O.A.; Balogun, I.O.; Ogunnowo, A.A. Phytochemical screening andin vitro bioactivity of *Cnidoscolus aconitifolius* (Euphorbiaceae). *J. Med. Plant Res.* **2007**, *1*, 63–65.
9. Johnston, M.I.M.; Nkeiruka, U.; Nsofor, A.; Anayo, L.; Ekwutosi, T.; Ugochi, J. An Evaluation of the Phytochemical and Nutritional Compositions of Fresh Leaves of *Cnidoscolus aconitifolius. Int. J. Res. Stud. Biosci.* **2016**, *4*, 21–28.
10. Phumsuay, R.; Muangnoi, C.; Dasuni Wasana, P.W.; Hasriadi; Vajragupta, O.; Rojsitthisak, P.; Towiwat, P. Molecular Insight into the Anti-Inflammatory Effects of the Curcumin Ester Prodrug Curcumin Diglutaric Acid In vitro and In vivo. *Int. J. Mol. Sci.* **2020**, *21*, 5700. [CrossRef] [PubMed]
11. Guoyong, X.S.J.; Hongting, L.; Mingkun, A.; Feng, H.; Yiqun D.; Wei, T.; Yan, Z.; Yucheng, Z.; Minjian, Q. Chemical constituents and antioxidative, anti-inflammatory and anti-proliferative activities of wild and cultivated *Corydalis saxicola. Ind. Crop. Prod.* **2021**, *169*, 113647.
12. Adeniran, O.; Abimbade, S. Characterization of Compounds from Leaf Extracts of Tree Spinach (*Cnidoscolus aconitifolius* (Miller) I. M. Johnston. *Int. J. Sci. Res. Chem. Eng.* **2014**, *1*, 82–86.
13. Godínez-Santillán, R.I.; Chávez-Servín, J.L.; García-Gasca, T.; Guzmán-Maldonado, S.H. Phenolic characterization and antioxidant capacity of alcoholic extracts from raw and boiled leaves of *Cnidoscolus aconitifolius* (Euphorbiaceae). *Acta Bot. Mex.* **2019**, *126*, e1493. [CrossRef]
14. Ajiboye, B.O.; Oyinloye, B.E.; Agboinghale, P.E.; Ojo, O.A. *Cnidoscolus aconitifolius* (Mill.) I. M. Johnst leaf extract prevents oxidative hepatic injury and improves muscle glucose uptake ex vivo. *J. Food Biochem.* **2019**, *43*, e13065. [CrossRef] [PubMed]
15. García-Rodríguez, R.V.; Gutiérrez-Rebolledo, G.A.; Méndez-Bolaina, E.; Sánchez-Medina, A.; Maldonado-Saavedra, O.; Domínguez-Ortiz, M.Á.; Cruz-Sánchez, J.S. *Cnidoscolus chayamansa* Mc Vaugh, an important antioxidant, anti-inflammatory and cardioprotective plant used in Mexico. *J. Ethnopharmacol.* **2014**, *151*, 937–943. [CrossRef]
16. Jiménez-Aguilar, D.M.; Grusak, M.A. Evaluation of Minerals, Phytochemical Compounds and Antioxidant Activity of Mexican, Central American, and African Green Leafy Vegetables. *Plant Foods Hum. Nutr.* **2015**, *70*, 357–364. [CrossRef]
17. Thavamoney, N.; Sivanadian, L.; Tee, L.H.; Khoo, H.E.; Prasad, K.N.; Kong, K.W. Extraction and recovery of phytochemical components and antioxidative properties in fruit parts of *Dacryodes rostrata* influenced by different solvents. *J. Food Sci. Technol.* **2018**, *55*, 2523–2532. [CrossRef]
18. Rodríguez-De Luna, S.L.; Ramírez-Garza, R.E.; Serna Saldívar, S.O. Environmentally Friendly Methods for Flavonoid Extraction from Plant Material: Impact of Their Operating Conditions on Yield and Antioxidant Properties. *Sci. World J.* **2020**, *2020*, 6792069. [CrossRef]
19. The United States Pharmacopeia. *Procedures of Thin Layer Chromatography*; The United States Pharmacopeia: Rockville, MD, USA, 1979; pp. 7524–7526.
20. Marinova, D.; Ribarova, F.; Atanassova, M. Total phenolics and flavonoids in Bulgarian fruits and vegetables. *J. Univ. Chem. Technol. Metall.* **2005**, *40*, 255–260.
21. Pękal, A. Evaluation of Aluminium Complexation Reaction for Flavonoid Content Assay. *Food Anal. Methods* **2014**, *7*, 1776–1782. [CrossRef]
22. Gobierno de México. *Norma Oficial Mexicana, Especificaciones Técnicas para la Producción, Cuidado y uso de los Animales de Laboratorio*; Nom-062-Zoo; Diario Oficial de la Federación: Mexico City, Mexico, 1999; pp. 1–58.
23. Santos, I.J.M.; Leite, G.O.; Costa, J.G.M.; Alves, R.R.N.; Campos, A.R.; Menezes, I.R.A.; Freita, F.R.V.; Nunes, M.J.H.; Almeida, W.O. Topical Anti-Inflammatory Activity of Oil from *Tropidurus hispidus* (Spix, 1825). *Evid.-Based Complement. Altern. Med.* **2015**, *2015*, 140247. [CrossRef]
24. Ginwala, R.; Bhavsar, R.; Chigbu, D.I.; Jain, P.; Khan, Z.K. Potential Role of Flavonoids in Treating Chronic Inflammatory Diseases with a Special Focus on the Anti-Inflammatory Activity of Apigenin. *Antioxidants* **2019**, *8*, 35. [CrossRef] [PubMed]

25. Us-Medina, U.; Millán-Linares, M.d.C.; Arana-Argaes, V.E.; Segura-Campos, M.R. Actividad antioxidante y antiinflamatoriain vitro de extractos de chaya (*Cnidoscolus aconitifolius* (Mill.) I.M. Johnst). *Nutr. Hosp.* **2020**, *37*, 46–55.
26. Onasanwo, S.A.; Oyagbemi, A.A.; Saba, A.B. Anti-inflammatory and analgesic properties of the ethanolic extract of (*Cnidoscolus aconitifolius* in rats and mice. *J. Basic Clin. Physiol. Pharmacol.* **2011**, *22*, 37–41. [CrossRef] [PubMed]
27. Prasad, J.D.; Gunn, K.C.; Davidson, J.O.; Galinsky, R.; Graham, S.E.; Berry, M.J.; Bennet, L.; Gunn, A.J.; Dean, J.M. Anti-Inflammatory Therapies for Treatment of Inflammation-Related Preterm Brain Injury. *Int. J. Mol. Sci.* **2021**, *22*, 4008. [CrossRef]
28. Castro, J.P.; Ocampo, Y.C.; Franco, L.A. In vivo andin vitroanti-inflammatory activity of *Cryptostegia grandiflora* Roxb. ex R. Br. leaves. *Biol. Res.* **2014**, *47*, 1–8. [CrossRef]
29. Kelly, G.S. Quercetin Monograph. *Alter. Med. Rev.* **2011**, *16*, 172–194.
30. Rajnarayana, K.; Reddy, M.S.; Chaluvadi, M.R.; Krishna, D.R. Bioflavonoids classification, pharmacological, biochemical effects and therapeutic potential. *Indian J. Pharmacol.* **2001**, *33*, 2–16.
31. Lee, K.M.; Hwang, M.K.; Lee, D.E.; Lee, K.W.; Lee, H.J. Protective effect of quercetin against arsenite-induced COX-2 expression by targeting PI3K in rat liver epithelial cells. *J. Agric. Food Chem.* **2010**, *58*, 5815–5820. [CrossRef]
32. Morikawa, K.; Nonaka, M.; Narahara, M.; Torii, I.; Kawaguchi, K.; Yoshikawa, T.; Kumazawa, Y.; Morikawa, S. Inhibitory effect of quercetin on carrageenan-induced inflammation in rats. *Life Sci.* **2003**, *74*, 709–721. [CrossRef] [PubMed]

Article

Very Fast RP–UHPLC–PDA Method for Identification and Quantification of the Cannabinoids from Hemp Oil

Nicoleta Mirela Blebea [1,2,*], **Dan Rambu** [3], **Teodor Costache** [3] **and Simona Negreș** [4]

[1] Faculty of Pharmacy, "Carol Davila" University of Medicine and Pharmacy, 6 Traian Vuia Street, 020956 Bucharest, Romania
[2] Pharmacology and Pharmacotherapy Department, Faculty of Pharmacy, Ovidius University, 900527 Constanta, Romania
[3] Research Center for Instrumental Analysis SCIENT, 1E Petre Ispirescu Street, Tancabesti, 077167 Ilfov, Romania; dan.rambu@scient.ro (D.R.); teodor.costache@scient.ro (T.C.)
[4] Department of Pharmacology, Faculty of Pharmacy, "Carol Davila" University of Medicine and Pharmacy, 6 Traian Vuia Street, 020956 Bucharest, Romania; simona_negres@yahoo.com
* Correspondence: nicoleta.blebea@gmail.com; Tel.: +40-735157107

Abstract: In recent years, hemp oils have become ubiquitous in health products on the European market. As the trend continues to grow and more cannabinoids are researched for their therapeutic benefits, more academic and industrial interests are drawn to this direction. Cannabidiol, Δ9-tetrahydrocannabinol, and their acidic forms remain the most examined cannabinoids in hemp and cannabis oils, in the case of cannabidiol due to its proven health implications in numerous articles, and in the case of Δ9-tetrahydrocannabinol, due to the legislation in the European area. These oils sold on the internet contain a wide range of cannabinoids that could demonstrate their effects and benefits. As a result of these claims, we developed a robust and rapid method that can identify and quantify 10 of the most common cannabinoids found in hemp oils: cannabivarin, cannabidiolic acid, cannabigerolic acid, cannabigerol, cannabidiol, cannabinol, Δ9-tetrahydrocannabinol, Δ8-tetrahydrocannabinol, cannabichromene, and tetrahydrocannabinolic acid in less than 11 min, with reverse-phase–high-performance liquid chromatography–photodiode matrix system (RP–UHPLC–PDA) equipped with C18 column, eluting in a gradient using water and acetonitrile with formic acid as mobile phases. The quantification of 9 sample products presented in different matrixes was performed using a calibration curve obtained by analyzing standard solutions from a 10-cannabinoid-mix-certified reference standard. The developed method demonstrated the ability to identify and quantify the main cannabinoids in hemp oil and is a useful tool for pharmaceutical professionals.

Keywords: *Cannabis sativa* L.; cannabinoids; hemp oil; UHPLC–PDA; validation

Citation: Blebea, N.M.; Rambu, D.; Costache, T.; Negreș, S. Very Fast RP–UHPLC–PDA Method for Identification and Quantification of the Cannabinoids from Hemp Oil. *Appl. Sci.* **2021**, *11*, 9414. https://doi.org/10.3390/app11209414

Academic Editor: Emanuel Vamanu

Received: 7 September 2021
Accepted: 7 October 2021
Published: 11 October 2021

Publisher's Note: MDPI stays neutral with regard to jurisdictional claims in published maps and institutional affiliations.

1. Introduction

Cannabis sativa L. (hemp), is an annual herbaceous plant belonging to the *Cannabaceae* family and has been used for ages to produce hemp fiber (for clothing, rope, and paper), seeds and also as a medicinal plant [1,2]. *Cannabis sativa* L. has also been used for recreational and medical purposes [3,4]. *Cannabis* has a complex chemical composition, with approximately 540 metabolites reported, such as (phyto) cannabinoids, terpenoids, flavonoids, and secondary metabolites used to treat epilepsy, Alzheimer's disease, Parkinson's disease, multiple sclerosis, pain and nausea in cancer patients, diabetes, and eating disorders [5,6]. The most active of these are cannabinoids, a class represented by over 150 terpenophenolic compounds that accumulate mainly in the resin secreted from the trichomes of female plants [3,6,7]. The therapeutic properties of cannabis are attributed to cannabinoids [8]. According to its use, cannabis is divided into two distinct groups: marijuana (medicinal and recreational) and industrial hemp. Hemp serves more as an agricultural commodity, being appreciated for its fibers and seeds and, recently, for the properties of cannabidiol

(CBD) [9,10]. The two major neuroactive components in *Cannabis* plants are the main psychoactive cannabinoid, Δ9-tetrahydrocannabinol (THC), and the non-psychoactive cannabinoid, CBD [11]. THC and CBD are neutral homologues of tetrahydrocannabinolic acid (THCA) and cannabidiolic acid (CBDA), respectively [12]. A classification of cannabinoids can be made due to their chemical structure by dividing them into 11 groups, including cannabigerol (CBG), THC, cannabichromene CBD, cannabichromene (CBC), cannabinol (CBN), Δ8-tetrahydrocannabinol (Δ8-THC), cannabicyclol (CBL), cannabinodiol (CBND), cannabielsoin (CBE), cannabitriol (CBT), and miscellaneous types [13]. CBD products have grown in popularity due to their low THC content, as well as due to the medical benefits attributed to CBD [14]. As a result, a multitude of products are marketed as supplements, improved formulas with CBD, the most marketed being CBD oils [14]. As CBD oil consumers have limited means to analyze their chemical composition, they may accidentally purchase products with undesirable properties given the different effects of cannabinoids [15]. As a result, it is important to implement quality control methods so that consumers are confident that CBD products have the desired effects [5,16,17]. The legal status of *Cannabis* compounds is different from one country to another. There are countries in which THC and CBD are classified in the same class of prohibited substances, while in other countries CBD products are legal [1,18]. For this reason, it is becoming increasingly important to have methods for quantifying the profile and cannabinoid content of CBD oils to ensure product uniformity and quality [19,20]. Cannabinoids may be detected by many and different analytical methods, including immunoassays (EMIT®, Elisa, fluorescent polarization, radioimmunotest). Are used techniques of flat chromatography; classic thin layer chromatography (TLC) [20], optimum performance laminar chromatography (OPLC) and multiple development automatization (AMD), gas chromatography–mass spectrometry (GC–MS) [21–23], high-performance liquid chromatography–mass spectrometry (HPLC–MS) [20,24–27]. TLC is useful in the laboratory for rapid screening of cannabinoid content in a sample [20]. Various analysis techniques are used in conjunction with HPLC to detect cannabinoids [20,25,27–31]. Methods include MS and ultraviolet (UV) [19,20,24–27,31]. UV detection is less expensive and simpler than detecting MS [31]; therefore, it is used for the quantification of major cannabinoids, and MS is used for the quantification of minor cannabinoids. In general LC–MS/MS techniques are considered more sensitive and selective analysis procedures than UHPLC analysis [20,27], but the UHPLC analysis is a more economical and accessible procedure for quantifying 10 cannabinoids from hemp oil [27,32]. Differences in cannabinoid concentrations justify the need to provide stronger regulation and control over the composition of hemp oil and CBD oil [20,28]. Individual doses throughout administration should be adjusted for CBD bioavailability [33]. This is of fundamental importance for the safety of consumers, as hemp oil preparations are also used for therapeutic purposes, regardless of whether they are registered as food supplements [20,28,34,35]. In addition to the studied methods, it is necessary to develop new procedures in the analysis of hemp oils for the identification and quantification of the 10 most relevant cannabinoids, which can be used in the pharmaceutical field. This study describes the development of the method and validation, the separation and detection of the main cannabinoids in hemp oil by reverse-phase ultra-high-performance liquid chromatography using the photodiode matrix detector (RP–UHPLC–PDA) with a single wavelength.

2. Materials and Methods

2.1. Chemicals and Materials

Analytical and chromatographic grade chemicals and solvents that were used for validation and analysis include methanol, purchased from VWR Chemicals (Radnor, PA, USA), acetonitrile from Merk (Darmstadt, Germany), and formic acid, purchased from Fisher Chemicals (Pittsburgh, UK). A mix standard of 10 phytocannabinoids in acetonitrile was provided at a concentration of 250 µg/mL for each component consisting of CBG, Δ9-THC, CBD, CBC, CBN, Δ8-THC, CBL, CBND, CBE, CBT and was obtained from Cayman Chemical (Ann Arbor, MI, USA). Ultrapure water was supplied by a Mili-Q water purification system from Millipore (Bedford, MA, USA).

2.2. Instrumentation

PerkinElmer Flexar FX-15 UHPLC system was used, equipped with a photodiode array detector (PDA), quaternary pump, column oven, and autosampler. Chromatographic separation was performed using a PerkinElmer Brownlee™ SPP C18, 2.7 μm, 3.0 mm × 150 mm column (PerkinElmer, Shelton, CT, USA). All data analysis, peak purity, and processing were performed using the PerkinElmer Chromera® CDS software.

2.3. Chromatographic Condition

The analysis was performed using as mobile phases 100% ultrapure water + 0.1% formic acid (solvent A) and 100% acetonitrile + 0.1% formic acid (solvent B) in a gradient elution mode, starting from 33% A and 67% B, increasing to 95% B in 5.5 min and maintaining at 95% B for 2 min. Equilibration time was 4.5 min at 33% A and 67% B before each injection, 10 μL sample volume + 5 μL air volume was injected in partial-loop mode. The flow rate was set at 1 mL/min, the temperature in the samples' compartment was 5 °C and 40 °C in the column oven. The acquisition was made at a wavelength of 228 nm.

2.4. Standard and Quality Control (QS) Solutions Preparation

A working stock solution of 50 μg/mL (also the first point on the calibration curve) was prepared by diluting 200 μL of the 10 phytocannabinoids standard mix with 800 μL acetonitrile. From this solution, serial dilutions were made at 25, 10, 5, and 2.5 μg/mL. The stock solution was also diluted to 16.7 μg/mL with acetonitrile, representing de QC solution.

2.5. Test Materials and Sample Preparation

Test materials consisting of 9 CBD (S1–S9) oils were obtained from reputable online retailers in order to validate the method on the most common matrixes used in the manufacturing process. Information about the composition and method employed in oil production was taken from the product label, and the resulting data were as follows: S1, S2, S5, and S6 were hemp extracts (HE) + hemp seed oil (HSO); S3 was produced by infusion under pressure in extra virgin hemp oil (INF + HO); S4 was produced by mixing HSO + CBD and terpenes (T); S7 by extracting with supercritical fluid CO_2 (SCO2); S8 was CBD isolate + HSO and S9 is CBD isolate + medium-chain triglyceride oil (MCT). The subsequent sample preparation was carried out as follows: 1000 mg of vegetal oil was accurately weighted on a calibrated semi-micro balance in a 15 mL centrifuge tube. The extraction was conducted by adding 10 mL of methanol over the sample and vortexing at high speed for 3 min. A full 2 mL syringe was collected from the extract and filtered through a 0.45 μm nylon filter into a glass tube, and 1 mL was transferred in a new tube and diluted with 2 mL of methanol; 1 mL was then transferred to a 10 mL volumetric flask, which was brought to volume with methanol (300-fold dilution); the last step was repeated for 3000-fold dilution.

2.6. Method Validation Parameters

The method was validated for specificity, precision, accuracy, linearity, the limit of detection (LOD), and limit of quantification (LOQ), according to the International Conference on Harmonization (ICH) guidelines (ICH Q2A 1994; ICH Q2B 1996) to assure the reliability of the results.

2.6.1. Specificity

The specificity was determined by matching the acquired spectra and elution order of the compounds from the reference solution with those from the sample solution considering the available standard chromatogram and data from literature [36]; for this purpose, 10 μL from blank, reference, and sample solutions were injected into the system. In order to asset the matrix effect and interferences from other coeluting analytes, peak purity analysis was performed.

2.6.2. Precision

The system precision or the agreement between the area values for each analyte was evaluated as repeatability at the same concentration level following 6 successive injections from individual 2.5 μg/mL reference solutions. It was expressed as relative standard deviation (RSD%) with an acceptance criterion of ≤5% RSD.

2.6.3. Accuracy

The accuracy, as the measurement of the closeness of experimental value to the actual amount, was established by injecting 3 standard solutions of known concentration (16.7 μg/mL) and reporting the practical concentration determined by the system to the theoretical one. The accuracy was expressed as percent recovery, where $100 \pm 10\%$ was considered acceptable.

2.6.4. Linearity

Linearity, defined as the ability of the method to obtain test results that are directly proportional to the analyte concentration within a specific range [37], was assessed in the range of 2.5–50 μg/mL (0.75–15 mg/g from 3000-fold dilution, or 7.5–150 mg/g for 300-fold dilution, in sample units) from three injections for each point on the curve using the least-squares method and by calculating the coefficient of determination R^2. An R^2 coefficient higher than 0.99 is considered an acceptable criterion of linearity.

2.6.5. Limit of Detection (LOD) and Limit of Quantification (LOQ)

LOD, described as the smallest concentrations of an analyte that can be reliably distinguished from zero, and LOQ, as the lowest concentration of the analyte that can be determined with acceptable repeatability and trueness [38], were estimated using the standard deviations of y-intercepts of the regression line (σ) and calibration curve slope (S) based on the following formulas:

$$LOD = 3.3 \, \sigma/S$$

$$LOQ = 10 \, \sigma/S$$

3. Results

3.1. Specificity

Spectra analysis and comparison confirmed the identity of each analyte in their eluting order on the reported method. As cannabis and hemp oil matrices contain a wide range of compounds that are extracted in the manufacturing process, good discrimination between matrix compounds and the target analytes was obtained with a suitable resolution that allows proper integration and quantification of each peak of interest, as shown in Figure 1. Retention times used for the identification of the compounds in the samples are displayed in Table 1.

At least 113 cannabinoids have been identified in cannabis and hemp plants. Due to the high probability of being coextracted and because of the close homology between the cannabinoid species, the risk of interferences and peak overlapping is increased. As shown in the table below, Table 2, the purity index provided for each sample shows that the method developed is capable of discriminating the target analytes from the sample matrix.

High values for CBC in S 2, CBN in S 4, and CBDV in S7 indicate a high probability that matrix-specific interferences are present at those particular retention times.

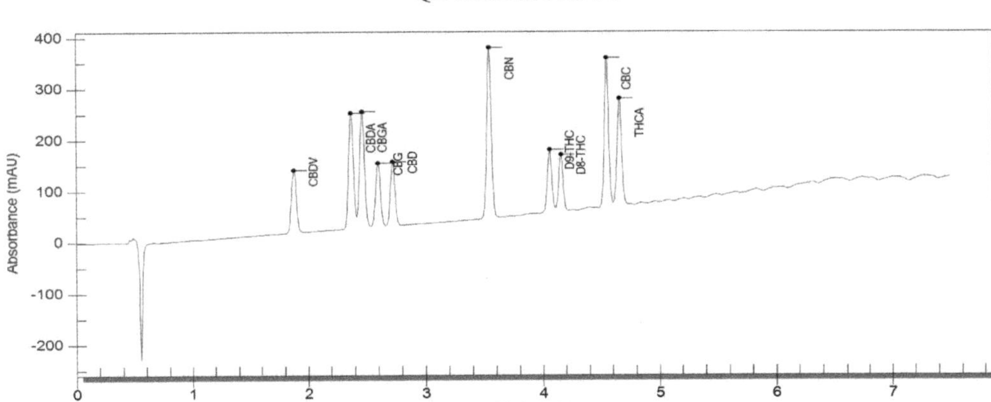

Figure 1. Chromatogram of 10 cannabinoids standard mix (16.7 µg/mL).

Table 1. Retention times (Rt) of cannabinoids identified in the standard mix.

Analyte	CBDV	CBDA	CBGA	CBG	CBD	CBN	Δ9-THC	Δ8-THC	CBC	THCA
Rt (min)	1.87	2.36	2.46	2.59	2.71	3.53	4.05	4.15	4.54	4.65

Table 2. Peak purity indexes of each peak from sample solutions (300 fold).

Analyte	S1	S2	S3	S4	S5	S6	S7	S8	S9
CBDV	1.29	1.02	N/A	1.62	1.26	N/A	4.2	1.82	1.96
CBDA	1.09	1.45	N/A	1.36	1.48	1.68	1.33	1.57	N/A
CBGA	2.07	N/A	N/A	N/A	1.46	1.52	1.67	N/A	N/A
CBG	1.23	1.05	N/A	N/A	1.09	N/A	1.41	N/A	N/A
CBD	1.29	1.06	1.12	2.21	1.13	1.09	1.29	1.33	1.07
CBN	N/A	1.68	1.42	3.37	N/A	N/A	1.19	N/A	N/A
Δ9-THC	1.29	2.06	1:19	N/A	1.54	1.15	1.40	N/A	N/A
Δ8-THC	N/A	N/A	N/A	1.27	N/A	N/A	N/A	N/A	N/A
CBC	1.14	7.23	2.15	1.39	1.05	1.05	N/A	N/A	N/A
THCA	1.42	1.19	N/A	N/A	1.15	1.21	N/A	N/A	N/A

3.2. Precision

Precision at the lowest point on the calibration curve (2.5 µg/mL) was evaluated based on reported concentration RSD%. The method was found to be reproducible, with an RSD value of 1.54% for CBDV; 0.45% for CBDA; 0.72% for CBGA; 1.88% for CBG; 1.63% for CBD%; 1.14% for CBN; 3.77% for Δ9-THC; 4.25% for Δ8-THC; 3.38% for CBC; 2.79% for THCA. These results indicate that the method is precise at the evaluated concentration.

3.3. Accuracy

To assess how close the values measured are from the true concentrations, accuracy at 16.7 µg/mL was calculated as percent recovery. The results indicated in Table 3 demonstrate a good accuracy of the proposed method at the evaluated concentration.

Table 3. Results for accuracy.

Analyte	Theoretical Concentration (µg/mL)	Amount Recovered (µg/mL)	Recovery (Mean ± %RSD)
THCA	16.69	16.69	100.12 ± 0.20
		16.75	
		16.69	
CBDV	16.71	16.88	101.38 ± 0.26
		16.97	
		16.94	
CBDA	16.70	16.92	101.38 ± 0.02
		16.92	
		16.93	
CBGA	16.71	16.92	101.29 ± 0.05
		16.92	
		16.91	
CBG	16.67	16.82	101.01 ± 0.12
		16.82	
		16.85	
CBD	16.71	16.83	100.89 ± 0.15
		16.86	
		16.88	
CBN	16.79	17.13	102.12 ± 0.15
		17.17	
		17.13	
Δ9-THC	16.65	17.01	102.00 ± 0.17
		16.96	
		16.97	
Δ8-THC	16.69	16.87	101.78 ± 0.80
		16.94	
		17.13	
CBC	16.67	16.88	101.05 ± 0.18
		16.82	
		16.83	

3.4. Linearity

The calibration curves were linear within the concentration range of 2.5–50 µg/mL (0.75–15 mg/g from 3000-fold dilution, or 7.5–150 mg/g for 300-fold dilution, in sample units) for each analyzed compound. The calibration curve exhibited a good linear regression, and a value higher than 0.999 for R^2 (coefficient of determination) was obtained for all cannabinoids, as shown in Table 4.

Given the variable range of concentrations at which cannabinoids from hemp and cannabis oils are found at different manufacturers, a two-dilution system was implemented that allows all the compounds analyzed to be included in the calibration curve domain thus validated.

Table 4. Results for linearity, regression equation, and R^2 for each analyte.

Analyte	Regression Equation	R^2
CBDV	y = 22,747x + 11,129	0.9999
CBDA	y = 37,845x + 26,076	0.9997
CBGA	y = 39,081x + 26,087	0.9996
CBG	y = 21,367x + 11,536	0.9999
CBD	y = 21,100x + 10,921	0.9998
CBN	y = 47,883x + 51,955	0.9990
Δ9-THC	y = 18,904x + 14,026	0.9997
Δ8-THC	y = 17,184x + 10,464	0.9998
CBC	y = 44,184x − 6791.7	0.9998
THCA	y = 34,511x − 7682.5	0.9997

3.5. Limit of Detection (LOD) and Limit of Quantification (LOQ)

LOD and LOQ were estimated from the data acquired for linearity. It was found that the main cannabinoids of interest can be detected at a concentration of 0.22 mg/g for Δ9-THC (0.022%) and 0.19 mg/g for CBD (0.019%) and can easy be quantified at 0.68 mg/g for Δ9-THC (0.068%) and 0.58 mg/g (0.058%) for CBD. The proposed method provides sufficient sensibility for Δ9-THC, around 10 times lower than the legal limit imposed in some countries (0.2%). Higher limits for CBN are probably due to different absorption at 228 nm. Limits for other cannabinoids that can be analyzed with this method are displayed in Table 5.

Table 5. Limit of detection (LOD) and limit of quantification (LOQ), determined by the calibration curves in standard units (µg/mL) and sample units (mg/g).

Analyte	LOD (µg/mL)	LOQ (µg/mL)	LOD (mg/g)	LOQ (mg/g)
CBDV	0.70	2.12	0.21	0.64
CBDA	1.69	5.13	0.51	1.54
CBGA	1.78	5.39	0.53	1.62
CBG	0.67	2.03	0.20	0.61
CBD	0.63	1.92	0.19	0.58
CBN	3.31	10.04	0.99	3.01
Δ9-THC	0.75	2.27	0.22	0.68
Δ8-THC	0.82	2.50	0.25	0.75
CBC	1.30	3.93	0.39	1.18
THCA	1.84	5.59	0.55	1.68

3.6. Cannabinoids in Hemp Oil

The method developed in this study was applied to the qualitative–quantitative analysis of the main cannabinoids from four samples of hemp oil. The amount of each cannabinoid was calculated using the equation obtained from a freshly prepared calibration curve. The following sequence was injected into the chromatographic system:

- 1× blank;
- 1× Calibration solution 2.5 µg/mL;
- 1× QC solution;
- 1× 3000-Fold samples;
- 1× 300-Fold samples;

- $1\times$ QC every 10 samples;
- $1\times$ QC after all samples;
- $1\times$ Blank;
- $1\times$ Wash.

As system suitability criteria, we chose to verify the accuracy of the 2.5 µg/mL calibration solution and QC solution; all the analyzed solutions passed the admissibility criteria (±10%).

Concentration in mg/g was calculated by following formula:

$$C\ (mg/g) = C\ (\mu g/mL) \times d/Swg \times 1000$$

where C (µg/mL) is the concentration extrapolated on the calibration curve, d is dilution factor (300 or 3000), Swg is sample weight in grams, and 1000 represents the transformation factor from µg to mg.

The main components CBD and CBDA were found in all samples, except two for CBDA in a concentration range of 1.42 to 166.32 mg/g for CBD and 1.62 to 18.80 mg/g for CBDA. Other cannabinoids also present in oils were CBDV (2.52 to 14.70 mg/g), CBGA, which was detected in two samples (1.19 to 2.64 mg/g) and as traces in the other two samples, CBG, which was quantified in only one sample (2.13 mg/g) and detected in other three, CBN (4.75 mg/g), which was found in one sample and identified in other three, Δ9-THC (1.06 to 1.35 mg/g), Δ8-THC, which was detected only in one sample in too low amount to be quantified, and CBC, which was present in six samples at a lower concentration, suggesting the need to decrease the dilution fold in order to be quantified and THCA (1.80–2.75 mg/g).

A more detailed view can be found in Table 6. Representative chromatograms of sample solutions S5 are shown in Figure 2a,b, and the rest of the chromatograms can be reviewed in the Supplementary Materials (Figures S1–S9).

Table 6. Results for hemp oils, analyzed with the proposed method and regression equations used for quantification.

Analyte	S1 HE + HSO		S2 HE + HSO		S3 INF + HO		S4 CBD + T + HSO		S5 HE + HSO		S6 HE + HSO		S7 SCO2		S8 CBD + HSO		S9 CBD + MCT	
	Rt	C	Rt	C	Rt	C	Rt	C	Rt	C	Rt	C	Rt	C	Rt	C	Rt	C
CBDV	1.84	<LOQ	1.89	14.70	N/A	<LOD	1.96	<LOQ	2.09	<LOQ	N/A	<LOD	1.81	2.52	1.94	<LOQ	2.03	<LOQ
CBDA	2.38	50.06	2.38	18.77	N/A	<LOD	2.38	1.62	2.54	17.81	2.54	14.65	2.38	6.79	2.34	<LOQ	N/A	<LOD
CBGA	2.48	1.19	N/A	<LOD	N/A	<LOD	N/A	<LOD	2.64	0.98	2.64	<LOQ	2.47	<LOQ	N/A	<LOD	N/A	<LOD
CBG	2.61	<LOQ < L	2.61	2.13	N/A	<LOD	N/A	<LOD	2.78	<LOQ	N/A	<LOD	2.60	<LOQ	N/A	<LOD	N/A	<LOD
CBD	2.73	2.34	2.73	70.09	2.73	2.53	2.73	<LOQ	2.90	4.37	2.90	1.42	2.73	166.32	2.80	89.88	2.90	50.67
CBN	N/A	<LOD	3.55	<LOQ	3.52	<LOQ	3.62	<LOQ	N/A	<LOD	N/A	<LOD	3.54	4.57	N/A	<LOD	N/A	<LOD
Δ9-THC	4.07	<LOQ	4.07	<LOQ	4.06	<LOQ	N/A	<LOD	4.22	1.35	4.22	<LOQ	4.06	1.06	N/A	<LOD	N/A	<LOD
Δ8-THC	N/A	<LOD	N/A	<LOD	N/A	<LOD	4.11	<LOQ	N/A	<LOD	N/A	<LOD	4.16	<LOD	N/A	<LOD	N/A	<LOD
CBC	4.55	<LOQ	4.55	<LOQ	4.55	<LOQ	4.55	<LOQ	4.69	<LOQ	4.70	<LOQ	4.55	<LOD	N/A	<LOD	N/A	<LOD
THCA	4.66	2.01	4.65	<LOQ	N/A	<LOD	4.67	<LOD	4.79	2.75	4.79	1.80	4.65	<LOD	N/A	<LOD	N/A	<LOD

Rt: retention time in min; S1–S10: sample 1–sample 10; C: concentration in mg/g; <LOQ: peak detected but too low to be quantified; <LOD: no peak detected; N/A: not applicable.

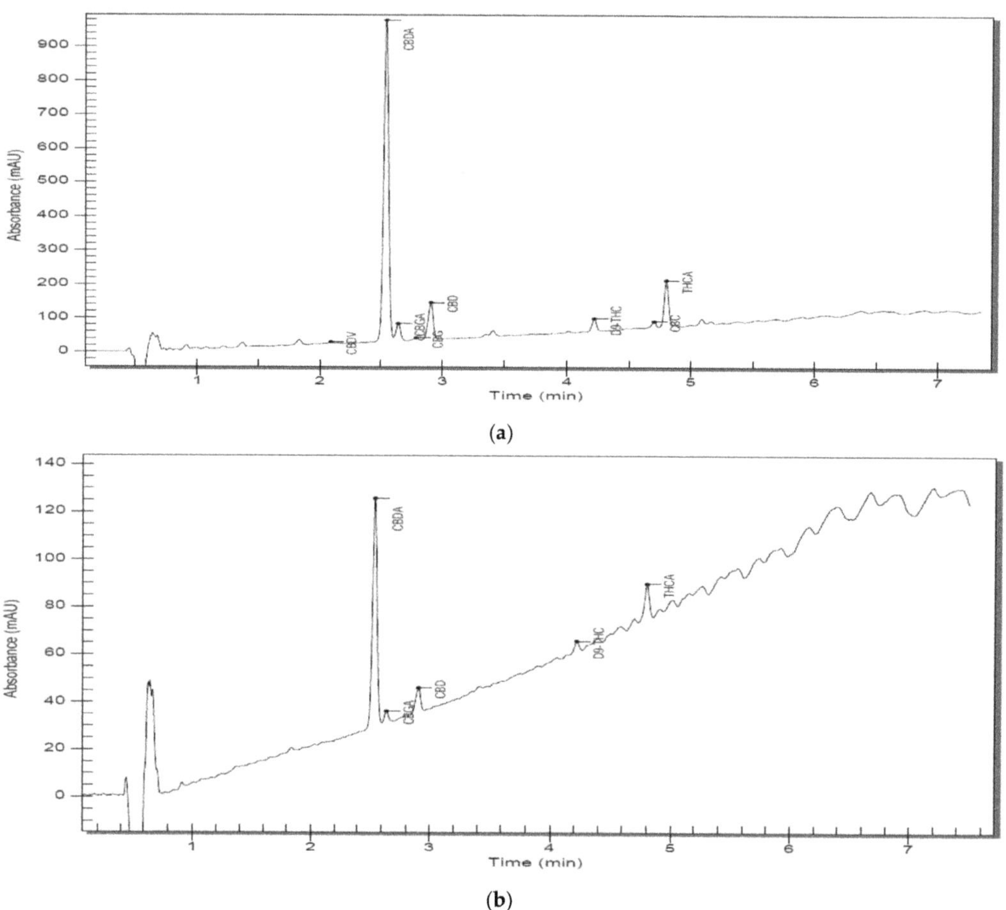

Figure 2. (**a**) Chromatographic trace of S5 D300 analyzed by RP–HPLC–PDA; (**b**) chromatographic trace of an S5 D3000 analyzed by RP–HPLC–PD.

4. Conclusions

The developed assay was validated by evaluating specificity, linearity, LOD, LOQ, accuracy, and precision and it was successfully applied to the analysis of hemp oils. The described method focuses on the quantification of major cannabinoids present in oils but can also be applied to check THC and other cannabinoids at levels >0.075% in order to assess the compliance with each country's own jurisdiction. Phytocannabinoids screening offers a unique view about the potency and health effects of hemp and cannabis oils that implies a synergic action of many compounds at various concentrations.

High throughput of samples can be analyzed in a working day, leading to economies of scale in regard to standards, solvents, energy, and analyst's time. Additionally, even if not all HPLC systems have the same performance, the proposed method can be operated on a wide variety of high-pressure liquids chromatographs due to the acceptable pressure generated by the system, solvents column, and temperature (around 5100 psi or 350 bar for our system). The two dilutions of 300-fold and 3000-fold allow a wide range of unknown samples to be analyzed within the method concentration range. If the proposed method is intended to be used for quality control of finished products, the sample dilution can also be optimized for the desired purpose; with regard to the various compositions of the extracts that can bring interferences, simple adjustments can be employed when validating the method for a specific product.

Supplementary Materials: The following are available online at https://www.mdpi.com/article/10.3390/app11209414/s1, Figure S1-a: Chromatogram of S1 D300 analyzed by RP-UHPLC-PDA; Figure S1-b: Chromatogram of S1 D3000 analyzed by RP-UHPLC-PDA; Figure S2-a: Chromatogram of S2 D300 analyzed by RP-UHPLC-PDA; Figure S2-b: Chromatogram of S2 D3000 analyzed by RP-UHPLC-PDA; Figure S3-a: Chromatogram of S3 D300 analyzed by RP-UHPLC-PDA; Figure S3-b: Chromatogram of S3 D3000 analyzed by RP-UHPLC-PDA; Figure S4-a Chromatogram of S4 D300 analyzed by RP-UHPLC-PDA; Figure S4-b: Chromatogram of S4 D3000 analyzed by RP-UHPLC-PDA; Figure S5-a: Chromatogram of S5 D300 analyzed by RP-UHPLC-PDA; Figure S5-b: Chromatogram of S5 D3000 analyzed by RP-UHPLC-PDA; Figure S6-a: Chromatogram of S6 D300 analyzed by RP-UHPLC-PDA; Figure S6-b: Chromatogram of S6 D3000 analyzed by RP-UHPLC-PDA; Figure S7-a: Chromatogram of S7 D3000 analyzed by RP-UHPLC-PDA; Figure S7-b; Chromatogram of S7 D300 analyzed by RP-UHPLC-PDA; Figure S8-a: Chromatogram of S8 D300 analyzed by RP-UHPLC-PDA; Figure S8-b: Chromatogram of S8 D3000 analyzed by RP-UHPLC-PDA; Figure S9-a: Chromatogram of S9 D300 analyzed by RP-HPLC-PDA; Figure S9-b: Chromatogram of S9 D3000 analyzed by RP-HPLC-PDA.

Author Contributions: Conceptualization, N.M.B. and D.R.; methodology, N.M.B.; software, T.C.; validation, D.R., T.C. and N.M.B.; formal analysis, D.R.; investigation, S.N.; resources, T.C.; data curation, D.R.; writing—original draft preparation, N.M.B.; writing—review and editing, N.M.B.; visualization, S.N.; supervision, T.C.; project administration, T.C.; funding acquisition, T.C. All authors have read and agreed to the published version of the manuscript.

Funding: This research received no external funding.

Institutional Review Board Statement: Not applicable.

Informed Consent Statement: Not applicable.

Data Availability Statement: Not applicable.

Acknowledgments: This study was performed in collaboration with the Research Center of Instrumental Analysis SCIENT, 1E Petre Ispirescu Street, 77167 Tancabesti, Ilfov, Romania.

Conflicts of Interest: All the authors declare that there is no conflict of interest with this study.

Abbreviations

CBDV	Cannabidivarin
CBDA	Cannabidiolic acid
CBGA	Cannabigerolic acid
CBG	Cannabigerol
CBD	Cannabidiol
CBND	Cannabinodiol
CBE	Cannabielsoin
CBN	Cannabinol
Δ9-THC	Δ9-Tetrahydrocannabinol
Δ8-THC	Δ8-Tetrahydrocannabinol
CBC	Cannabichromene
THCA	Tetrahydrocannabinolic acid
CBL	Cannabicyclol
CBT	Cannabitriol

References

1. Lafaye, G.; Karila, L.; Blecha, L.; Benyamina, A. Cannabis, cannabinoids, and health. *Dialogues Clin. Neurosci.* **2017**, *19*, 309–316. [PubMed]
2. Vlad, R.A.; Hancu, G.; Ciurba, A.; Antonoaea, P.; Rédai, E.M.; Todoran, N.; Muntean, D.L. Cannabidiol—Therapeutic and legal aspects. *Pharmazie* **2020**, *75*, 463–469. [PubMed]

3. Bonini, S.A.; Premoli, M.; Tambaro, S.; Kumar, A.; Maccarinelli, G.; Memo, M.; Mastinu, A. *Cannabis sativa*: A comprehensive ethnopharmacological review of a medicinal plant with a long history. *J. Ethnopharmacol.* **2018**, *5*, 300–315. [CrossRef]

4. Schluttenhofer, C.; Yuan, L. Challenges towards Revitalizing Hemp: A Multifaceted Crop. *Trends Plant. Sci.* **2017**, *22*, 917–929. [CrossRef] [PubMed]

5. Namdar, D.; Mazuz, M.; Ion, A.; Koltai, H. Variation in the compositions of cannabinoid and terpenoids in Cannabis sativa derived from inflorescence position along the stem and extraction methods. *Ind. Crop. Prod.* **2018**, *113*, 376–382. [CrossRef]

6. Eržen, M.; Košir, I.J.; Ocvirk, M.; Kreft, S.; Čerenak, A. Metabolomic Analysis of Cannabinoid and Essential Oil Profiles in Different Hemp (*Cannabis sativa* L.) Phenotypes. *Plants* **2021**, *10*, 966. [CrossRef] [PubMed]

7. Hanuš, L.O.; Meyer, S.M.; Muñoz, E.; Taglialatela-Scafati, O.; Appendino, G. Phytocannabinoids: A unified critical inventory. *Nat. Prod. Rep.* **2016**, *33*, 1357–1392. [CrossRef]

8. Hazekamp, A.; Choi, Y.H.; Verpoorte, R. Quantitative analysis of cannabinoids from *Cannabis sativa* using 1H-NMR. *Chem. Pharm. Bull.* **2004**, *52*, 718–721. [CrossRef]

9. Sandler, L.N.; Beckerman, J.L.; Whitford, F.; Gibsonet, K.A. Cannabis as conundrum. *Crop. Prot.* **2019**, *117*, 37–44. [CrossRef]

10. Blebea, N.M.; Bucur, L.A. Pharmacotherapeutic options in neoplastic diseases-part IV. *Farmacist.ro* **2021**, *4*, 15–18. [CrossRef]

11. Ibarra-Lecue, I.; Mollinedo-Gajate, I.; Meana, J.J.; Callado, L.F.; Diez-Alarcia, R.; Uriguen, L. Chronic cannabis promotes pro-hallucinogenic signaling of 5-HT2A receptors through Akt/mTOR pathway. *Neuropsychopharmacology* **2018**, *43*, 2028–2035. [CrossRef]

12. Aizpurua-Olaizola, O.; Soydaner, U.; Öztürk, E.; Schibano, D.; Simsir, Y.; Navarro, P.; Etxebarria, N.; Usobiaga, A. Evolution of the Cannabinoid and Terpene Content during the Growth of Cannabis sativa Plants from Different Chemotypes. *J. Nat. Prod.* **2016**, *79*, 324–331. [CrossRef]

13. Berman, P.; Futoran, K.; Lewitus, G.M.; Mukha, D.; Benami, M.; Shlomi, T.; Meiri, D. A new ESI-LC/MS approach for comprehensive metabolic profiling of phytocannabinoids in Cannabis. *Sci. Rep.* **2018**, *8*, 14280. [CrossRef]

14. VanDolah, H.J.; Bauer, B.; Mauck, K.F. Clinicians' Guide to Cannabidiol and Hemp Oils. *Mayo Clin. Proc.* **2019**, *94*, 1840–1851. [CrossRef]

15. Fischedick, J.T.; Hazekamp, A.; Erkelens, T.; Choi, Y.H.; Verpoorte, R. Metabolic fingerprinting of *Cannabis sativa* L., cannabinoids and terpenoids for chemotaxonomic and drug standardization purposes. *Phytochemistry* **2010**, *71*, 2058–2073.

16. Dussy, F.E.; Hamberg, C.; Luginbühl, M.; Schwerzmann, T. Isolation of Delta9-THCA-A from hemp and analytical aspects concerning the determination of Delta9-THC in cannabis products. *Forensic Sci. Int.* **2005**, *149*, 3–10. [CrossRef]

17. Fischedick, J.; Kooy, F.V.D.; Verpoorte, R. Cannabinoid receptor 1 binding activity and quantitative analysis of *Cannabis sativa* L. smoke and vapor. *Chem. Pharm. Bull.* **2010**, *58*, 201–207. [CrossRef] [PubMed]

18. Corroon, J.; Kight, R. Regulatory Status of Cannabidiol in the United States: A Perspective. *Cannabis Cannabinoid Res.* **2018**, *3*, 190–194. [CrossRef] [PubMed]

19. Aizpurua-Olaizola, O.; Omar, J.; Navarro, P.; Olivares, M.; Etxebarria, N.; Usobiaga, A. Identification and quantification of cannabinoids in Cannabis sativa L. plants by high performance liquid chromatography-mass spectrometry. *Anal. Bioanal. Chem.* **2014**, *406*, 7549–7560. [CrossRef] [PubMed]

20. Blebea, N.M.; Negreș, S. Methods for quantification of the main cannabinoids in CBD oil. In Proceedings of the Geolinks International Conference on Environmental Sciences 2021, Online Conference on Environmental Sciences, Burgas, Bulgaria, 17–18 May 2021; pp. 57–64.

21. Rodrigues, A.; Yegles, M.; Van Elsué, N.; Schneider, S. Determination of cannabinoids in hair of CBD rich extracts consumers using gas chromatography with tandem mass spectrometry (GC/MS-MS). *Forensic Sci. Int.* **2018**, *292*, 163–166. [CrossRef]

22. Cardenia, V.; Gallina Toschi, T.; Scappini, S.; Rubino, R.C.; Rodriguez-Estrada, M.T. Development and validation of a Fast gas chromatography/mass spectrometry method for the determination of cannabinoids in *Cannabis sativa* L. *J. Food Drug Anal.* **2018**, *26*, 1283–1292. [CrossRef]

23. Leghissa, A.; Hildenbrand, Z.L.; Foss, F.W.; Schug, K.A. Determination of cannabinoids from a surrogate hops matrix using multiple reaction monitoring gas chromatography with triple quadrupole mass spectrometry. *J. Sep. Sci.* **2018**, *41*, 459–468. [CrossRef]

24. Patel, B.; Wene, D.; Fan, Z.T. Qualitative and quantitative measurement of cannabinoids in cannabis using modified HPLC/DAD method. *J. Pharm Biomed. Anal.* **2017**, *146*, 15–23. [CrossRef]

25. Pellati, F.; Brighenti, V.; Sperlea, J.; Marchetti, L.; Bertelli, D. New Methods for the Comprehensive Analysis of Bioactive Compounds in *Cannabis sativa* L. (hemp). *Molecules* **2018**, *23*, 2639. [CrossRef] [PubMed]

26. Ciolino, L.A.; Ranieri, T.L.; Taylor, A.M. Commercial cannabis consumer products part 2: HPLC-DAD quantitative analysis of cannabis cannabinoids. *Forensic Sci. Int.* **2018**, *289*, 438–447. [CrossRef] [PubMed]

27. Rodríguez, E.; Szijj, J.; Inglott, A.; Azzopardi, L.M. Analytical Techniques Used for Analysis of Cannabinoids. *Cannabis Sci. Technol.* **2021**, *4*, 34–46.

28. Blebea, N.M.; Costache, T.; Negreș, S. The qualitative and quantitative analysis of CBD in hemp oils by UHPLC with PDA and applications. *Sci. Pap. Ser. D Anim. Sci.* **2019**, *62*, 138–142.

29. Raharjo, T.J.; Verpoorte, R. Methods for the analysis of cannabinoids in biological materials: A review. *Phytochem. Anal.* **2004**, *15*, 79–94. [CrossRef]

30. Citti, C.; Braghiroli, D.; Vandelli, M.A.; Cannazza, G. Pharmaceutical and biomedical analysis of cannabinoids: A critical review. *J. Pharm. Biomed. Anal.* **2018**, *147*, 565–579. [CrossRef]
31. Leghissa, A.; Hildenbrand, Z.L.; Schug, K.A. A review of methods for the chemical characterization of cannabis natural products. *J. Sep. Sci.* **2018**, *41*, 398–415. [CrossRef]
32. Mandrioli, M.; Tura, M.; Scotti, S.; Toschi, T.G. Fast Detection of 10 Cannabinoids by RP-HPLC-UV Method in *Cannabis sativa* L. *Molecules* **2019**, *4*, 2113. [CrossRef]
33. Pavlovic, R.; Nenna, G.; Calvi, L.; Panseri, S.; Borgonovo, G.; Giupponi, L.; Cannazza, G.; Giorgi, A. Quality Traits of "Cannabidiol Oils": Cannabinoids Content, Terpene Fingerprint and Oxidation Stability of European Commercially Available Preparations. *Molecules* **2018**, *23*, 1230. [CrossRef] [PubMed]
34. Hazekamp, A. The Trouble with CBD Oil. *Med. Cannabis Cannabinoids* **2018**, *1*, 65–72. [CrossRef]
35. Winkler, P. The analysis of pesticides and cannabinoids in cannabis using LC-MS/MS. In *Comprehensive Analytical Chemistry*; Ferrer, I., Thurman, E.M., Eds.; Elsevier: Amsterdam, The Netherlands, 2020; Volume 90, pp. 277–313.
36. Hazekamp, A.; Peltenburg, A.; Verpoorte, R.; Giroud, C. Chromatographic and Spectroscopic Data of Cannabinoids from *Cannabis sativa* L. *J. Liq. Chromatogr. Relat. Technol.* **2005**, *28*, 2361–2382. [CrossRef]
37. Le, T.H.H.; Phung, T.H.; Le, D.C. Development and Validation of an HPLC Method for Simultaneous Assay of Potassium Guaiacolsulfonate and Sodium Benzoate in Pediatric Oral Powder. *J. Anal. Methods Chem.* **2019**, *2019*, 1–9. [CrossRef]
38. Thompson, M.; Ellison, S.L.R.; Wood, R. Harmonized Guidelines for single-laboratory validation of method of analyses (IUPAC Technical Report). *Pure Appl. Chem.* **2002**, *74*, 835–855. [CrossRef]

Article

Influence of Drying Type of Selected Fermented Vegetables Pomace on the Natural Colorants and Concentration of Lactic Acid Bacteria

Emilia Janiszewska-Turak [1,*], Weronika Kołakowska [1], Katarzyna Pobiega [2] and Anna Gramza-Michałowska [3]

[1] Department of Food Engineering and Process Management, Institute of Food Sciences, Warsaw University of Life Sciences—SGGW, 02-787 Warsaw, Poland; weronika_kolakowska@sggw.edu.pl

[2] Department of Food Biotechnology and Microbiology, Institute of Food Sciences, Warsaw University of Life Sciences—SGGW, 02-787 Warsaw, Poland; katarzyna_pobiega@sggw.edu.pl

[3] Department of Gastronomy Science and Functional Foods, Faculty of Food Science and Nutrition, Poznań University of Life Sciences, Wojska Polskiego 31, 60-624 Poznań, Poland; anna.gramza@up.poznan.pl

* Correspondence: emilia_janiszewska_turak@sggw.edu.pl; Tel.: +48-22-5937366

Featured Application: Powders obtained from freeze-dried fermented pomace can be used as a source of pigment and LAB. After analysis, it could be stated that future tests for drying conditions, of convective drying as well as freeze-drying, are needed to obtain higher concentrations of bacteria.

Abstract: Nowadays, foods with probiotic bacteria are valuable and desired, because of their influence on human gut and health. Currently, in the era of zero waste, the food industry is interested in managing its waste. Therefore, the aim of the study was to determine the influence of drying process on the physicochemical properties of fermented vegetable pomace. The work included examining the influence of the lactic acid bacteria (*Levilactobacillus brevis*, *Lactiplantibacillus plantarum*, *Limosilactobacillus fermentum* and its mixture in the ratio 1:1:1) used for vegetable fermentation (beetroot, red pepper, carrot), obtaining pomace from fermented vegetables, and then selection of drying technique using the following methods: convection drying (CD) or freeze-drying (FD) on the physical and chemical properties of pomace. In the obtained pomace and its dried form, dry substance, water activity, color, and active substances such as betalains and carotenoids by spectrophotometric method and also bacteria concentration were evaluated. After fermentation of pomace from the same vegetable, a similar concentration of lactic acid bacteria was found as well as dry substances, color and colorants. Results of physico-chemical properties were related to the used vegetable type. After drying of pomace, it could be seen a high decrease in bacteria and colorant concentration (betalains, carotenoids) independently from drying and vegetable type as well as used starter cultures. The smallest change was observed for spontaneously fermented vegetables compared to those in which the starter culture was used.

Keywords: lactic acid fermentation; betalain; carotenoids; red pepper; beetroot; carrot; drying

Citation: Janiszewska-Turak, E.; Kołakowska, W.; Pobiega, K.; Gramza-Michałowska, A. Influence of Drying Type of Selected Fermented Vegetables Pomace on the Natural Colorants and Concentration of Lactic Acid Bacteria. *Appl. Sci.* **2021**, *11*, 7864. https://doi.org/10.3390/app11177864

Academic Editors: Emanuel Vamanu and Catarina Guerreiro Pereira

Received: 29 June 2021
Accepted: 24 August 2021
Published: 26 August 2021

Publisher's Note: MDPI stays neutral with regard to jurisdictional claims in published maps and institutional affiliations.

1. Introduction

The current trend to promote healthy eating is towards fermented products with probiotic properties. For this reason, fermented fruits and vegetables have become increasingly popular. The reason for this is the increase in the number of people who limit their lactose intake due to food intolerances or allergies to milk proteins. Fermented vegetable or fruit and their juices are therefore an interesting alternative for those people [1,2]. Fermentation technologies are classified as long-lasting food preservation processes, they depend mainly on: salt concentration in the brine and temperature, as well as the addition of starter strains and the degree of multiplication of lactic acid bacteria [3]. Fermented food has a special quality, texture and taste, and has a beneficial effect on health [4,5].

In the lactic acid fermentation process, the use of starter cultures increases the chance for the desired microbiota to dominate over the other microorganisms, and to carry out a proper and controlled lactic acid fermentation process. A properly selected starter culture may also fulfill a protective and technological function, which may lead to the elimination or reduction of chemical or thermal preservation methods [6]. Traditional fermentation of vegetables depends on the microorganisms found in the raw material and is carried out spontaneously. The utilization of the LAB strain depends on the fermented matrix. Using lactic acid bacteria could introduce a potential method to improve the sensory and nutritional quality of fermented food [7]. Plant-derived LAB strains such as *Lactiplantibacillus plantarum* are most commonly used for the fermentation of the plant matrix [8–10]. LAB strains such as *Levilactobacillus brevis* or *Limosilactobacillus fermentum* can also be used for the fermentation of fruits and vegetables due to the possibility of fermentation of sugars present in these raw materials [11].

The commonly used in the fermentation process raw materials are: carrot, tomato, red bell pepper, cabbage, cucumber and beetroot. The pickled products can be eaten whole (fermented vegetables) or transferred into juices. During the production of fermented juices, one of the technologies involves fermenting whole vegetables and then pressing them for juice. During pressing, pomace is produced, which is classified as a waste from the fruit and vegetable industry. Currently, waste management in the world is heading towards transforming as much as possible into useful products, using processes involving microorganisms [12,13]. What's more, pomaces are rich in active substances such as polyphenols, natural pigments, fiber, vitamins and minerals [14–16] and LAB. One of the ways of using pomace is adding it to food, most often in a dried powdered form [17].

Drying is a basic process in the food industry [18,19]. The quality of dried products depends on the application of the appropriate drying method. With the use of suitable parameters of the drying process, such as air velocity, temperature, humidity, and by using methods that do not aerate the dried material, each of the methods can be effective [20,21]. However, dehydrated foods may still undergo adverse changes, e.g., auto-oxidation of fat, oxidation of vitamin C, discoloration, or retrogradation of starch [22]. One of the simplest methods of drying raw materials is convection/air drying, it is characterized by simplicity of construction and easy operation of devices. Convection drying (CD) is a process in which the mass and heat move simultaneously. This method is not ideal because of the nutritional value, color, appearance and taste of the dried product deteriorate during drying [20,23].

The second popular method is freeze-drying (FD). The color of freeze-dried products is more similar to the raw material from which they are made than the droughts produced with the use of other drying methods [24–26]. Freeze-drying is the most conservative way of drying vegetables and fruits as well as LAB, because the resulting product is of fairly good quality, suitable for long-term storage at ambient temperature without losing its nutritional value, as well as properties such as color, texture and aroma [27,28].

To the best of our knowledge, drying of carrot pomace [29–31], using beetroot pomace or its extracts [32–34] were investigated, however, no information about the drying of fermented vegetable pomaces are available in the literature. The aim of the study was testing the possibility to manage pomace generated during the production of juices based on fermented vegetables as a potential source of pigments and LAB. The range of the work included (1) carrying out the fermentation process of selected vegetables, using the microbiota of each vegetable or starter cultures; (2) obtaining pomace; (3) in order to obtain higher stability, drying of the pomace by convective drying or freeze-drying; (4) testing the influence of different drying type (convective and freeze-drying) on the stability of pigments and lactic acid bacteria in fermented vegetable pomaces by determination of bacteria and pigment content before and after the drying process, determination of color, dry matter and water activity in vegetable pomace.

2. Materials and Methods

2.1. Raw Materials and Microorganisms

Beetroot (*Beta vulgaris*), carrot (*Dacus carota*), and red pepper (*Capsicum annuum* L.) were purchased from a local supermarket (Bronisze, Poland) and stored in a temperature range 4–6 °C maximum for 2 days before used. The strains were obtained from the American Type Culture Collection (ATCC, Manassas, VA, USA) and Collection of Industrial Microorganisms (KKP, Warsaw, Poland).

2.2. Technological Treatment

2.2.1. Fermentation

Preparation of Inocula

Different probiotic bacterial strains (LAB), chosen after the literature survey were used: (*Levilactobacillus brevis* KKP 804, *Limosilactobacillus fermentum* KKP 811, *Lactiplantibacillus plantarum* ATCC 4080). The bacterial strains were cultured on de Man Rogosa and Sharpe Agar (MRS; Biomaxima, Poland) and incubated at 37 °C $\pm$ 1 °C for 48 h. Bacterial inocula were prepared in sterile 0.85% NaCl (w/v) solution to reach a population of approximately 1×10^9 CFU/mL. The study was carried out by separately inoculating *L. brevis* (LB), *L. plantarum* (LP), and *L. fermentum* (LF) strains and a combination of tested strains in the ratio of 1:1:1 (L MIX). For spontaneous fermentation (SF) no lactic acid bacteria were added, only microbiota from vegetables was present.

Fermentation Process

The fermentation process was conducted in accordance with Janiszewska-Turak et al. [35] protocol. Vegetables were washed, peeled and sliced. Then were placed into glass jars to which 2% solution of NaCl in a proportion 1:1 w/w was added. A concentration of 1% v/v of the inoculum was added to the water. Jars were closed for the creation of anaerobic conditions. All experiments were left for 5–7 days depending on the vegetable (the pH in jars was tested from the 5th to 7th day until achieving a minimum pH of 3.9). After obtaining pH level, the fermentation process was stopped by placing jars into the refrigerator for at least 12 h. After a maximum of 2 days from the end, all samples were analyzed. All experiments were done in duplicate. Experiments were done in parallel.

2.2.2. Juice Pressing

The fermented vegetables were used to obtain pomaces. The process was made with a juicer model NS-621CES (Kuvings, Daegu, Korea). Separately, juice and pomace were collected. Pomace was used in this research.

2.2.3. Freeze-Drying

Obtained pomaces were placed on a petri dish and frozen at −40 °C (Shock Freezer HCM 51.20, Irinox, Treviso, Italy) for 5 h. Freeze drying was carried out in ALPHA 1–4 freeze dryer (Christ, Osterode, Germany) for 24 h at a heating shelf temperature of 30 °C and the constant pressure of 63 Pa, a safety pressure was set up at 103 Pa.

2.2.4. Convective Drying

Convective air drying (CD: forced air circulation at the level of 2 m s^{-1}) was performed in a laboratory convective dryer (made in our department, Warsaw, Poland) with an electronic scale (AXIS, Gdańsk, Poland). Vegetables were placed on the sieves and simultaneously heated by air with a set temperature of 45 °C.

2.3. Analytical Method

2.3.1. Dry Matter and Water Activity

Dry matter (d.m.) was evaluated in vegetables at each stage of the process. For all samples, gravimetric method was used. About 0.6–1 g of sample was placed in a dish and dried by vacuum drying method (Memmert VO400, Schwabach, Germany) under the pressure of 10 mPa in 75 °C for 24 h until constant weight according to information from Rybak et al. [28] and Janiszewska-Turak et al. [14]. Measurements were made in triplicate.

The water activity (a_w) was measured at a temperature of 25 °C by a Rotronic Hygroscop DT hygrometer (Rotronic, Zurich, Switzerland). All measurements were done in triplicate.

2.3.2. Color Analysis

The color components were measured using a colorimeter CR-5 (Konica Minolta, Sakai Osaka, Japan) in CIE L*a*b* system. The protocol of measurement and calculations was described by Rybak et al. [18]. Calibration was made with a white pattern (L* 92.49, a* 1.25, b* 1.92). The measurement was done on a glass transparent petri dish on which the pomace was placed at a 5-mm layer, with standard illumination C, illuminant D65, angle of observation 2° was used. All measurements were made in five repetitions.

2.3.3. Betalain Analysis

Spectrophotometric method described by Janiszewska and Włodarczyk [36] and Janiszewska-Turak et al. [14] with some modifications was used to measure betalain content. For this, a spectrophotometer (Spectronic 200; Thermo Fisher Scientific Inc., Waltham, MA, USA) was used. A phosphate buffer with pH 6.5 was used for pigment extraction. Sample of approximately of 0.7 g for raw pomace or 0.05 g for dried was mixed with 50 mL of phosphate buffer for 10 min. All measurements were made in tree repetition for each sample.

The determination of betalain concentration was calculated in terms of betanin (mg betanin/100 g d.m.) and vulgaxanthin-I (mg vulgaxanthin-I/100 g d.m.). Pigment content calculations were based on the absorption values measured at 538 nm for betanin, for vulgaxanthin-I at 476 nm. The absorbance at 600 nm was measured and used to correct the amount of impurities

2.3.4. Carotenoids Analysis

Carotenoids were measured in accordance to methodology presented by Janiszewska-Turak and Witrowa-Rajchert [37] based on spectrophotometric measurements: 0.3 g of the grounded raw or dried sample was weighed into a centrifuge tube with the addition of distilled water (20 mL) and Carrez I and II solutions (each of 1 mL) (VWR Chemicals BDH Prolabo, Leuven, Belgium). The absorbance of the colored solutions was measured at 450 nm (Spectronic 200; Thermo Fisher Scientific Inc., Waltham, MA, USA). The total carotenoid content (TCC) was determined on the basis of the equation presented in methodology [37]. The analysis was conducted in triplicate.

2.3.5. Microbiological Analysis

The total count of bacteria by pour plate method was used for enumerating the viable cell LAB. In sterile conditions, 10 g of the fermented pomace samples were homogenized (Stomacher 400 Circulator, Seward, UK) for 1 min with 90 mL of 0.85% sterile sodium chloride (NaCl) [35]. Serial dilutions of the homogenates were poured into plates with MRS agar counts at 30 ± 1 °C for 48 h for lactic acid bacteria counts. The concentration of LAB was recorded as log CFU per g dry substance. The samples were analyzed in triplicate.

2.4. Statistical Analysis

The obtained results were subjected to a statistical analysis using the Statistica 13 software (StatSoft, Warsaw, Poland), using one-way analysis of variance with Tukey HSD test at a significance level of $\alpha = 0.05$. The other parameters were determined using MS Excel 16.

3. Results and Discussion

Physical (Table 1) and chemical properties (Tables 1 and 2), as well as microbiological analysis (Table 2) of the tested pomaces, were presented.

Table 1. General properties of beetroot, carrot and red pepper pomace.

Sample Name	Dry Matter d.m. (%)	Water Activity a_w (-)	Color		
			L*	a*	b*
B	12.1 ± 0.2 b	-	13.1 ± 1.2 a	17.7 ± 2.9 efg	1.9 ± 0.8 d
B_CD	88.5 ± 1.4 c	0.42 ± 0.01 f	21.0 ± 0.7 c	11.9 ± 1.1 abc	1.9 ± 0.4 d
B_FD	97.5 ± 0.3 f	0.22 ± 0.00 b	34.0 ± 0.7 f	16.1 ± 0.3 def	5.0 ± 0.4 fgh
B_SF	5.2 ± 0.6 a	-	27.7 ± 1.7 d	23.9 ± 0.7 h	−2.8 ± 0.1 b
B_SF_CD	92.6 ± 0.8 de	0.37 ± 0.00 d	16.1 ± 2.0 ab	16.5 ± 2.1 def	0.1 ± 0.3 c
B_SF_FD	96. ± 0.13 ef	0.23 ± 0.01 b	27.7 ± 1.7 d	23.9 ± 0.7 h	−2.8 ± 0.1 b
B_LB	2.9 ± 0.6 a	-	18.3 ± 0.4 bc	14.3 ± 1.2 cde	3.9 ± 0.5 ef
B_LB_CD	90.5 ± 0.0 cd	0.41 ± 0.00 e	13. ± 1.15 a	11.6 ± 1.3 abc	2.3 ± 0.4 d
B_LB_FD	95.2 ± 1.3 ef	0.24 ± 0.01 b	31.3 ± 0.4 def	17.6 ± 0.3 def	3.3 ± 0.3 de
B_LP	9.3 ± 0.1 b	-	14.3 ± 1.8 ab	21.1 ± 2.2 gh	−1.7 ± 0.9 b
B_LP_CD	90.2 ± 2.4 cd	0.46 ± 0.01 g	16.3 ± 1.2 ab	9.2 ± 1.2 a	−2.5 ± 0.2 b
B_LP_FD	94.9 ± 0.1 ef	0.14 ± 0.00 a	29.4 ± 0.6 de	18.4 ± 0.3 fg	−5.6 ± 0.2 a
B_LF	4.0 ± 0.0 a	-	21.1 ± 0.5 c	13.1 ± 0.9 bcd	6.0 ± 0.4 h
B_LF_CD	95.3 ± 0.1 ef	0.27 ± 0.00 c	15.0 ± 7.3 ab	9.5 ± 4.7 a	3.0 ± 1.5 d
B_LF_FD	95.6 ± 0.4 ef	0.24 ± 0.00 b	32.8 ± 1.2 ef	13.3 ± 1.1 bcd	4.4 ± 0.5 efg
B_LMIX	4.6 ± 0.0 a	-	16.2 ± 0.3 ab	18.0 ± 0.7 efg	5.8 ± 0.4 gh
B_LMIX_CD	90.8 ± 1.6 cd	0.45 ± 0.01 g	14.8 ± 1.5 ab	11.5 ± 1.6 abc	1.6 ± 0.7 d
B_LMIX_FD	96.4 ± 0.2 ef	0.24 ± 0.00 b	33.6 ± 1.1 ef	17.6 ± 1.0 efg	3.2 ± 0.4 de
C	16.8 ± 0.1 a	-	45.3 ± 1.3 b	26.5 ± 0.4 bcd	34.9 ± 0.4 ghi
C_CD	89.8 ± 0.1 b	0.49 ± 0.02 c	38.1 ± 1.3 a	23.2 ± 2.0 a	23.9 ± 2.2 a
C_FD	96.7 ± 0.1 cd	0.21 ± 0.00 a	49.2 ± 0.9 c	23.3 ± 0.4 a	26.1 ± 0.6 a–d
C_SF	11.6 ± 1.1 a	-	44.2 ± 1.2 b	29.0 ± 0.2 de	32.3 ± 0.3 fgh
C_SF_CD	91.6 ± 0.1 bc	0.32 ± 0.00 b	49.5 ± 0.3 c	30.2 ± 1.2 e	29.3 ± 0.7 def
C_SF_FD	94.4 ± 0.6 bcd	0.22 ± 0.01 a	58.0 ± 1.5 d	24.4 ± 1.0 ab	25.6 ± 1.1 ab
C_LB	14.5 ± 0.6 a	-	45.2 ± 2.2 b	28.4 ± 1.8 de	34.9 ± 2.5 ghi
C_LB_CD	93.4 ± 0.4	0.35 ± 0.00 b	42.3 ± 1.9 b	30.6 ± 1.2 e	29.5 ± 1.2 def
C_LB_FD	97.9 ± 0.4 d	0.17 ± 0.01 a	58.0 ± 1.6 d	26.7 ± 0.6 b-d	30.3 ± 0.6 ef
C_LP	14.7 ± 0.0 a	-	44.3 ± 1.0 b	30.7 ± 0.3 e	36.2 ± 0.3 hi
C_LP_CD	95.0 ± 0.1 bcd	0.45 ± 0.07 c	38.7 ± 2.7 a	24.0 ± 3.7 ab	25.3 ± 5.5 ab
C_LP_FD	97.3 ± 0.7 cd	0.22 ± 0.01 a	58.1 ± 2.5 d	26.8 ± 1.7 b–e	30.5 ± 1.0 b–f
C_LF	12.9 ± 0.5 a	-	45.6 ± 1.1 b	29.2 ± 0.4 de	35.3 ± 1.0 ghi
C_LF_CD	91.6 ± 0.1 bc	0.46 ± 0.01 c	43.0 ± 1.9 b	27.8 ± 1.4 cde	27.4 ± 1.0 a–e
C_LF_FD	93.6 ± 2.7 bdc	0.14 ± 0.01 a	58.3 ± 0.5 d	24.4 ± 1.2 ab	27.6 ± 0.6 a–e
C_LMIX	11.0 ± 0.4 a	-	44.9 ± 1.2 b	29.8 ± 1.7 e	37.3 ± 1.6 i
C_LMIX_CD	89.5 ± 2.4 b	0.35 ± 0.00 b	43.9 ± 2.4 b	28.3 ± 1.9 de	29.4 ± 1.8 def
C_LMIX_FD	97.4 ± 0.1 cd	0.16 ± 0.00 a	55.6 ± 0.1 d	25.5 ± 1.0 abc	28.7 ± 0.0 b–e
P	13.0 ± 0.5 a	-	30.7 ± 1.8 abc	27.0 ± 2.7 bcd	27.9 ± 2.3 e–h
P_CD	81.9 ± 0.1 b	0.48 ± 0.07 bc	34.9 ± 1.9 cd	24.2 ± 0.6 abc	20.3 ± 1.7 abc
P_FD	91.7 ± 0.3 cd	0.19 ± 0.01 a	38.8 ± 4.8 de	20.5 ± 3.9 a	19.0 ± 2.2 a
P_SF	9.9 ± 0.5 a	-	31.0 ± 0.6 abc	31.3 ± 3.8 def	22.9 ± 2.6 a–d

Table 1. *Cont.*

Sample Name	Dry Matter d.m. (%)	Water Activity a$_w$ (-)	Color		
			L*	a*	b*
P_SF_CD	82.9 ± 0.4 b	0.50 ± 0.00 d	38.3 ± 1.7 de	32.2 ± 1.7 ef	25.7 ± 1.8 d–g
P_SF_FD	93.2 ± 0.0 d	0.31 ± 0.00 b	37.0 ± 3.3 de	24.5 ± 3.3 abc	26.8 ± 1.2 d–g
P_LB	8.6 ± 0.4 a	-	26.9 ± 0.8 a	30.7 ± 1.9 def	30.0 ± 1.5 ghi
P_LB_CD	83.9 ± 3.6 b	0.48 ± 0.00 c	38.3 ± 1.4 de	31.5 ± 1.5 def	23.9 ± 1.9 b–e
P_LB_FD	95.0 ± 0.1 d	0.12 ± 0.00 a	40.7 ± 4.5 ef	21.9 ± 3.5 ab	22.9 ± 3.4 a–d
P_LP	8.6 ± 0.2 a	-	28.9 ± 0.6 a	35.8 ± 1.0 f	33.1 ± 2.3 i
P_LP_CD	86.6 ± 1.7 bc	0.42 ± 0.00 c	44.6 ± 2.1 f	27.5 ± 3.1 cde	29.1 ± 2.5 e–h
P_LP_FD	96.3 ± 0.3 d	0.31 ± 0.00 b	30.4 ± 3.0 abc	26.7 ± 4.9 bcd	19.2 ± 4.8 ab
P_LF	8.4 ± 0.0 a	-	26.9 ± 1.6 a	31.5 ± 1.0 def	29.9 ± 2.3 f–i
P_LF_CD	82.5 ± 0.2 b	0.42 ± 0.00 c	37.5 ± 1.2 de	31.8 ± 0.6 def	27.8 ± 1.1 e–h
P_LF_FD	96.7 ± 0.0 d	0.31 ± 0.00 a	40.6 ± 1.8 ef	25.3 ± 1.3 abc	31.4 ± 1.8 hi
P_LMIX	9.0 ± 0.2 a	-	29.9 ± 0.8 ab	34.3 ± 1.4 f	33.4 ± 2.0 i
P_LMIX_CD	94.1 ± 0.2 d	0.44 ± 0.00 c	31.5 ± 1.1 abc	24.7 ± 0.6 abc	20.6 ± 1.2 abc
P_LMIX_FD	94.3 ± 0.0 d	0.33 ± 0.00 b	34.5 ± 5.6 bcd	23.4 ± 2.2 abc	24.6 ± 4.5 c–f

Abbreviations of sample name: B—Beet root or C—carrot or P—pepper; SF—spontaneously fermented, LB—*Levilactobacillus brevis*, LF—*Limosilactobacillus fermentum*, LP—*Lactiplantibacillus plantarum*, LMIX—*Lactobacillus* mixture of those 3 in proportion 1:1:1; CD—convective drying, FD—freeze drying; a, b … —different indexes for individual series mean statistically significant differences for each vegetable separately in a column at the level of $p < 0.05$.

Table 2. Concentration of LAB in raw or fermented pomace.

	Raw	FD	CD
	log CFU/g$_{s.d.}$ ± SD		
B_LB	6.70 ± 0.02	2.93 ± 0.06	<1
B_LP	6.61 ± 0.02	2.46 ± 0.12	<1
B_LF	6.63 ± 0.03	2.82 ± 0.06	<1
B_LMIX	5.06 ± 0.08	2.04 ± 0.03	<1
C_LB	6.29 ± 0.20	2.56 ± 0.06	<1
C_LP	6.05 ± 0.06	3.06 ± 0.05	<1
C_LF	6.72 ± 0.06	3.34 ± 0.04	<1
C_LMIX	5.65 ± 0.01	2.86 ± 0.03	<1
P_LB	6.65 ± 0.04	2.95 ± 0.04	<1
P_LP	7.40 ± 0.02	3.48 ± 0.04	<1
P_LF	7.16 ± 0.06	3.62 ± 0.03	<1
P_LMIX	6.07 ± 0.09	2.94 ± 0.02	<1

Abbreviations of sample name: B—beetroot or C—carrot or P—pepper; LB—*Levilactobacillus brevis*, LF—*Limosilactobacillus fermentum*, LP—*Lactiplantibacillus plantarum*, LMIX—*Lactobacillus* mixture of those 3 in proportion 1:1:1; CD—convective drying, FD—freeze drying.

3.1. Physical Properties of Pomace

3.1.1. Pomace Water Activity (a$_w$) and Dry Matter (d.m.)

From the physical properties in pomaces dry mater, water activity and color were measured. That information is needed to define the basic properties of the pomace obtained by convection drying or freeze-drying. Both the dry matter content and water activity indicate the storage properties of dried substances [38,39] while color is the main feature by which usually consumers rate the quality of the product [14,40]. Higher dry matter content and lower water activity may indicate good storage properties of the obtained product [20,41]. In pomace, dry matter ranged from 2.9–12.1 for beetroot pomace to 11.0–16.8 for raw carrot pomace. Dry matter for red pepper was placed in the middle between those two and ranged from 8.4 to 13.0. The highest content of dry matter in all non-dried pomace was observed for pomaces obtained from raw vegetables, non-used for fermentation. The fermentation process caused a decrease in dry matter, independently from vegetable or lactobacillus strains during the fermentation process. Similar values (7.4–8.1) for dry matter of fermented pepper were observed in the research of Hallmann et al. [42] who have fermented different cultivars of pepper. It is related to the fermentation process in which the activity of LAB linked to the mineral nitrogen availability can cause intensive growth of bacteria and lead to the tissue structure loosening. That loosening structure in the final step can cause releasing from the vegetable tissue non-dissolved substances [42]. Moreover, that the loose structure of vegetables during pressing the juice can transfer more substances into the juice than leave it inside the pomace.

After the drying process, the dry matter increased in all samples. The dry matter was above the value of 89, which makes the product dry and allows the product to be included in the drought. The exception was convection-dried red pepper pomace, which, as a result of convection drying, obtained only about 82% of the dry weight. This may be related to a high sugar content in the red pepper vegetable [43], which does not allow the removal of sufficient water under the presented experimental conditions. In all experiments, dry matter obtained after freeze-drying was statistically significantly higher than those from CD, the exceptions were samples from fermented by *Limosilactobacillus fermentum* for beetroot and carrot pomace and for fermented by a mixture of *Lactobacillus* red pepper pomace for which no statistically significant difference was seen.

Moisture content (100 minus dry matter) determines how much water is in food, while water activity (a_w) shows how that water will respond to microorganisms. The higher the water activity, the faster the growth of microorganisms (e.g., bacteria, yeasts or molds) [44,45]. It is stated that most of the microorganism activity starts when a_w is above 0.6 [46]. In the present research, a higher decrease was observed for beetroot and carrot pomaces dried in freeze dryer (Table 1). Water activity did not exceed 0.5 for the presented samples, in samples for FD water avidity was for beetroot pomaces at level 0.14–0.25, for carrot pomaces at level 0.14–0.22 and the highest for FD sample range was observed for red pepper pomaces 0.12–0.33. A water activity below 0.3 is beneficial for the stability of dried vegetables as it reduces the amount of water available for microbial growth and therefore powders can be stored longer [47]. Analyzing the samples, it could be stated that obtained after freeze-drying, dried pomace of beetroot and carrot can be treated as a stable material.

3.1.2. Pomace Color

Analyzing the color coefficients, it was seen that the drying process has changed the color coefficient of pomaces. For lightness, of all samples, raw or fermented with different types of LAB, a decrease after convective drying and an increase after freeze-drying were noted, an exception was raw beetroot for which after both dryings an increase was noted (Table 1). It can be related to the relatively faster degradation of the pomace top layer during convection drying, and more rapid evaporation of water from the sample surface, which could cause crust formation due to collapsing tissue walls by shrinkage [48] and partial degradation of pigments [49]. In the case of freeze-dried pomace—water is rapidly removed from the entire surface of the sample—which does not change the structure [18]. For all vegetables, beetroot, carrot and red pepper pomace drying has caused a significant decrease in coefficient a* after CD and no changes after FD. Exception samples of red pepper pomace obtained after fermentation with *Lactiplantibacillus plantarum* and LAB mixture for them also after FD decrease was seen. For coefficient b* of beetroot pomace, no clear correlation between drying type or used LAB in the fermentation process was noted, while for carrot and red pepper decrease in this parameter was observed. An exception was the samples obtained after spontaneous fermentation and with application of *Limosilactobacillus fermentum* for which increase in this coefficient was observed. It could be related to the low value in raw material in comparison to other raw samples. A similar observation was made for red pepper by Pinar et al. [50], who have used convective and freeze-drying processes. They mentioned that the chromatic parameters were influenced by the drying type, which is related to the discoloring effect during the drying process [50].

In all analyzed samples, no relation to the fermentation process and LAB in raw samples, independently from the vegetable, was observed.

3.2. Pigment Content

3.2.1. Pigment Content in Beetroot Pomace

Pigment content in beetroot pomaces was divided in measurement into red-violet betalain (Figure 1a) and yellow vulgaxanthin-I (Figure 1b). In the tested samples after drying process, two behaviors were observed in the pigment content. In raw beetroot pomace, pomaces fermented with *Limosilactobacillus fermentum* or *Lactiplantibacillus plantarum* no changes in red/violet pigment content were found while for the rest of the samples a significant decrease was observed. Almost the same observation was made for yellow pigment content in beetroot pomace, in raw beetroot pomace, pomaces fermented with *Lactiplantibacillus plantarum* or *Levilactobacillus brevis* where no changes were observed. For the rest of the samples, a statistically significant decrease was noted. The normal situation for the drying process is that the pigments are exposed to hot air in the CD and may also be degraded during the freeze-drying process. However, the pigments in the raw material are protected by the tissue that is not damaged during fermentation.

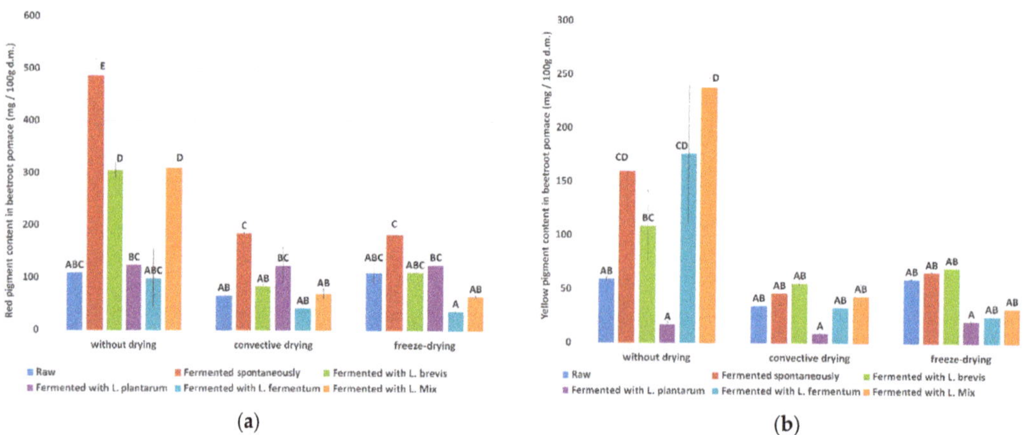

(a)　　　(b)

Figure 1. Pigment content in tested fermented beetroot pomace (a) Red pigment (betanin) content in beetroot pomace; (b) Yellow pigment content in fermented beetroot pomace. A, B . . . —different indexes for individual series mean statistically significant differences for given values at the level of $p < 0.05$.

The highest degradation of red/violet pigment after drying was observed for samples with fermentation by *Lactobacillus Mix* (77% and 79% for CD and FD) and *Levilactobacillus brevis* (73% and 64% for CD and FD), while for the yellow pigment it was observed for samples *Lactobacillus Mix* (82% and 86% for CD and FD) and *Limosilactobacillus fermentum* (81% and 86% for CD and FD). The smallest degradation after drying was observed in raw beetroot samples (about 40% and 0% for CD and FD), independently from the drying type and analyzed pigment. Furthermore, a small change was observed for *Lactiplantibacillus plantarum* (1% for red/violet pigment and 48% for yellow pigment), however, the amount of red and yellow pigment was the lowest in all analyzed samples.

After the fermentation process in raw beetroot pomace, a statistically significant increase in red/violet and yellow pigment content was observed, the exception was sample fermented by *Lactiplantibacillus plantarum* in which a decrease trend in yellow and no change in red pigment was evaluated.

In fermented samples, the pH during fermentation has changed and decreased below 4. In the case of beetroot fermentation, the obtained pH below 4 favors the maintenance of betalain pigments protection for which the optimal pH is 3–6 [51–53].

3.2.2. Pigment Content in Carrot and Red Pepper Pomace

In carrot and in red pepper, pomace carotenoid content as an equivalent of β-carotene was measured. Drying of pomace of carrot and red pepper, independently from drying technique, caused a decrease in carotenoid content, the exception was a sample of carrot pomace fermented with *Lactobacillus Mix* for which no statistically significant change was found. Comparison between carrot and red pepper pomace carotenoid content showed similarity in both vegetables (Figure 2).

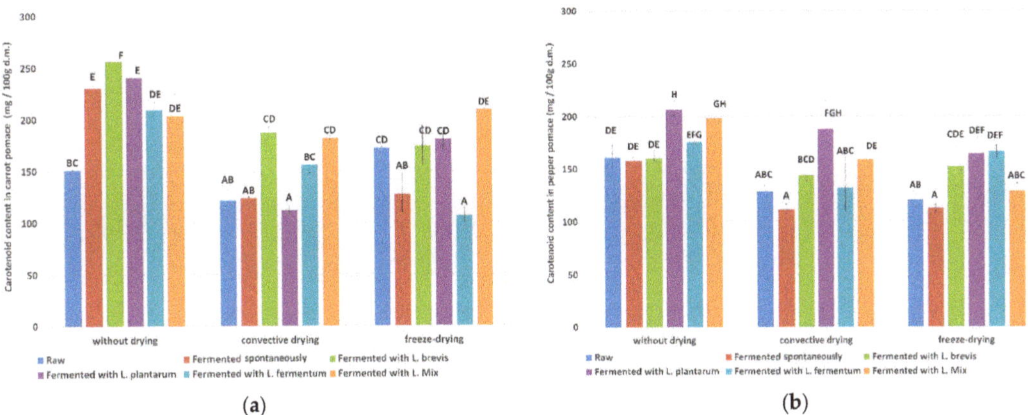

Figure 2. Pigment content in tested fermented (**a**) carrot pomace; (**b**) pepper pomace. A, B … —different indexes for individual series mean statistically significant differences for given values at the level of $p < 0.05$.

The highest degradation of β-carotene pigment after drying was observed for carrot samples with spontaneous fermentation (46% and 45% for CD and FD), *Lactiplantibacillus plantarum* (about 53% for CD) and *L. fermentum* (about 49% for FD). In red pepper samples degradation was at level of 30–35% for raw and spontaneously fermented red pepper pomace independently from drying type and for freeze-dried pomaces with *Lactobacillus Mix*.

After the fermentation process in carrot, a statistically significant increase in pomace pigment content was observed, while for red pepper only for samples fermented with *Lactiplantibacillus plantarum* and *L. Mix* a significant increase was seen. A similar observation was mentioned by Lee et al. [54] for fermented red pepper. After fermentation process, they observed increase in the content of carotenoids (especially the volatile one) and Bartkiene et al. [55] observed the same for fermented tomato pulp by selected starter cultures, e.g., *Lactobacillus sakei* or *Pediococcus acidilactici*, in which an increase in lycopene was evaluated. Mapelli-Brahm et al. [5] have concluded that behavior of carotenoids after fermentation varies depending on the plant material and conditions of the fermentation process. Moreover, when planning the fermentation process, it should be remembered that due to enzymatic activity, bacteria can increase the extraction of carotenoids from the tissue, and what is more, the fermentation process is not related to the degradation of carotenoids [5,56].

3.3. Effect of Convection Drying and Freeze-Drying on the Concentration of Lactic Acid Bacteria in Pomace

Table 2 shows the effect of the drying method on the concentration of lactic acid bacteria in fermented beetroot, carrot and pepper.

In raw pomace, the concentration of LAB ranged from 5.06 to 7.40. The LAB concentration depended both on the fermented raw material (fermented pepper pomace had more LAB than beetroot and carrot) and on the bacterial species used in fermentation (in the case of fermentation using a combination of strains, the LAB concentration was lower by 1 log cycle compared to fermentation using single strains of bacteria). After drying the pomace using convective drying, the LAB concentration dropped below the detection limit. Better results were obtained after lyophilization of fermented pomace, as the decrease in the concentration of LAB was at the level of 3–4 log cycles compared to raw pomace.

Until now, fermented fruit and vegetable pomace has been investigated as a dietary component, but no attempt has been made to test the LAB concentration after the drying. It has been shown that supplementation of blueberry pomace with carbon can contribute to increasing the viability of LAB during fermentation, which in turn leads to the creation of a new type of probiotic food [57]. Moreover, fermented mulberry pomace has demonstrated the possibility of creating functional foods from LAB [58]. Unfortunately, such food cannot be stored for a long time, therefore drying the fermented pomace can extend its shelf life. However, it is necessary to select an appropriate drying method to limit the decrease in the viability of lactic acid bacteria.

4. Conclusions

The food market is strongly interested to offer tasty and affordable products which would bring new functionality and health value. Therefore, a targeted fermentation and drying process of vegetable pomace offers an easy way to deliver probiotic bacteria into food products.

Our research showed that after fermentation of pomace from the same vegetable, independently from starter culture or spontaneous fermentation, a similar concentration of lactic acid bacteria was found as well as dry substances, color and colorants content. The drying process of pomace, however, caused a decrease in bacteria and colorant concentration (betalains, carotenoids) independently from drying and vegetable type as well as the used starter cultures. The slightest change was observed for spontaneously fermented vegetables in comparison to those in which the starter culture was used. Analysis of pomace physical properties indicated that freeze-dried pomaces of beetroot and carrot can be treated as a stable material with extended shelf life.

The results of the experiment showed that the fermented vegetable pomace is a promising approach towards the improvement of the functionality of food products. Obtained dried vegetable pomaces can be used in food formulations offering enrichment with pigments and probiotic bacteria with acceptable color attributes and stable quality during storage. Therefore, our findings provide a possible new direction for the design of functional products, however, further research considering the changes in sensory and polyphenolic compounds is required.

Author Contributions: Conceptualization, E.J.-T.; methodology, E.J.-T. and K.P.; validation, E.J.-T. and K.P.; formal analysis, E.J.-T.; investigation, W.K. and K.P.; data curation, E.J.-T., K.P. and A.G.-M.; writing—original draft preparation, E.J.-T., K.P. and A.G.-M.; writing—review and editing, E.J.-T., K.P. and A.G.-M.; visualization, E.J.-T.; supervision, E.J.-T. and A.G.-M. All authors have read and agreed to the published version of the manuscript.

Funding: This research received no external funding.

Data Availability Statement: Data available from corresponding author.

Conflicts of Interest: The authors declare no conflict of interest.

References

1. Zaręba, D.; Ziarno, M. Alternatywne probiotyczne napoje warzywne i owocowe. *Bromatol. Chem. Toksykol.* **2011**, *2*, 160–168.
2. Pimentel, T.C.; da Costa, W.K.A.; Barão, C.E.; Rosset, M.; Magnani, M. Vegan probiotic products: A modern tendency or the newest challenge in functional foods. *Food Res. Int.* **2021**, *140*, 110033. [CrossRef]

3. Kieliszek, M.; Pobiega, K.; Piwowarek, K.; Kot, A.M. Characteristics of the proteolytic enzymes produced by lactic acid bacteria. *Molecules* **2021**, *26*, 1858. [CrossRef]

4. Nair, M.R.; Chouhan, D.; Sen Gupta, S.; Chattopadhyay, S. Fermented foods: Are they tasty medicines for *Helicobacter pylori* associated peptic ulcer and gastric cancer? *Front. Microbiol.* **2016**, *7*, 1148. [CrossRef]

5. Mapelli-Brahm, P.; Barba, F.J.; Remize, F.; Garcia, C.; Fessard, A.; Mousavi Khaneghah, A.; Sant'Ana, A.S.; Lorenzo, J.M.; Montesano, D.; Meléndez-Martínez, A.J. The impact of fermentation processes on the production, retention and bioavailability of carotenoids: An overview. *Trends Food Sci. Technol.* **2020**, *99*, 389–401. [CrossRef]

6. de Souza, J.V.; Dias, F.S. Protective, technological, and functional properties of select autochthonous lactic acid bacteria from goat dairy products. *Curr. Opin. Food Sci.* **2017**, *13*, 1–9. [CrossRef]

7. Chen, Y.; Ouyang, X.; Laaksonen, O.; Liu, X.; Shao, Y.; Zhao, H.; Zhang, B.; Zhu, B. Effect of *Lactobacillus acidophilus*, *Oenococcus oeni*, and *Lactobacillus brevis* on composition of bog bilberry juice. *Foods* **2019**, *8*, 430. [CrossRef]

8. Ilango, S.; Antony, U. Probiotic microorganisms from non-dairy traditional fermented foods. *Trends Food Sci. Technol.* **2021**. [CrossRef]

9. Malik, M.; Bora, J.; Sharma, V. Growth studies of potentially probiotic lactic acid bacteria (*Lactobacillus plantarum*, *Lactobacillus acidophilus*, *and Lactobacillus casei*) in carrot and beetroot juice substrates. *J. Food Process. Preserv.* **2019**, *43*, e14214. [CrossRef]

10. Bontsidis, C.; Mallouchos, A.; Terpou, A.; Nikolaou, A.; Batra, G.; Mantzourani, I.; Alexopoulos, A.; Plessas, S. Microbiological and chemical properties of chokeberry juice fermented by novel lactic acid bacteria with potential probiotic properties during fermentation at 4 degrees C for 4 weeks. *Foods* **2021**, *10*, 768. [CrossRef]

11. Garcia, C.; Guerin, M.; Souidi, K.; Remize, F. Lactic fermented fruit or vegetable juices: Past, present and future. *Beverages* **2020**, *6*, 8. [CrossRef]

12. Okonko, I.O.; Adeola, O.; Aloysius, F.; Damilola, A.; Adewale, O. Utilization of food wastes for sustainable development. *Electron. J. Environ. Agric. Food Chem.* **2009**, *8*, 263–286.

13. Khubber, S.; Marti-Quijal, F.J.; Tomasevic, I.; Remize, F.; Barba, F.J. Application of fermentation to recover high-added value compounds from food by-products: Antifungals and Antioxidants. *Ferment. Process. Emerg. Conv. Technol.* **2021**, 195–219. [CrossRef]

14. Janiszewska-Turak, E.; Rybak, K.; Grzybowska, E.; Konopka, E.; Witrowa-Rajchert, D. The influence of different pretreatment methods on color and pigment change in beetroot products. *Molecules* **2021**, *26*, 3683. [CrossRef] [PubMed]

15. Grasso, S. Extruded snacks from industrial by-products: A review. *Trends Food Sci. Technol.* **2020**, *99*, 284–294. [CrossRef]

16. Michalska, A.; Wojdyło, A.; Lech, K.; Łysiak, G.P.; Figiel, A. Effect of different drying techniques on physical properties, total polyphenols and antioxidant capacity of blackcurrant pomace powders. *LWT* **2017**, *78*, 114–121. [CrossRef]

17. Šeregelj, V.; Ćetković, G.; Čanadanović-Brunet, J.; Šaponjac, V.T.; Vulić, J.; Lević, S.; Nedović, V.; Brandolini, A.; Hidalgo, A. Encapsulation of carrot waste extract by freeze and spray drying techniques: An optimization study. *LWT-Food Sci. Technol.* **2021**, *138*, 110696. [CrossRef]

18. Rybak, K.; Wiktor, A.; Witrowa-Rajchert, D.; Parniakov, O.; Nowacka, M. The quality of red bell pepper subjected to freeze-drying preceded by traditional and novel pretreatment. *Foods* **2021**, *10*, 226. [CrossRef]

19. Sidor, A.; Drożdżyńska, A.; Brzozowska, A.; Gramza-Michałowska, A. The effect of plant additives on the stability of polyphenols in dried black chokeberry (*Aronia melanocarpa*) fruit. *Foods* **2021**, *10*, 44. [CrossRef]

20. Janowicz, M.; Lenart, A. Selected physical properties of convection dried apples after HHP treatment. *LWT-Food Sci. Technol.* **2015**, *63*, 828–836. [CrossRef]

21. Piotrowski, D.; Kostyra, E.; Grzegory, P.; Janiszewska-Turak, E. Influence of drying methods on the structure, mechanical and sensory properties of strawberries. *Eur. Food Res. Technol.* **2021**, *247*, 1859–1867. [CrossRef]

22. Tylewicz, U.; Nowacka, M.; Rybak, K.; Drozdzal, K.; Dalla Rosa, M.; Mozzon, M. Design of Healthy Snack Based on Kiwifruit. *Molecules* **2020**, *25*, 3309. [CrossRef]

23. Różyło, R. Recent trends in methods used to obtain natural food colorants by freeze-drying. *Trends Food Sci. Technol.* **2020**, *102*, 39–50. [CrossRef]

24. Krzykowski, A.; Dziki, D.; Rudy, S.; Gawlik-Dziki, U.; Polak, R.; Biernacka, B. Effect of pre-treatment conditions and freeze-drying temperature on the process kinetics and physicochemical properties of pepper. *LWT* **2018**, *98*, 25–30. [CrossRef]

25. LaTorre-Snyder, M. Lyophilization: The basics. *Pharm. Process.* **2017**, *32*, 24–25.

26. Kulczyński, B.; Sidor, A.; Gramza-Michałowska, A. Antioxidant potential of phytochemicals in pumpkin varieties belonging to Cucurbita moschata and Cucurbita pepo species. *CyTA-J. Food* **2020**, *18*, 472–484. [CrossRef]

27. Nowak, D.; Jakubczyk, E. The freeze-drying of foods—The characteristic of the process course and the effect of its parameters on the physical properties of food materials. *Foods* **2020**, *9*, 1488. [CrossRef]

28. Rybak, K.; Parniakov, O.; Samborska, K.; Wiktor, A.; Witrowa-Rajchert, D.; Nowacka, M. Energy and quality aspects of freeze-drying preceded by traditional and novel pre-treatment methods as exemplified by red bell pepper. *Sustainability* **2021**, *13*, 2035. [CrossRef]

29. Drobot, V.; Kovbasa, V.; Bondarenko, Y.; Bilyk, O.; Hryshchenko, A. Use of dried carrot pomace in the technology of wheat bread for elderly people. *Food Sci. Technol.* **2019**, *13*, 98–105. [CrossRef]

30. Sharma, K.D.; Karki, S.; Thakur, N.S.; Attri, S. Chemical composition, functional properties and processing of carrot—A review. *J. Food Sci. Technol.* **2012**, *49*, 22–32. [CrossRef]

31. Kumar, N.; Sarkar, B.C.; Sharma, H.K. Mathematical modelling of thin layer hot air drying of carrot pomace. *J. Food Sci. Technol.* **2012**, *49*, 33–41. [CrossRef]

32. Hidalgo, A.; Brandolini, A.; Canadanovic-Brunet, J.; Cetkovic, G.; Tumbas Saponjac, V. Microencapsulates and extracts from red beetroot pomace modify antioxidant capacity, heat damage and colour of pseudocereals-enriched einkorn water biscuits. *Food Chem.* **2018**, *268*, 40–48. [CrossRef]

33. Tumbas Saponjac, V.; Canadanovic-Brunet, J.; Cetkovic, G.; Jakisic, M.; Djilas, S.; Vulic, J.; Stajcic, S. Encapsulation of Beetroot Pomace Extract: RSM Optimization, Storage and Gastrointestinal Stability. *Molecules* **2016**, *21*, 584. [CrossRef]

34. Vulić, J.J.; Ćebović, T.N.; Čanadanović-Brunet, J.M.; Ćetković, G.S.; Čanadanović, V.M.; Djilas, S.M.; Tumbas Šaponjac, V.T. In vivo and in vitro antioxidant effects of beetroot pomace extracts. *J. Funct. Foods* **2014**, *6*, 168–175. [CrossRef]

35. Janiszewska-Turak, E.; Hornowska, Ł.; Pobiega, K.; Gniewosz, M.; Witrowa-Rajchert, D. The influence of *Lactobacillus* bacteria type and kind of carrier on the properties of spray-dried microencapsules of fermented beetroot powders. *Int. J. Food Sci. Technol.* **2021**, *56*, 2166–2174. [CrossRef]

36. Janiszewska, E.; Włodarczyk, J. Influence of spray drying conditions on the beetroot pigments retention after microencapsulation process. *Acta Agrophysica* **2013**, *20*, 343–356.

37. Janiszewska-Turak, E.; Witrowa-Rajchert, D. The influence of carrot pretreatment, type of carrier and disc speed on the physical and chemical properties of spray-dried carrot juice microcapsules. *Dry. Technol.* **2021**, *39*, 439–449. [CrossRef]

38. Nowacka, M.; Fijalkowska, A.; Wiktor, A.; Rybak, K.; Dadan, M.; Witrowa-Rajchert, D. Changes of mechanical and thermal properties of cranberries subjected to ultrasound treatment. *Int. J. Food Eng.* **2017**, *13*, 20160306. [CrossRef]

39. Sledz, M.; Wiktor, A.; Nowacka, M.; Witrowa-Rajchert, D. Drying kinetics, microstructure and antioxidant properties of basil treated by ultrasound. *J. Food Process. Eng.* **2017**, *40*, e12271. [CrossRef]

40. Ribeiro, J.S.; Veloso, C.M. Microencapsulation of natural dyes with biopolymers for application in food: A review. *Food Hydrocoll.* **2021**, *112*, 106374. [CrossRef]

41. Janiszewska, E.; Witrowa-Rajchert, D.; Kidon, M.; Czapski, J. Effect of the applied drying method on the physical properties of purple carrot pomace. *Int. Agrophys.* **2013**, *27*, 143–149. [CrossRef]

42. Hallmann, E.; Marszałek, K.; Lipowski, J.; Jasińska, U.; Kazimierczak, R.; Średnicka-Tober, D.; Rembiałkowska, E. Polyphenols and carotenoids in pickled bell pepper from organic and conventional production. *Food Chem.* **2019**, *278*, 254–260. [CrossRef] [PubMed]

43. Di Scala, K.; Crapiste, G. Drying kinetics and quality changes during drying of red pepper. *LWT-Food Sci. Technol.* **2008**, *41*, 789–795. [CrossRef]

44. Labuza, T.P.; Altunakar, B.; Barbosa-Canovas, G.V.; Fontana, A.J.; Schmidt, S.J. *Water Activity in Foods: Fundamentals and Applications*; Wiley-Blackwell: Isengard, HD, USA, 2007; p. 109.

45. Tonon, R.V.; Baroni, A.F.; Brabet, C.; Gibert, O.; Pallet, D.; Hubinger, M.D. Water sorption and glass transition temperature of spray dried açai (*Euterpe oleracea* Mart.) juice. *J. Food Eng.* **2009**, *94*, 215–221. [CrossRef]

46. Barbosa, J.; Brandão, T.R.S.; Teixeira, P. Spray drying conditions for orange juice incorporated with lactic acid bacteria. *Int. J. Food Sci. Technol.* **2017**, *52*, 1951–1958. [CrossRef]

47. Tonon, R.V.; Brabet, C.; Hubinger, M.D. Anthocyanin stability and antioxidant activity of spray-dried açai (*Euterpe oleracea* Mart.) juice produced with different carrier agents. *Food Res. Int.* **2010**, *43*, 907–914. [CrossRef]

48. Ozturk, O.K.; Takhar, P.S. Physical and viscoelastic properties of carrots during drying. *J. Texture Stud.* **2020**, *51*, 532–541. [CrossRef] [PubMed]

49. Doymaz, İ. Drying kinetics, rehydration and colour characteristics of convective hot-air drying of carrot slices. *Heat Mass Transf.* **2016**, *53*, 25–35. [CrossRef]

50. Pinar, H.; Çetin, N.; Ciftci, B.; Karaman, K.; Kaplan, M. Biochemical composition, drying kinetics and chromatic parameters of red pepper as affected by cultivars and drying methods. *J. Food Compos. Anal.* **2021**, *102*, 103976. [CrossRef]

51. Fernández-López, J.A.; Fernández-Lledó, V.; Angosto, J.M. New insights into red plant pigments: More than just natural colorants. *RSC Adv.* **2020**, *10*, 24669–24682. [CrossRef]

52. Fu, Y.; Shi, J.; Xie, S.Y.; Zhang, T.Y.; Soladoye, O.P.; Aluko, R.E. Red beetroot betalains: Perspectives on extraction, processing, and potential health benefits. *J. Agric. Food Chem.* **2020**, *68*, 11595–11611. [CrossRef] [PubMed]

53. Chhikara, N.; Kushwaha, K.; Sharma, P.; Gat, Y.; Panghal, A. Bioactive compounds of beetroot and utilization in food processing industry: A critical review. *Food Chem.* **2019**, *272*, 192–200. [CrossRef]

54. Lee, S.M.; Lee, J.Y.; Cho, Y.J.; Kim, M.S.; Kim, Y.S. Determination of volatiles and carotenoid degradation compounds in red pepper fermented by Lactobacillus parabuchneri. *J. Food Sci.* **2018**, *83*, 2083–2091. [CrossRef] [PubMed]

55. Bartkiene, E.; Vidmantiene, D.; Juodeikiene, G.; Viskelis, P.; Urbonaviciene, D. Lactic acid fermentation of tomato: Effects on cis/trans lycopene isomer ratio, β-carotene mass fraction and formation of L (+)-and D (−)-lactic acid. *Food Technol. Biotechnol.* **2013**, *51*, 471–478.

56. Rodriguez-Amaya, D.B. *A Guide to Carotenoid Analysis in Foods*; ILSI Press: Washington, DC, USA, 2001.

57. Yan, Y.; Zhang, F.; Chai, Z.; Liu, M.; Battino, M.; Meng, X. Mixed fermentation of blueberry pomace with *L. rhamnosus* GG and *L. plantarum*-1: Enhance the active ingredient, antioxidant activity and health-promoting benefits. *Food Chem. Toxicol.* **2019**, *131*, 110541.
58. Tang, S.; Cheng, Y.; Wu, T.; Hu, F.; Pan, S.; Xu, X. Effect of *Lactobacillus plantarum*-fermented mulberry pomace on antioxidant properties and fecal microbial community. *LWT* **2021**, *147*, 111651. [CrossRef]

applied sciences

Article

Flavonoids from Fig (*Ficus carica* Linn.) Leaves: The Development of a New Extraction Method and Identification by UPLC-QTOF-MS/MS

Chunjian Zhao [1,2,3,4], Shen Li [1,2], Chunying Li [1,2,3,*], Tingting Wang [1,2], Yao Tian [1,2] and Xin Li [1,2]

[1]   Key Laboratory of Forest Plant Ecology, Ministry of Education, Northeast Forestry University, Harbin 150040, China; zcj@nefu.edu.cn (C.Z.); klp18ls@nefu.edu.cn (S.L.); klp18wtt@nefu.edu.cn (T.W.); klp18ty@nefu.edu.cn (Y.T.); klp19lx@nefu.edu.cn (X.L.)
[2]   College of Chemistry, Chemical Engineering and Resource Utilization, Northeast Forestry University, Harbin 150040, China
[3]   State Engineering Laboratory for Bio-Resource Eco-Utilization, Northeast Forestry University, Harbin 150040, China
[4]   Collaborative Innovation Center for Development and Utilization of Forest Resources, Harbin 150040, China
*   Correspondence: nefujane@aliyun.com; Tel./Fax: +86-451-8219-0848

Citation: Zhao, C.; Li, S.; Li, C.; Wang, T.; Tian, Y.; Li, X. Flavonoids from Fig (*Ficus carica* Linn.) Leaves: The Development of a New Extraction Method and Identification by UPLC-QTOF-MS/MS. *Appl. Sci.* **2021**, *11*, 7718. https://doi.org/10.3390/app11167718

Academic Editor: Emanuel Vamanu

Received: 23 June 2021
Accepted: 20 August 2021
Published: 22 August 2021

Abstract: Flavonoid-rich leaves of the *Ficus carica* L. plant are usually discarded as waste. In this work, ultrasonic enzyme-assisted aqueous two-phase extraction (UEAATPE) was proposed as an innovative method to estimate the total flavonoids present in *F. carica* L. leaves. Total flavonoids were analyzed qualitatively and quantitatively by UPLC-QTOF-MS. At 38% (*w/w*) ethanol/18% (*w/w*) ammonium sulfate, we achieved the optimum conditions in which to establish an easy-to-form aqueous two-phase extraction (ATPE) as the final system. The optimal UEAATPE conditions were set at an enzymatic concentration of 0.4 U/g, 150 min enzymolysis time, an enzymolysis temperature of 50 °C, a liquid–solid ratio of 20:1 (mL/g), and 30 min ultrasonic time. The yields of the total flavonoids, i.e., 60.22 mg/g, obtained by UEAATPE were found to be 1.13-fold, 1.21-fold, 1.27-fold, and 2.43-fold higher than those obtained by enzyme-assisted ATPE (EAATPE), ultrasonic-assisted ATPE (UAATPE), ATPE, and soxhlet extraction (SE) methods, respectively. Furthermore, eleven flavonoids from the leaves of the *F. carica* L. plant were completely identified and fully characterized. Among them, ten flavonoids have been identified for the first time from the leaves of the *F. carica* L. plant. These flavonoids are quercetin 3-*O*-hexobioside-7-*O*-hexoside, 2-carboxyl-1,4-naphthohydroquinone-4-*O*-hexoside, luteolin 6-*C*-hexoside, 8-*C*-pentoside, kaempferol 6-*C*-hexoside-8-*C*-hexoside, quercetin 6-*C*-hexobioside, kaempferol 6-*C*-hexoside-8-*C*-hexoside, apigenin 2″-*O*-pentoside, apigenin 6-*C*-hexoside, quercetin 3-*O*-hexoside, and kaempferol 3-*O*-hexobioside. Therefore, *F. carica* L. leaves contain new kinds of unidentified natural flavonoids and are a rich source of biological activity. Therefore, this research has potential applications and great value in waste handling and utilization.

Keywords: *Ficus carica* L.; flavonoids; ultrasonic enzyme co-assisted; aqueous two-phase extraction; UPLC-QTOF-MS/MS; identification

1. Introduction

Plant-derived organic waste mainly includes crop stalks, leftover branches and wood strips, fallen leaves, dry vines, weeds, and nut shells from the production process. Among the large traditional agricultural countries, agricultural organic waste has the following four characteristics: large quantity, poor quality, low price, and harmful properties [1]. In most agricultural organic waste treatment processes, the treatment efficiency is not high, and the environmental damage caused by improper treatment methods is relatively serious [2]. Most agricultural organic waste, due to its relatively abundance, can help protect the environment and save energy while improving comprehensive utilization of agricultural organic wastes.

109

Ficus carica L., a fig plant, has a long history as a Moraceae [3]. Figs are native to the Mediterranean coast, from Turkey to Afghanistan, having been established in the region since ancient times [4]. In China, figs were introduced from Persia during the Tang Dynasty and have been cultivated in both the south and the north, especially in Xinjiang and Shandong provinces [5]. As a crop with a long history and due to its cultivation and high nutritional value, figs have always been a source of food for human survival. Regarding their nutritional value, figs have recently been used in food processing [6,7]. Figs also have extensive medicinal value, as well as having functions in the nourishment of the stomach, the clearance of the intestines, the reduction of swelling, and detoxification [8–10].

Currently, fig leaves with a high biomass and many bioactive compounds are usually discarded, resulting in a waste of resources [11]. Fig leaves also contain flavonoids, sugars, pectin, tannins, vitamin C, trace elements, and other bioactive components [12,13]. They have many pharmacological effects due to the large amounts of flavonoids contained in fig leaves. They can prevent cardiovascular diseases and are anti-osteoporotic and are used in the treatment of diarrhea, for scavenging oxidative free radicals and for blood lipid reduction, sore throats, and immune regulation [14–17]. However, most of the flavonoids have not been identified or characterized. Therefore, in order to develop and utilize fig resources reasonably, extraction methods of flavonoids should necessarily be developed.

At present, traditional extraction methods of flavonoids are commonly used. However, the traditional methods are inefficient and environmentally unfriendly [18]. In order to overcome these shortcomings, the aqueous two-phase extraction (ATPE) method could be used to replace conventional extraction methods. As an economic, mild, and simple separation method, it has been widely used in the field of natural product separation [19]. Additionally, enzyme-assisted extraction (EAE) is a better pretreatment method for separation. Cellulase can remove the pectin in the cell wall so that the material in the cell can be dissolved quickly and fully. This can improve the extraction yield of effective ingredients while reducing the consumption volume of solvents without destroying the structure of compounds. An enzymatic reaction is widely used in extracting various compounds from natural products due to its mildness, economical nature, and environmental protection [20]. Recently, UAE has been widely used in natural product extraction because of its time efficiency and reduced solvent usage [21–23]. For purposes of increasing the extraction efficiency of ATPE, a combination of the UAE and ATPE methods was developed [24,25].

In this study, on the basis of the advantages of UAE, EAE and ATPE [26,27], an ultrasonic enzyme-assisted ATPE (UEAATPE) method was developed for its environmental protection characteristics. Furthermore, the identification and characterization of eleven flavonoids was achieved using UPLC-QTOF-MS/MS from fig leaves. This research is of great significance for improving the development of fig leaves and promoting the rational utilization of this resource. This can be used not only to develop new resources, but also to effectively use green waste and further achieve recycling to promote agricultural development. In the current situation of environmental protection and sustainable development, it is also an effective waste management method.

2. Materials and Methods

2.1. Materials and Chemicals

We collected fig leaves in Chengshan Town, Rongcheng City. The fig leaves were thoroughly dried in a cool and dark place. For further study, we used a disintegrator to crush the dried leaves into powder (60 mesh). All chemicals used, unless stated otherwise, were of analytical grade and purchased from Sigma-Aldrich (St. Louis, MO, USA). The standard solution was stored at $-20\,^{\circ}$C and used for the subsequent experiment.

2.2. Apparatus

Acquity^{TM} ultra-high-performance liquid chromatography (Waters, Milford, MA, USA); Triple TOF 5600+ time-of-flight mass spectrometer with electrospray ion source (AB SCIEX, Framingham, MA, USA); ASE350 Rapid Solvent Extraction Apparatus (Dionex,

Sunnyvale, CA, USA); AG135 Precision electronic balance (Mettler Toledo, Greifensee, Switzerland); KQ-100e Ultrasonic cleaning instrument (Kunshan Ultrasonic Instrument Co., LTD., Kunshan, China).

2.3. Ultrasonic Enzyme-Assisted Aqueous Two-Phase Extraction (UEAATPE)

ATPSs were screened on the basis of the formation described in reference [28]. Each of the salts tested (ammonium sulfate, dipotassium hydrogen phosphate, sodium carbonate, sodium sulfate, calcium chloride, potassium dihydrogen phosphate, and sodium chloride) was dissolved in deionized water. The salt solution was mixed with ethanol by a vortex stirrer. ATPS was formed when the mixture showed two-phase separation at the cloud point. Due to rapid phase formation and stratification, an ethanol/$(NH_4)_2SO_4$ system was chosen [29]. A diagram of UEAATPE is shown in Figure 1.

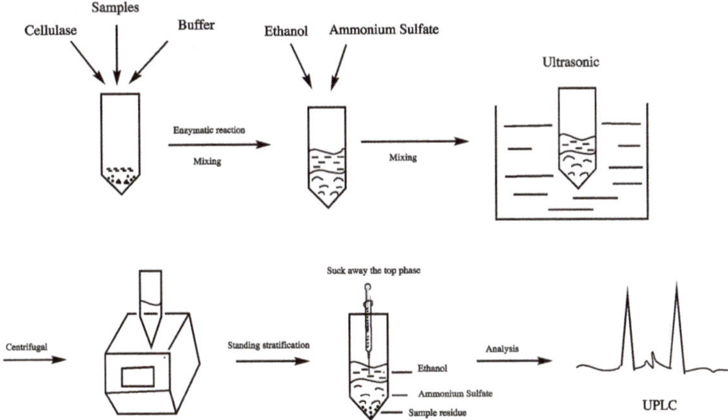

Figure 1. Diagram of UEAATPE.

To a total of 0.2 g fig leaf powder, 4.0 g disodium hydrogen phosphate–citric acid buffer solution and 0.3 U/g cellulase were first added in a 10 mL graduated test tube and then mixed evenly by a vortex mixer and placed in a water bath at a certain constant temperature. Then, ammonium sulfate and ethanol were added into the enzymatic slurry and vibrated for 10 min by a vortex mixer to completely dissolve the salt. The suspension was given ultrasonic (100 w) treatment for 30 min. After ultrasonic treatment, the mixture was mixed well and then set at room temperature for 30 min to form an aqueous two-phase system.

2.4. Determination of Total Flavonoids

The total flavonoids were determined by the method described in [28]. The extracted solutions (0.3 mL) were transferred to a 10 mL test tube, to which sodium nitrite solution (5%; 0.3 mL) was added. The mixture was allowed to stand for 6 min, and then 0.3 mL of 10% aluminum nitrate solution was added. After another 6 min, this was followed by the addition of 4 mL of 4% sodium hydroxide solution. The absorbance of the mixture was measured at 510 nm using a UV-Vis spectrophotometer (Perkin–Elmer Lambda 25, Waltham, MA, USA). After 15 min, the flavonoid contents in the extracts were determined in comparison to a standard curve that was plotted using rutin. The results were the averages of triplicate analyses. The calibration curve was obtained using rutin as the standard as shown in Table 1. Then, the extraction yield of total flavonoids was calculated according to Equation (1):

$$Y = C_t V_t / M_t \tag{1}$$

where Y (mg/g) represents the yield of total flavonoids; C_t (mg/mL) represents total flavonoid concentrations in the top phases; and V_t (mL) is the volume of the top phases. M_t (g) is the total mass of the fig leaf powder.

Table 1. Calibration curve, correlation coefficient, LOD, and LOQ of the total flavonoids.

Analytes	Calibration Curve	R^2	Linear Range (μg/mL)	LOD (μg/mL)	LOQ (μg/mL)
rutin	y = 5.5778 x + 0.0635	0.9929	3.901–250	2.196	7.032

2.5. Experiment Design of UEAATPE

After testing in a single-factor experiment, four factors were selected and combined in the proposed methods to assess the main role of BBD and its interactions. These were X_1: 36–40% ethanol concentration; X_2: 16–20% ammonium sulfate concentration; X_3: 10:1–30:1 (mL/g) liquid–solid ratio; X_4: 20–40 min ultrasonic time. Y represents the yield of total flavonoids in different ranges.

2.6. Comparison of Different Extraction Methods

The extraction yield of total flavonoids was compared with UEAATPE, EAATPE, UAATPE, ATPE, and SE. The mass fractions of ethanol and ammonium sulfate were 38% (w/w) and 18% (w/w), respectively. The other fixed extraction conditions under optimization were enzymatic hydrolysis for 180 min at a 0.4 U/g cellulase concentration, a fixed temperature of 50 °C, an extraction time of 30 min, and a liquid–solid ratio of 20:1 mL/g.

2.7. The Analysis of UPLC-QTOF-MS/MS

The supernatant was evaporated until dry with a rotary evaporator (RE-52AA, Shanghai Huxi Instrument, Shanghai, China) under reduced pressure in a 60 °C water bath. The suspended sample was re-dissolved in a methanol solution. Then, before the UPLC-QTOF-MS analysis, the solution was filtered through a 0.22 μm microporous membrane.

The samples were separated on an ACQUITY UPLC HSS T3 (150 mm × 2.1 mm i.d., 1.8 μm, Waters). The column temperature was maintained at 50 °C, and the injection volume was set as 2 μL with 0.3 mL/min as a fixed flow rate. The mobile phase was composed of acetonitrile (A) and 0.1% (v/v) formic acid in aqueous solution (B). The gradient elution conditions were as follows: 0–2 min, 5% A; 2–25 min, 5–40% A; 25–32 min, 40–95% A. The chromatogram was obtained at 254 nm, and semi-quantitative calculations were performed for each compound based on the relative peak area and rutin standard.

In the negative ion mode, m/z 100–1500 was used as the mass spectrum data acquisition condition. Atomizing air (GS1): 55 psi; atomizing air (GS2): 55 psi; source temperature (TEM): 550 °C; source voltage (IS): −4500 V. Level 1 scan: de-cluster voltage (DP) and focusing voltage (CE): 100 V and 10 V. Secondary scan: TOF MS~Product Ion~IDA mode was used to collect mass spectrum data. The CID energy was −20, −40, and −60 V. Before sample injection, a CDS pump was used for mass axis correction to make the mass axis error less than 2 ppm.

2.8. Statistical Analysis

The data are presented as mean ± standard deviation (SD). All data for this study were adopted for analysis of variance (ANOVA) to determine significant differences. The significance of such differences between mean values was determined using Duncan's test ($p < 0.05$). ANOVA and Duncan's multiple range tests were performed with SPSS19 (SPSS, Chicago, IL, USA).

3. Results and Discussion

3.1. Screening of Phase Ratio in ATPS System

3.1.1. Selection of Ethanol Mass Fraction

Single-factor conditions were fixed to allow analysis of other factors. As shown in Table 2, the effect of the ethanol mass fraction on the yield of total flavonoids was studied. Before the ethanol mass fraction reached 38% (w/w), the yield of total flavonoids was positively correlated with it; then, the yield of flavonoids showed a downward trend with increasing ethanol concentration. The reason for this is that as the mass fraction of the ethanol in the system increased, the concentration of the ethanol in the top phase increased, and the polarity decreased. Under these conditions, it was more conducive to perform the extraction of flavonoids. As the mass fraction of ethanol exceeded 38% (w/w), the polarity of the upper phase decreased. This resulted in the immiscibility of the flavonoids and the salt of the precipitation [30]. At this time, the amount of some fat-soluble organic compounds also increased, which inhibited the leaching of total flavonoids. Therefore, the optimal parameter for the ethanol concentration was 38% (w/w).

Table 2. Effects of different influencing factors on the yield of total flavonoids.

No.	Mass Fraction of Ethanol (%)	Mass Fraction of Ammonium Sulfate (%)	Concentration of Enzymes (U/g)	Enzymolysis Time (min)	Enzymolysis Temperature (°C)	Ultrasonic Time (min)	Liquid–Solid Ratio (mL/g)	Yield of Total Flavonoids (mg/g)
1	32	18	0.3	180	50	30	20	56.83
2	34	18	0.3	180	50	30	20	58.92
3	36	18	0.3	180	50	30	20	59.46
4	38	18	0.3	180	50	30	20	60.02
5	40	18	0.3	180	50	30	20	58.18
6	38	14	0.3	180	50	30	20	47.79
7	38	16	0.3	180	50	30	20	57.03
8	38	18	0.3	180	50	30	20	60.19
9	38	20	0.3	180	50	30	20	58.41
10	38	22	0.3	180	50	30	20	55.40
11	38	18	0.3	180	50	30	20	47.13
12	38	18	0.4	180	50	30	20	60.00
13	38	18	0.5	180	50	30	20	59.03
14	38	18	0.6	180	50	30	20	57.25
15	38	18	0.7	180	50	30	20	57.17
16	38	18	0.3	90	50	30	20	46.13
17	38	18	0.3	120	50	30	20	50.24
18	38	18	0.3	150	50	30	20	53.20
19	38	18	0.3	180	50	30	20	59.89
20	38	18	0.3	210	50	30	20	50.00
21	38	18	0.3	180	35	30	20	41.55
22	38	18	0.3	180	40	30	20	44.99
23	38	18	0.3	180	45	30	20	58.09
24	38	18	0.3	180	50	30	20	60.12
25	38	18	0.3	180	55	30	20	52.53
26	38	18	0.3	180	50	10	20	56.11
27	38	18	0.3	180	50	20	20	57.99
28	38	18	0.3	180	50	30	20	60.91
29	38	18	0.3	180	50	40	20	59.23
30	38	18	0.3	180	50	50	20	57.10
31	38	18	0.3	180	50	30	10	55.01
32	38	18	0.3	180	50	30	20	60.01
33	38	18	0.3	180	50	30	30	58.32
34	38	18	0.3	180	50	30	40	58.29
35	38	18	0.3	180	50	30	50	58.10

3.1.2. Selection of Ammonium Sulfate Mass Fraction

It can be seen in Table 2 that as the ammonium sulfate mass fraction increased, the yield of flavonoids showed a trend of rising first and then falling. However, the yield of flavonoids changed little in the range 17~22% of ammonium sulfate. When the ammonium sulfate mass fraction reached 18% (w/w), the highest yield of total flavonoids was obtained.

This occurred in view of the mass fraction of ammonium sulfate directly affecting the phase ratio in the ATPS system. The volume ratio of the upper and lower phases and the total extraction capacity of the solvent were the factors that affected the extraction rate of total flavonoids [31]. Therefore, we selected 18% (w/w) of ammonium sulfate concentration as the optimal parameter.

3.2. Univariate Analysis of UEAATPE

3.2.1. Effects of Enzyme Concentration on Flavonoid Yield

When determining the enzyme concentration, economic effects should also be considered. The idea was to attain complete extraction and avoid excessive use of enzymes. It can be seen from Table 2 that the total flavonoid content changed under different cellulase concentrations of 0.3, 0.4, 0.5, 0.6, and 0.7 U/g. In the range of cellulase concentration from 0.30 to 0.4 U/g, the total flavonoid yield showed a positive correlation. However, the same change trend between 0.4 and 0.7 U/g was not clear. The reason for this is that cellulase destroyed the cell wall and released bioactive ingredients [32]. However, a superfluous enzyme concentration can saturate the substrate. Excess enzymes should not be combined with substances that cause waste [33]. In brief, a 0.4 U/g concentration of cellulase was chosen for further experimental optimization.

3.2.2. Effects of Enzymolysis Time on Flavonoid Yield

It can be seen from Table 2 that an obvious trend of flavonoid yield was observed between 90 and 180 min, and then the yield of flavonoids began to decrease after 210 min. Enzymatic digestion of the cell walls appeared to have occurred in the sample and the maximum amount of flavonoids was released in 180 min. Enzymatic hydrolysis time affected the yield of target components, and the enzymatic hydrolysis time was short, which did not allow the target components to fully dissolve. The long enzymatic hydrolysis time not only increased the extraction cost, but also led to an increase in impurities. It may be that as the time increases, some flavonoids are oxidized to form quinone compounds and reduce the yield of the target compounds of the top phase. This indicates that the enzymatic hydrolysis of 180 min is the optimal time to catalyze cell wall hydrolysis.

3.2.3. Effects of Enzymolysis Temperature on Flavonoid Yield

Table 2 shows the changing trend of total flavonoids. As the temperature reached between 35 °C and 50 °C, the extraction yield of flavonoids showed a positive correlation. Generally speaking, enzyme activity is closely related to temperature. Enzyme activity and reaction rate can be increased by raising the temperature of enzyme hydrolysis. However, at temperatures above 50 °C, the extraction yield of flavonoids decreased significantly. It may be that excessive temperatures denature the enzyme. It can be seen that the optimal enzymolysis temperature parameter was set to 50 °C.

3.2.4. Effects of Ultrasonic Time on Flavonoid Yield

Ultrasound is a crucial parameter influencing total flavonoid yield. As shown in Table 2, the parameter range of the extraction time was set from 10 to 50 min. Under the condition of a fixed power of 200 w, the yield of total flavonoids was studied. Before the ultrasound time reached 30 min, the extraction yield of total flavonoids was positively correlated with the ultrasound time. At 30 min, it reached the highest yield of 60.07 mg/g, then the yields of flavonoids decreased with the further increase in ultrasonic time [34]. This was mainly due to the large amount of time taken by the ultrasonic reaction, leading to some flavonoid oxidation or degradation, so the yield of flavonoids decreased. Thus, 30 min of extraction time was selected for the subsequent experiments.

3.2.5. Effects of the Liquid–Solid Ratio on Flavonoid Yield

The effect of different liquid–solid ratios on the yield of flavonoids is shown in Table 2. Economically speaking, an appropriate liquid–solid ratio is very important for flavonoid extraction. The extraction yield of total flavonoids showed an upward trend as the liquid–solid ratio increased from 10:1 to 20:1 mL/g, and then began to decline after it exceeded 20:1 mL/g. This is because the larger the liquid–solid ratio, the more adverse the penetration of the solvent and solute diffusion, resulting in less sufficient flavone dissolution, and the extraction yield of total flavonoids will thus become lower. However, the addition of dried fig leaf powder can absorb water from the aqueous phase and increase the ethanol concentration. The total amount of leaching decreased with the increase in the amount of powder added due to the decrease in permeability and diffusion capacity. Considering this factor economically, in order to avoid significant wastage of the solvent, the liquid–solid ratio in the optimal test design was 20:1 mL/g.

3.3. Optimization of UEAATPE

As shown in Table 3, the test resulted in 29 uncontrolled runs. The correlation between the response and the independent variable can be visualized with a 3D surface plot. Figure 2 indicates the influence of the ethanol concentration %, (X_1), ammonium sulfate concentration %, (X_2), liquid–solid ratio mL/g, (X_3), and ultrasound time min, (X_4), on the yield of total flavonoids and their interaction. Furthermore, it is possible to predict the optimal value of the response and the corresponding experimental conditions through the F value (>7.84) and p value (<0.01), as shown in Table 4. This result shows that the model can describe the total flavonoid yield of UEAATPE well. The equation of the response variables and independent variables is as follows:

$$Y = 59.60 - 0.03X_1 + 0.48X_2 - 4.66X_3 - 1.63X_4 + 1.18\,X_1X_2 + 2.95\,X_1X_3 + 3.24\,X_1X_4 + 2.00\,X_2X_3 - 0.59\,X_2X_4 - 4.48X_3\,X_4 - 5.19\,X_1{}^2 - 3.37\,X_2{}^2 - 9.60\,X_3{}^2 - 8.41\,X_4{}^2 \tag{2}$$

where Y represents yields of total flavonoids (mg/g); X_1, X_2, X_3, and X_4, respectively, represent the ethanol concentration (%), ammonium sulfate concentration (%), liquid–solid ratio (mL/g), and ultrasonic time (min).

Table 3. Experimental data and total flavonoid extraction analyzed by the Box–Behnken approach.

Runs	Ethanol Concentration (X_1, %)	Ammonium Sulfate Concentration (X_2, %)	Liquid—Solid Ratio (X_3, mL/g)	Ultrasonic Time (X_4, min)	Yield of Total Flavonoids (mg/g)
	Independent Variables				
1	38	18	30:1	20	43.71
2	38	18	10:1	20	44.62
3	36	18	10:1	30	52.80
4	38	16	20:1	40	45.44
5	36	18	30:1	30	35.25
6	38	16	10:1	30	49.99
7	40	18	30:1	30	42.99
8	40	18	10:1	30	48.75
9	38	20	10:1	30	48.13
10	36	18	20:1	20	48.74
11	38	18	20:1	30	60.39
12	40	18	20:1	20	42.87
13	40	18	20:1	40	48.16
14	38	18	10:1	40	49.70
15	38	20	20:1	40	45.84
16	38	16	30:1	30	39.54
17	38	20	30:1	30	45.68
18	38	18	30:1	40	30.87
19	36	18	20:1	40	41.04
20	40	16	20:1	30	49.59

Table 3. *Cont.*

Runs	Independent Variables				Yield of Total Flavonoids (mg/g)
	Ethanol Concentration (X_1, %)	Ammonium Sulfate Concentration (X_2, %)	Liquid—Solid Ratio (X_3, mL/g)	Ultrasonic Time (X_4, min)	
21	38	18	20:1	30	58.94
22	38	18	20:1	30	59.11
23	36	20	20:1	30	51.40
24	40	20	20:1	30	51.11
25	38	16	20:1	20	48.90
26	38	18	20:1	30	61.08
27	36	16	20:1	30	54.60
28	38	18	20:1	30	58.50
29	38	20	20:1	20	51.70

Table 4. Analysis of variance for the extraction yield of total flavonoids by quadratic model.

Source	Sum of Squares	Mean Square	F-Value	*p*-Value	Significance
Model	1.40×10^3	100.32	28.36	<0.0001	significant
X_1	1.05×10^{-2}	1.05×10^{-2}	2.97×10^{-3}	0.9573	not significant
X_2	2.81	2.81	0.79	0.3882	not significant
X_3	260.89	260.89	73.76	<0.0001	significant
X_4	31.69	31.69	8.96	0.0097	not significant
$X_1 X_2$	5.57	5.57	1.58	0.2301	not significant
$X_1 X_3$	34.76	34.76	9.83	0.0073	not significant
$X_1 X_4$	42.11	42.11	11.91	0.0039	not significant
$X_2 X_3$	16.00	16.00	4.52	0.0517	not significant
$X_2 X_4$	1.44	1.44	0.41	0.5342	not significant
$X_3 X_4$	80.28	80.28	22.70	0.0003	not significant
X_1^2	175.31	1.75×10^2	49.57	<0.0001	significant
X_2^2	73.68	73.68	20.83	0.0004	not significant
X_3^2	598.54	5.99×10^2	1.70×10^2	<0.0001	significant
X_4^2	459.01	4.60×10^2	1.30×10^2	<0.0001	significant
Lack of Fit	44.82	4.48	3.82	0.1042	not significant
R^2			0.9659		

According to appropriate extraction conditions (independent variables) and actual operation analysis by Design Expert software, all these conditions were modified as follows: 38% (*w/w*) ethanol/18% (*w/w*) ammonium sulfate; liquid–solid ratio of 20:1 mL/g, and ultrasonic time of 30 min. Under these conditions, it was possible to obtain 60.22 mg/g flavonoids by UEAATPE. According to the RSM prediction model, the above experimental value matches the fitted value (RSD < 1.72%).

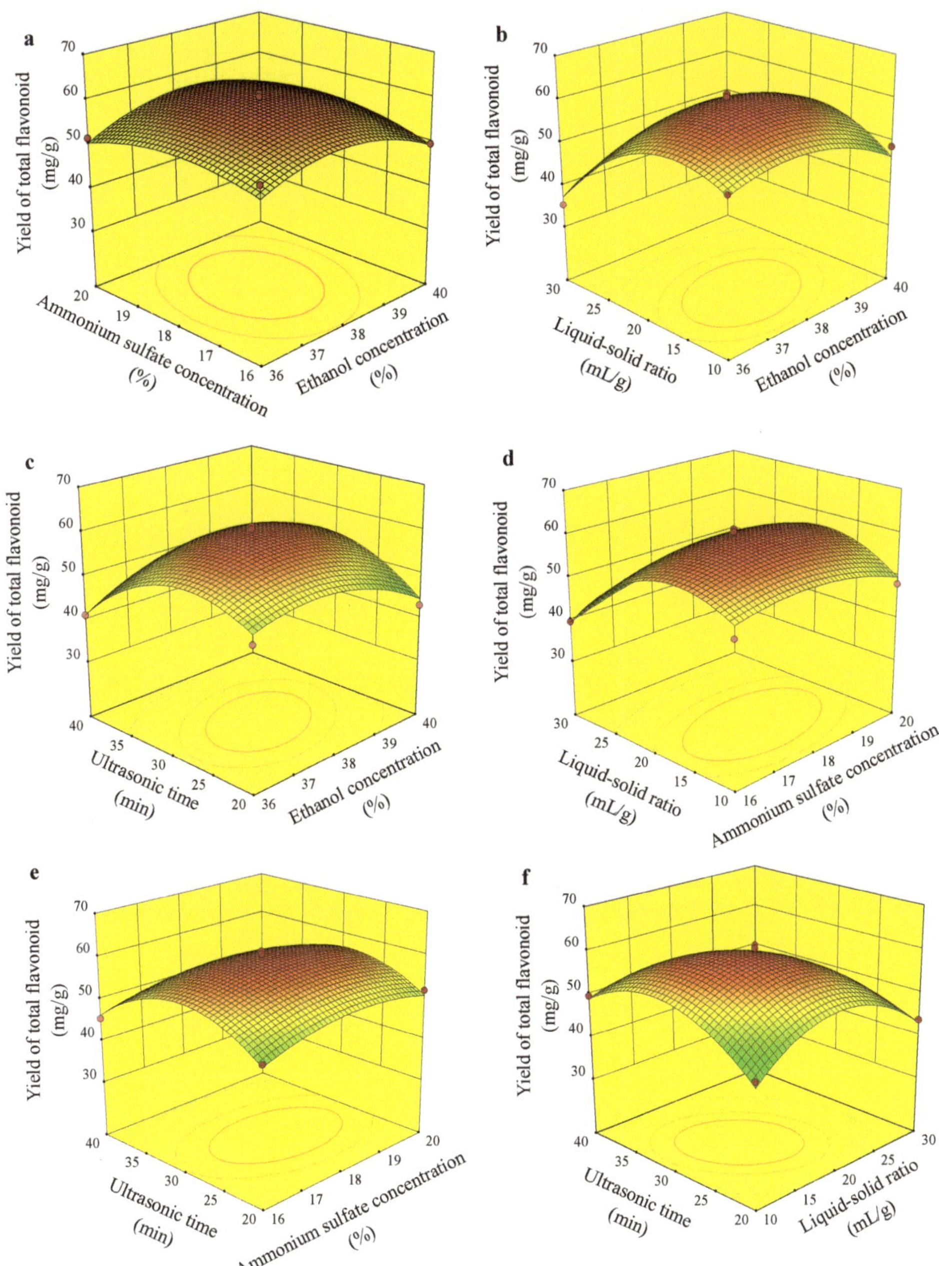

Figure 2. Response surface representations for total flavonoids in fig leaves (the variables are as follows: (**a**) interaction of ethanol concentration, % and ammonium sulfate concentration, %; (**b**) interaction of liquid–solid ratio, mL/g and ammonium sulfate concentration, %; (**c**) interaction of ultrasonic time, min and ethanol concentration, %; (**d**) interaction of liquid–solid ratio, mL/g and ammonium sulfate concentration, %; (**e**) interaction of ultrasonic time, min and ammonium sulfate concentration, %; (**f**) interaction of ultrasonic time, min and liquid–solid ratio, mL/g.

3.4. Comparison of Different Methods

The results shown in Figure 3 indicate the yields of flavonoids that were obtained by UEAATPE, EAATPE, UAATPE, ATPE, and SE. Among the above five methods, the highest yield of total flavonoids was extracted by UEAATPE. At the same time, the remaining four methods in order of extraction yield were UAE > EAE > ATPE > SE (53.48, 49.59, 47.46, and 24.79 mg/g, respectively). The extraction yield of total flavonoids with SE was lower than that with the other four methods. Therefore, UEAATPE is a prospective method for the extraction of flavonoids.

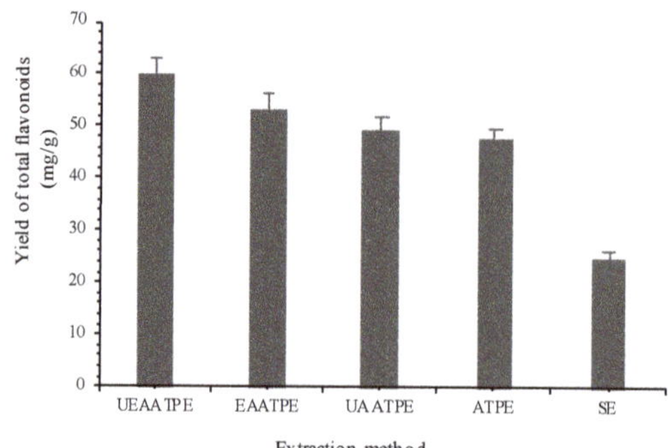

Figure 3. Effects of different methods on the yield of total flavonoids.

3.5. Identification of Flavonoids in Fig Leaves

The UPLC chromatography of flavonoid extraction by UEAATPE was detected at 254 nm, as shown in Figure 4. The identification and structure elucidation of the compounds from the leaves was completed by UPLC-QTOF-MS/MS in negative ion mode. Approximately 11 peaks were separated in the extract. Subsequently, the ESI-MS^1 and MS^2 were used to identify and characterize the flavonoids from the fig leaves. The mass spectra and fragmentation pathways can be seen in Figures S1–S11 (see the Supplementary Material). The identification of peaks was performed using reference data such as retention time and mass spectrum, as shown in Table 5. According to the structural characteristics, these compounds are all flavonoids. The chemical structure compounds identified are shown in Figure 5.

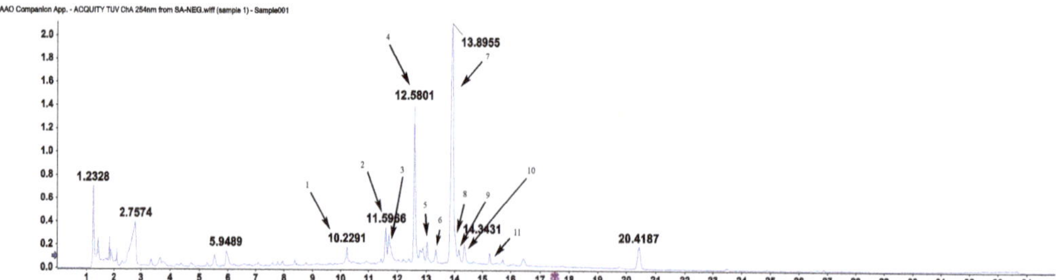

Figure 4. The UPLC chromatogram of flavonoid extraction from fig leaves detected at 254 nm.

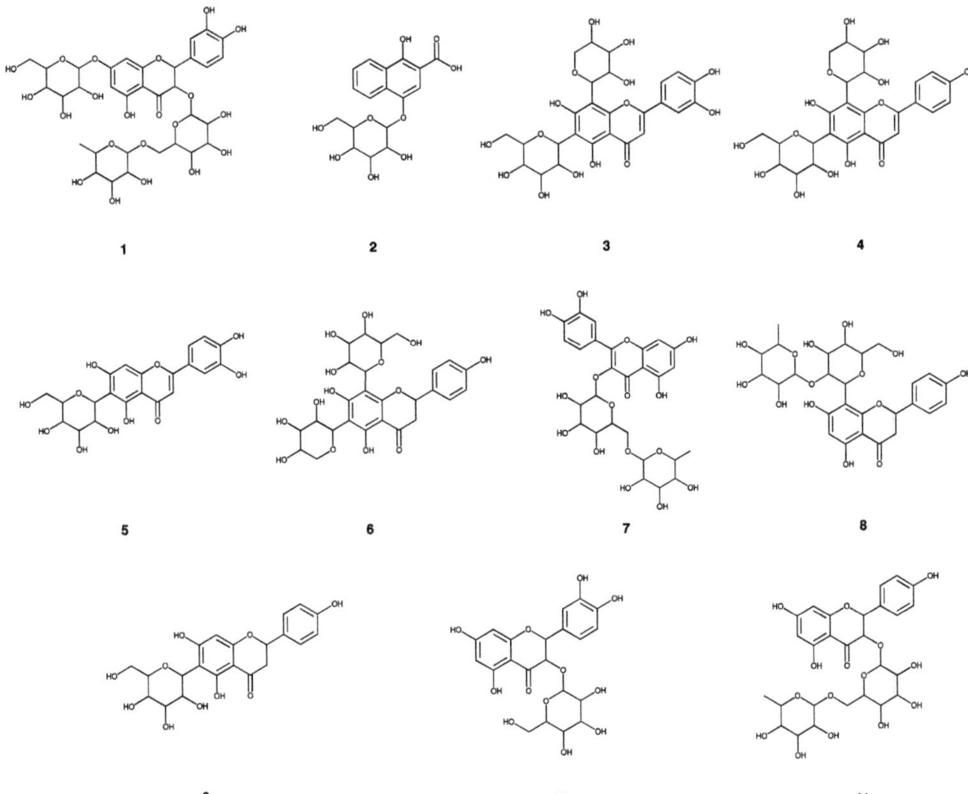

Figure 5. Chemical structures of the compounds identified from fig leaves (refer to Table 5): Compounds 1 through 11 are 3-*O*-(rhamnopyranosyl-glucopyranosyl)-7-*O*-(glucopyrnosyl)-quercetin; 2-carboxyl-1,4-naphthohydroquinone-4-*O*-hexoside; luteolin 6-*C*-hexoside, 8-*C*-pentoside; kaempferol 6-*C*-hexoside-8-*C*-hexoside; quercetin 6-*C*-hexobioside; kaempferol 6-*C*-hexoside-8-*C*-hexoside; quercetin 3-*O*-hexobioside; apigenin 2''-*O*-pentoside; apigenin 6-*C*-hexoside; quercetin 3-*O*-hexoside; kaempferol 3-*O*-hexobioside.

Table 5. Flavonoids identified in the fig leaf extracts using UPLC-QTOF-MS/MS.

Peak No.	t_R	MS	MS/MS	Molecular Weight	Molecular Formula	Identification
1	10.22	771.2027	609.1530, 462.0838, 301.0357	772.20621	$C_{33}H_{40}O_{21}$	3-*O*-(rhamnopyranosyl-glucopyranosyl)-7-*O*-(glucopyrnosyl)-quercetin
2	11.59	365.0881	203.0352, 159.0454, 130.0422	366.09508	$C_{17}H_{18}O_9$	2-carboxyl-1,4-naphthohydroquinone-4-*O*-hexoside
3	11.7	579.1366	519.1194, 489.1083, 429.0856, 369.0635	580.14282	$C_{26}H_{28}O_{15}$	luteolin 6-*C*-hexoside, 8-*C*-pentoside
4	12.58	563.1414	473.1115, 443.1001, 353.0670	564.14791	$C_{26}H_{28}O_{14}$	kaempferol 6-*C*-hexoside-8-*C*-hexoside
5	12.87	447.0934	369.0615, 357.0622, 327.0511, 297.0397, 285.0396, 133.0280	448.10050	$C_{21}H_{20}O_{11}$	quercetin 6-*C*-hexobioside
6	13.02	563.1412	443.1001, 353.0670	564.14791	$C_{26}H_{28}O_{14}$	kaempferol 6-*C*-hexoside-8-*C*-hexoside
7	13.89	609.1470	301.0362, 151.0031, 257.0450, 273.0477	610.15338	$C_{27}H_{30}O_{16}$	quercetin 3-*O*-hexobioside
8	14.02	577.1577	457.1164, 293.0454	578.16356	$C_{27}H_{30}O_{14}$	apigenin 2''-*O*-pentoside
9	14.14	432.1056	341.0673, 311.0564, 283.0612	432.10565	$C_{21}H_{20}O_{10}$	apigenin 6-*C*-hexoside
10	14.34	463.0890	301.0357	464.09548	$C_{21}H_{20}O_{12}$	quercetin 3-*O*-hexoside
11	15.25	593.1524	285.0403	594.15847	$C_{27}H_{30}O_{15}$	kaempferol 3-*O*-hexobioside

The 11 compounds were divided into flavonoid oxygen glycoside compounds and flavonoid carbon glycoside compounds [35]. As shown in Figure 6, flavonoid glycosides mainly underwent Y^- type cleavage; that is, the glycosyl is removed, and the hydroxy is retained, which is represented by Y^-. The right subscript of the ion indicates the type of glycosyl. For example, Y_{H^-} means the cleavage of the hexose to remove the glycosyl. Flavonoid glycosides mainly underwent the cleavage of the sugar ring (X cleavage). When cleavage occurs, the position of the broken bond on the sugar ring is indicated by the left superscript, and the type of the cleavage glycosyl is indicated by the right subscript. This included hexose (H), pentose (*p*), and deoxyhexacarbonose (D). For example, $^{0,2}X_H$ represents a ring-opening cleavage caused by the breakage of the 0, 2 bond on the hexose [36,37].

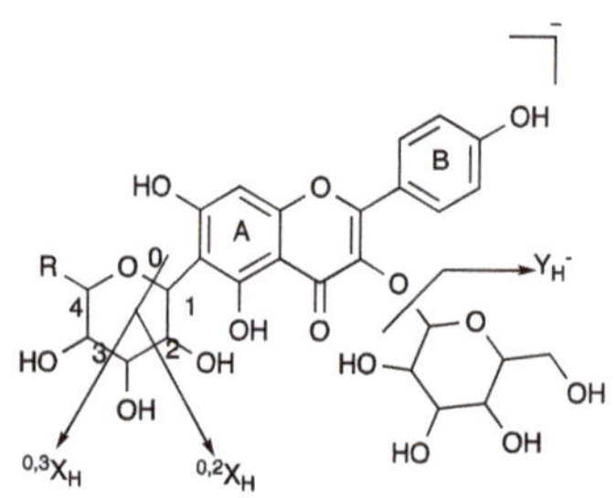

Figure 6. Mass spectrometric analysis of flavonoid glycosides in negative ion mode.

Compounds **1**, **2**, **7**, **10**, and **11** were all flavonoid oxygen glycoside compounds [38]. Quasi-molecular ion peaks can appear in negative ion modes of flavonoid oxygen glycoside compounds to cause glycosidic bond cleavage. This is characterized by a neutral loss of the glucosyl $C_6H_{10}O_5$ (162), rhamnosyl $C_6H_{10}O_4$ (146), or xylose $C_5H_8O_4$ (132). The aglycon structure can continue to lose $-CH_3$ and $-CO_2$ or RDA cleavage can occur in the C ring of the flavonoid aglycon.

The MS spectrum of Compound **1** (t_R = 10.22 min) showed an $[M - H]^-$ ion at m/z 771.2027, and the fragment ions of oxygen glycosides: m/z 609.1530, 462.0838, and 301.0357 were generated. Compound **1** was identified as quercetin 3-*O*-hexobioside-7-*O*-hexoside [39].

Compound **10** (t_R = 14.34 min) showed an $[M - H]^-$ ion at m/z 463.0890. According to m/z 301 $[M-162-H]^-$ secondary mass spectrometry, we speculated that there was glucose in the structure of the compound, and the molecular weight of the parent nucleus was 301, which was quercetin. Based on Scifinder and Reaxy database retrieval, Compound **10** was presumed to be quercetin 3-*O*-hexoside [40]. Compound **1** and Compound **10** had the same mass spectrum fragmentation pathway. The relative molecular mass of Compound **10** was 464. The fragment ion peak (m/z 301) was caused by the loss of a quasi-molecular ion peak and a neutral fragment of m/z 162. According to the relative molecular mass, the neutral fragment of m/z 162 was preliminarily judged to be a glucose group. Since $[M-H-162]^-$ (m/z 301) is a fragment ion of quercetin aglycone, this indicates that there was a glucosyl group in the compound. The fragmentation of m/z 301 ion generated two main characteristic ions: m/z 179 and m/z 151. The m/z 179 may be obtained by the fragmentation of m/z 301 and the transfer of 2 H ions. The m/z 151 fragment ion was generated by the RDA reaction of m/z 301. In addition, the loss of one molecule of CO at m/z 179 also generated m/z 151 ions. This is consistent with the cleavage pathway for quercetin 3-*O*-hexobioside quercetin aglycon fragment ions, and the cleavage law conformed to the structural characteristics of quercetin 3-*O*-hexoside.

Compound **2** (t_R = 11.59 min) showed an $[M - H]^-$ at m/z 365.0881 and product ions at m/z 203.0352, 159.0454, and 130.0422. Among them, m/z 203.0352 represented $[M-H-C_6H_{10}O_5]^-$ obtained after the precursor ion lost glucose, which is the aglycon of the compound. The m/z 159.0454 represents the fragment ion obtained after the precursor ion had lost 206 Da $[M-C_{11}H_{10}O_4-H]^-$. The m/z 130.0422 represents the fragment ion obtained after the precursor ion had lost 235 Da $[M-C_{12}H_{11}O_5-H]^-$. According to the fragment ions in the secondary mass spectrum and the related literature, we speculated that the compound was 2-carboxyl-1,4-naphthohydroquinone-4-O-hexoside [38].

Compound **7** (t_R = 13.89 min) was the main flavonoid in fig leaves. This showed the MS spectrum $[M - H]^-$ ion at m/z 609.1407, and the product ions at m/z 301.0362, 151.0031, 257.0450, and 273.0477 were generated. Under the bombardment of 25% energy, the quercetin 3-O-hexobioside aglycone was lost to form fragments of m/z 301.0362. In the tandem mass spectrometry of the m/z 301.0362 ion, there were three main ways of fragmentation. The first was fragmentation of m/z 301.0362 through RDA to form fragments of m/z 151. The second type was fragmentation of m/z 301. The loss of the carbonyl group from the C ring formed fragments of m/z 273.0477. The compound was identified as quercetin 3-O-hexobioside [41].

Compound **11** was eluted at 15.25 min. It showed the $[M - H]^-$ ion at m/z 593.1524 and the product ions at m/z 285 $[M-162-146-H]^-$ were generated. The quasi-molecular ion peak lost a neutral fragment of m/z 309, resulting in fragment ions $[M-H-308]^-$ (m/z 285). According to the relative molecular mass, the neutral fragment of m/z 308 was initially judged to be rutin. According to the literature, the substance was kaempferol 3-O-hexobioside [42].

Compounds **3**, **4**, **5**, **6**, **8**, and **9** are all flavonoid glycosides [38]. Quasi-molecular ion peaks appeared in negative ion modes for the flavonoid carboglycosides. The negative ion mode showed higher abundance for fragment ions. We observed the ring-opening and cleavage of the sugar ring and the subsequent neutral loss of sugar residues, CO, aglycon loss, CH_3, and other fragment ion peaks. The ring-opening cleavage of the sugar ring is the characteristic cleavage form of carbon glycosides. This mainly occurs in the sugar ring 0, 2 bond and 0, 3 bond, hexose neutral loss of $C_4H_8O_4$ or $C_3H_6O_3$, and the neutral loss of pentose $C_3H_6O_3$ or $C_2H_4O_2$.

Compound **3** (t_R = 11.7 min) had the $[M - H]^-$ ion at m/z 579.1366, and the product ions at m/z 519.1194, 489.1083, 429.0856, 399.0748, 369.0635, and 339.0521 were generated. At m/z 519.1194, a fragment ion was obtained after the precursor ion lost a $C_2H_4O_2$ fragment $[M-H-C_2H_4O_2]^-$. The m/z 489.1083 represents the fragment ion obtained after the precursor ion had lost 90 Da $[M-C_3H_6O_3-H]^-$. The m/z 429.0856 represents the fragment ion obtained after the precursor ion had lost 150 Da $[M-C_3H_6O_3-C_2H_4O_2-H]^-$. The m/z 399.0748 represents the fragment ion obtained by discarding 180 Da of the precursor ion $[M-C_2H_4O_2-C_4H_8O_4-H]^-$. The m/z 369.0635 represents the fragment ion obtained by losing 120 Da of the precursor ion $[M-C_7H_{14}O_7-H]^-$. The m/z 339.0521 represents the fragment ion obtained by discarding 240 Da of the precursor ion $[M-C_7H_{14}O_7-CH_2O-H]^-$. According to the fragment ions in the secondary mass spectrum and related literature, we speculated that the compound was luteolin 6-C-hexoside, 8-C-pentoside [43].

The retention time of Compound **4** was 12.58 min. Compound **4** had the $[M - H]^-$ ion at m/z 563.1414, and the product ions at m/z 473.1115, 443.1001, and 353.0670 were generated. It was identified as kaempferol 6-C-hexoside-8-C-hexoside [44]. The m/z 563 $[M - H]^-$ is its quasi-molecular ion peak, and m/z 443.1001, 473.1115, and 353.0670 are the main fragments produced by its cracking. The most abundant fragment ion in MS_2 was m/z 443. There was a difference of 120 mass units between it and m/z 563. We speculated that it was produced by 0-2 cracking of the hexosyl part of $[M - H]^-$. Due to the difference of 90 mass units between m/z 563 → 473, we speculated that m/z 473 was caused by 0-3 cracking of the hexosyl part of $[M - H]^-$. The difference between m/z 563 → 353 was 120 + 90 mass units. We speculated that $[M - H]^-$ 0-2 cracking of hexose and pentose occurred simultaneously in the ESI source. The difference between m/z 563 → 383 was 120 + 60 mass units. We speculated that $[M - H]^-$ 0-2 cracking of pentose and 0-3 cracking of hexose may have occurred simultaneously in the ESI source. The difference between m/z 563 → 503 was 60 mass units. Presumably, $[M - H]^-$ 0-3 cracking of the pentose occurred in the ESI source. The main ion fragments appearing in the secondary mass spectrum were similar to those in the description of the cleavage behavior of the sugar moiety of flavonoid glycosides in the literature. Compound **6** (t_R = 13.02 min) showed the [M − H]$^-$ ion at m/z 473.1115, and the MS/MS mainly produced characteristic fragment ions of carbon glycosides: m/z 443.1001, 353.0670. It was identified as kaempferol 6-C-hexoside-8-C-hexoside [44]. The cleavage rules for isochafotaside and kaempferol 6-C-hexoside-8-C-hexoside were the same, but the peak order was different. Kaempferol 6-C-hexoside-8-C-hexoside was in the front; isochafotaside was in the back.

Compound **5** (t_R = 12.87 min) showed an MS spectrum for the [M − H]$^-$ ion at m/z 447.0934 and the product ions at m/z 369.0615, 357.0622, 327.0511, 297.0397, 285.0396, and 133.0280. It was identified as quercetin 6-C-hexobioside [45]. The available molecular formula is $C_{21}H_{20}O_{11}$. Among these, m/z 369.0615 represents the fragment ion [M-H-$C_2H_6O_3$]$^-$ after the precursor ion had lost 78 Da. The m/z 357.0622 represents the fragment ion [M-H-$C_3H_6O_3$]$^-$ after the precursor ion had lost 90 Da. The m/z 327.0511 represents the fragment ion after the precursor ion had lost 120 Da, [M-H-$C_4H_8O_4$]$^-$. The m/z 297.0397 represents the fragment ion after the precursor ion had lost 180 Da, [M-H-$C_5H_{10}O_5$]$^-$. The m/z 285.0396 represents the fragment ion [M-H-$C_6H_{10}O_5$]$^-$ after the precursor ion had lost 162 Da. It is the aglycon of the compound after removing glucose. The m/z 133.0280 represents the fragment ion of this compound after RDA fragmentation, [M-H-$C_7H_4O_4$]$^-$.

Compound **8** (t_R = 14.02 min) showed the [M − H]$^-$ ion at m/z 577.1577, and the product ions at m/z 457.1164 and 293.0454 were detected. We speculated that there was glucose in the structure of the compound and that the link between sugar and the parent nucleus was carbon glycoside. The molecular ion peak was m/z 577 [M − H]$^-$. Then, a fragment ion peak of [M-H-120]$^-$ appeared. We observed that it cleaved off the ion $C_4H_8O_4$, and the original structure may have been replaced by hexose. In addition, the position of the pentose phosphate pathway easily merged with the parent ring, and the hydroxyl group was dehydrated and broken and lost 164, forming m/z 413 [M-H-164]$^-$. When position 120 of hexose was closed again, m/z 293 [M-H-120]$^-$ was formed. Thus, Compound **8** was identified as apigenin 2″-O-pentoside [46].

Compound **9** (t_R = 14.14 min) showed the [M − H]$^-$ ion at m/z 432.1029, and the product ions at m/z 341.0673, 311.0564, and 283.0612 were generated. In the process of sugar chain cleavage, one molecule of $C_3H_6O_3$ was lost to generate fragment ions [M-H-90]$^-$. The m/z 311.0564 represents the fragment ion generated by losing one molecule of $C_4H_8O_4$ during the rupture of the glucosyl ring with the parent ion of m/z 431.0981 [M-H-120]$^-$. It continued to lose one molecule of CO to generate fragments of m/z 283.0612 [M-H-$C_4H_8O_4$-CO]$^-$. In addition, m/z 311.0564 lost one molecule of CH_2O to produce fragment ions m/z 281.0450 [M-H-$C_4H_8O_4$-CH_2O]$^-$. It was identified as apigenin 6-C-hexoside [47].

4. Conclusions

This study indicated that the proposed method could be successfully used for the pretreatment and identification of flavonoids in discarded fig leaves. Overall, an innovative extraction method was developed to extract flavonoids from discarded fig leaves by UEAATPE. The ATPE system of 38% (w/w) ethanol/18% (w/w) ammonium sulfate was established as the final system. Simultaneously, an enzymatic concentration of 0.4 U/g, 150 min enzymolysis, an enzymolysis temperature of 50 °C, a liquid–solid ratio of 20:1 (mL/g), and an extraction time of 30 min were obtained as the optimal UEAATPE conditions. Under these conditions, we obtained 60.22 mg/g flavonoids by UEAATPE. The yield of total flavonoids obtained by UEAATPE was 1.13-fold, 1.21-fold, 1.27-fold, and 2.43-fold higher than the yields obtained by the other four methods (EAATPE, UAATPE, ATPE, and SE), respectively. UEAATPE has been shown to be a promising method in the field of bioactive ingredient extraction. Among the eleven compounds characterized, ten flavonoids were first reported in fig leaves. The major flavonoid with the highest content was quercetin 3-O-hexobioside, at about 26.93%. The reuse of fig leaf waste and the development of biologically active ingredients can help to minimize the impact of agricultural waste on the environment. All of this is attributed to the potential medicinal properties of fig leaf flavonoids. In the future, this method may be increasingly applied in the development and utilization of agricultural waste, acting as a bond between promoting economic balance and environmental protection.

Supplementary Materials: The following are available online at https://www.mdpi.com/article/10.3390/app11167718/s1. Figure S1: The first order mass spectrometry (a), the secondary mass spectrometry (b) and the cleavage pathway (c) of compound 1; Figure S2: The first order mass spectrometry (a), the secondary mass spectrometry (b) and the cleavage pathway (c) of compound 2; Figure S3: The first order mass spectrometry (a), the secondary mass spectrometry (b) and the cleavage pathway (c) of compound 3; Figure S4: The first order mass spectrometry (a), the secondary mass spectrometry (b) and the cleavage pathway (c) of compound 4; Figure S5: The first order mass spectrometry (a), the secondary mass spectrometry (b) and the cleavage pathway (c) of compound 5; Figure S6: The first order mass spectrometry (a), the secondary mass spectrometry (b) and the cleavage pathway (c) of compound 6; Figure S7: The first order mass spectrometry (a), the secondary mass spectrometry (b) and the cleavage pathway (c) of compound 7; Figure S8: The first order mass spectrometry (a), the secondary mass spectrometry (b) and the cleavage pathway (c) of compound 8; Figure S9: The first order mass spectrometry (a), the secondary mass spectrometry (b) and the cleavage pathway (c) of compound 9; Figure S10: The first order mass spectrometry (a), the secondary mass spectrometry (b) and the cleavage pathway (c) of compound 10; Figure S11: The first order mass spectrometry (a), the secondary mass spectrometry (b) and the cleavage pathway (c) of compound 11.

Author Contributions: Conceptualization, C.Z. and S.L.; methodology, C.L.; software and validation, Y.T., X.L., and S.L.; formal analysis, C.Z.; investigation, S.L.; resources, S.L.; data curation, X.L.; writing—original draft preparation, S.L.; writing—review and editing, S.L.; visualization, C.Z.; supervision, C.L.; project administration, T.W.; funding acquisition, Y.T. All authors have read and agreed to the published version of the manuscript.

Funding: This work was financially supported by the Fundamental Research Fund for Central Universities (2572019CZ01), the National Natural Science Foundation (31870609), and Heilongjiang Touyan Innovation Team Program (Tree Genetics and Breeding Innovation Team).

Institutional Review Board Statement: Not applicable.

Informed Consent Statement: Not applicable.

Data Availability Statement: Not applicable.

Conflicts of Interest: The authors declare no conflict of interest.

References

1. Du, C.; Abdullah, J.J.; Greetham, D.; Fu, D.; Yu, M.; Ren, L.; Li, S.; Lu, D. Valorization of food waste into biofertiliser and its field application. *J. Clean. Prod.* **2018**, *187*, 273–284. [CrossRef]
2. Christensen, T.H.; Damgaard, A.; Levis, J.; Zhao, Y.; Bisinella, V. Application of LCA modelling in integrated waste management. *Waste Manag.* **2020**, *118*, 313–322. [CrossRef] [PubMed]
3. Khadivi, A.; Anjam, R.; Anjam, K. Morphological and pomological characterization of edible fig (*Ficus carica* L.) to select the superior trees. *Sci. Hortic.* **2018**, *238*, 66–74. [CrossRef]
4. Barolo, M.I.; Ruiz Mostacero, N.; López, S.N. *Ficus carica* L. (Moraceae): An ancient source of food and health. *Food Chem.* **2014**, *164*, 119–127. [CrossRef] [PubMed]
5. Chang, X.H.; Wu, C.Y.; Cao, Z.Y. *Flora of China*; Science Press: Beijing, China, 1998; Volume 23, p. 124.
6. Backes, E.; Leichtweis, M.G.; Pereira, C.; Carocho, M.; Ferreira, I.C.F.R. *Ficus carica* L. and *Prunus spinosa* L. extracts as new anthocyanin-based food colorants: A thorough study in confectionery products. *Food Chem.* **2020**, *333*, 127457. [CrossRef] [PubMed]
7. Palmeira, L.; Pereira, C.; Dias, M.I.; Abreu, R.M.V.; Ferreira, I.C.F.R. Nutritional, chemical and bioactive profiles of different parts of a Portuguese common fig (*Ficus carica* L.) variety. *Food Res. Int.* **2019**, *126*, 108572. [CrossRef] [PubMed]
8. Canal, J.R.; Torres, M.D.; Romero, A.; Pérez, C. A chloroform extract obtained from a decoction of *Ficus carica* leaves improves the cholesterolaemic status of rats with streptozotocin-induced diabetes. *Acta Physi-Ologica Hung.* **2000**, *87*, 71–76. [CrossRef]
9. Fouad, D.; Alhatem, H.; Abdel-Gaber, R.; Ataya, F. Hepatotoxicity and renal toxicity induced by gamma-radiation and the modulatory protective effect of *Ficus carica* in male albino rats. *Res. Vet. Sci.* **2019**, *125*, 24–35. [CrossRef]

10. Kais, R.; Dhekra, G.; Dalanda, W.; Slimen, S.; Mohamed, A.; Hichem, S.; Lamjed, M. *Ficus carica* aqueous extract alleviates delayed gastric emptying and recovers ulcerative colitis-enhanced acute functional gastrointestinal disorders in rats. *J. Ethnopharmacol.* **2018**, *224*, 242–249.

11. Alexandre, E.M.C.; Araújo, P.; Duarte, M.F.; De Freitas, V.; Pintado, M.; Saraiva, J.A. High-pressure assisted extraction of bioactive compounds from industrial fermented fig by-product. *J. Food Sci. Technol.* **2017**, *54*, 2519–2531. [CrossRef]

12. Abbasi, S.; Kamalinejad, M.; Babaie, D.; Shams, S.M.; Sadr, Z.; Gheysari, M.; Askari, V.R.; Rakhshandeh, H. A new topical treatment of atopic dermatitis in pediatric patients based on *Ficus carica* L. (Fig): A randomized, placebo-controlled clinical trial. *Complementary Ther. Med.* **2017**, *35*, 85–91. [CrossRef]

13. Mustafa, K.; Yu, S.; Zhang, W.; Mohamed, H.; Song, Y. Screening, characterization, and in vitro-ROS dependent cytotoxic potential of extract from *Ficus carica* against hepatocellular (HepG2) carcinoma cells. *S. Afr. J. Bot.* **2021**, *138*, 217–226. [CrossRef]

14. Patil, V.V.; Patil, V.R. Evaluation of anti-inflammatory activity of *Ficus carica* Linn. leaves. *Indian J. Nat. Prod. Resour.* **2011**, *2*, 151–155.

15. Pérez, C.; Canal, J.R.; Campillo, J.E.; Romero, A.; Torres, M.D. Hypotriglyceridaemic activity of *Ficus carica* leaves in experimental hypertriglyceridaemic rats. *Phytother. Res.* **2015**, *13*, 188–191. [CrossRef]

16. Pérez, C.; Domínguez, E.; Ramiro, J.M.; Romero, A.; Campillo, J.E.; Torres, M.D. A study on the glycaemic balance in streptozotocin iabetic rats treated with an aqueous extract of *Ficus carica* (fig tree) leaves. *Phytother. Res.* **1996**, *10*, 82–83. [CrossRef]

17. Vikas, P. Evaluation of anti-pyretic potential of *Ficus carica* leaves. *Int. J. Pharm. Sci. Rev. Res.* **2010**, *2*, 010.

18. Caleja, C.; Barros, L.; Prieto, M.A.; Bento, A.; Oliveira, M.B.P.P.; Ferreira, I.C.F.R. Development of a natural preservative obtained from male chestnut flowers: Optimization of a heat-assisted extraction technique. *Food Funct.* **2019**, *10*, 1352–1363. [CrossRef]

19. Soares, R.R.G.; Azevedo, A.M.; Alstine, J.M.V.; Aires-Barros, M.R. Partitioning in aqueous two-phase systems: Analysis of strengths, weaknesses, opportunities and threats. *Biotechnol. J.* **2015**, *10*, 1158–1169. [CrossRef]

20. Liu, Y.; Gong, G.L.; Zhang, J.; Jia, S.Y.; Li, F.; Wang, Y.Y.; Wu, S.H. Response surface optimization of ultrasound-assisted enzymatic extraction polysaccharides from *Lycium barbarum*. *Carbohyd. Polym.* **2014**, *110*, 278–284. [CrossRef]

21. Chmelová, D.; Škulcová, D.; Legerská, B.; Horník, M.; Ondrejovič, M. Ultrasonic-assisted extraction of polyphenols and antioxidants from *Picea abies* bark. *J. Biotechnol.* **2020**, *314–315*, 25–33. [CrossRef] [PubMed]

22. Lei, X.; Hu, W.-B.; Yang, Z.-W.; Hui, C.; Wang, N.; Liu, X.; Wang, W.-J. Enzymolysis-ultrasonic assisted extraction of flavanoid from *Cyclocarya paliurus* (Batal) Iljinskaja:HPLC profile, antimicrobial and antioxidant activity. *Ind. Crop. Prod.* **2019**, *130*, 615–626.

23. Chunying, L.; Yukun, Z.; Chunjian, Z.; Yujiao, N.; Kaiting, W.; Jingjing, Z.; Wenyan, Z. Ultrasonic Assisted-Reflux Synergistic Extraction of Camptothecin and Betulinic Acid from *Camptotheca acuminata* Decne. Fruits. *Molecules* **2017**, *22*, 1076.

24. Luo, X.; Cui, J.; Zhang, H.; Duan, Y.; Zhang, D.; Cai, M.; Chen, G. Ultrasound assisted extraction of polyphenolic compounds from red sorghum (*Sorghum bicolor* L.) bran and their biological activities and polyphenolic compositions. *Ind. Crop. Prod.* **2018**, *112*, 296–304. [CrossRef]

25. Orevi, T.; Antov, M. Ultrasound assisted extraction in aqueous two-phase system for the integrated extraction and separation of antioxidants from wheat chaff—ScienceDirect. *Sep. Purif. Technol.* **2017**, *182*, 52–58.

26. Jiao, Y.; Zuo, Y. Ultrasonic extraction and HPLC determination of anthraquinones, aloe-emodine, emodine, rheine, chrysophanol and physcione, in roots of *Polygoni multiflori*. *Phytochem. Anal.* **2009**, *20*, 272–278. [CrossRef] [PubMed]

27. Wang, C.; Zuo, Y. Ultrasound-assisted hydrolysis and gas chromatography–mass spectrometric determination of phenolic compounds in cranberry products. *Food Chem.* **2011**, *128*, 562–568. [CrossRef]

28. Song, H.; Yang, R.; Zhao, W.; Katiyo, W.; Hua, X.; Zhang, W. Innovative assistant extraction of flavonoids from pine (*Larix olgensis* Henry) needles by high-density steam flash-explosion. *J. Agric. Food Chem.* **2014**, *62*, 3806. [CrossRef]

29. Zhang, W.; Zhu, D.; Fan, H.; Liu, X.; Wan, Q.; Wu, X.; Liu, P.; Tang, J.Z. Simultaneous extraction and purification of alkaloids from *Sophora flavescens* Ait. by microwave-assisted aqueous two-phase extraction with ethanol/ammonia sulfate system. *Sep. Purif. Technol.* **2015**, *141*, 113–123. [CrossRef]

30. Xie, X.; Zhu, D.; Zhang, W.; Huai, W.; Wang, K.; Huang, X.; Zhou, L.; Fan, H. Microwave-assisted aqueous two-phase extraction coupled with high performance liquid chromatography for simultaneous extraction and determination of four flavonoids in *Crotalaria sessiliflora* L. *Ind. Crop. Prod.* **2017**, *95*, 632–642. [CrossRef]

31. Ma, F.Y.; Gu, C.B.; Li, C.Y.; Luo, M.; Wang, W.; Zu, Y.G.; Li, J.; Fu, Y.J. Microwave-assisted aqueous two-phase extraction of isoflavonoids from *Dalbergia odorifera* T. Chen leaves. *Sep. Purif. Technol.* **2013**, *115*, 136–144. [CrossRef]

32. Jung, S.; Lamsal, B.P.; Stepien, V.; Johnson, L.A.; Murphy, P.A. Functionality of soy protein produced by enzyme-assisted extraction. *J. Am. Oil Chem. Soc.* **2006**, *83*, 71–78. [CrossRef]

33. Huang, M.; Zhang, Y.; Xu, S.; Xu, W.; Chu, K.; Xu, W.; Zhao, H.; Lu, J. Identification and quantification of phenolic compounds in *Vitex negundo* L. var. cannabifolia (Siebold et Zucc.) Hand.-Mazz. using liquid chromatography combined with quadrupole time-of-flight and triple quadrupole mass spectrometers. *J. Pharm. Biomed. Anal.* **2015**, *108*, 11–20. [CrossRef] [PubMed]

34. Jovanovic-Malinovska, R.; Kuzmanova, S.; Winkelhausen, E. Application of ultrasound for enhanced extraction of prebiotic oligosaccharides from selected fruits and vegetables. *Ultrason. Sonochem.* **2015**, *22*, 446–453. [CrossRef] [PubMed]

35. Fabre, N.; Rustan, I.; De, H.E.; Quetin-Leclercq, J. Determination of flavone, flavonol, and flavanone aglycones by negative ion liquid chromatography electrospray ion trap mass spectrometry. *J. Am. Soc. Mass Spectrom.* **2001**, *12*, 707–715. [CrossRef]

36. Chen, H.; Zuo, Y. Identification of flavonol glycosides in American cranberry fruit. *Food Chem.* **2007**, *101*, 1357–1364. [CrossRef]

37. Zuo, Y.; Lu, X.; Anwar, F.; Hameed, S. Characterization of free and conjugated phenolic compounds in fruits of selected wild plants. *Food Chem.* **2016**, *190*, 80–89.
38. Komatsu, M.; Tomimori, T.; Hatayama, K.; Makiguchi, Y.; Mikuriya, N. Studies on the constituents of *Sophora species*. II. Constituents of Sophora subprostrata CHUN et T. CHEN. (2). Isolation and structure of new flavonoids, sophoradochromene and sophoranochromene. *Chem. Pharm. Bull.* **1970**, *18*, 741–745. [CrossRef]
39. Adjé, F.A.; Lozano, Y.F.; Gernevé, C.L.; Lozano, P.R.; Meudec, E.; Adima, A.A.; Gaydou, E.M. Phenolic acid and flavonol water extracts of Delonix regia red flowers. *Ind. Crop. Prod.* **2012**, *37*, 303–310. [CrossRef]
40. He, Z.; Xia, W. Analysis of phenolic compounds in chinese olive (*Canarium album* L.) fruit by RPHPLC-DAD-ESI-MS. *Food Chem.* **2007**, *105*, 1307–1311. [CrossRef]
41. Zhou, C.H.; Luo, Y.Y.; Lei, Z.X.; Wei, G. UHPLC-ESI-MS Analysis of Purified Flavonoids Fraction from Stem of *Dendrobium denneaum* Paxt. and Its Preliminary Study in Inducing Apoptosis of HepG2 Cells. *Evid. -Based Complementary Altern. Med.* **2018**, *2018*, 8936307. [CrossRef]
42. Sousa, A.D.; Maia, A.I.V.; Rodrigues, T.H.S.; Canuto, K.M.; Ribeiro, P.R.V.; de Cassia Alves Pereira, R.; Vieira, R.F.; de Brito, E.S. Ultrasound-assisted and pressurized liquid extraction of phenolic compounds from Phyllanthus amarus and its composition evaluation by UPLC-QTOF. *Ind. Crop. Prod.* **2016**, *79*, 91–103. [CrossRef]
43. He, L.; Zhang, Z.; Lu, L.; Liu, Y.; Li, S.; Wang, J.; Song, Z.; Yan, Z.; Miao, J. Rapid identification and quantitative analysis of the chemical constituents in *Scutellaria indica* L. by UHPLC-QTOF-MS and UHPLC-MS/MS. *J. Pharm. Biomed. Anal.* **2016**, *117*, 125–139. [CrossRef] [PubMed]
44. Batista, M.T.; Gomes, E.T. C-glycosylflavones from Ceratonia siliqua cotyledons. *Phytochemistry* **1993**, *34*, 1191–1193. [CrossRef]
45. Norbaek, R.; Brandt, K.; Kondo, T. Identification of flavone C-glycosides including a new flavonoid chromophore from barley leaves (*Hordeum vulgare* L.) by improved NMR techniques. *J. Agric. Food Chem.* **2000**, *48*, 1703–1707. [CrossRef] [PubMed]
46. Muth, D.; Marsden-Edwards, E.; Kachlicki, P.; Stobiecki, M. Differentiation of isomeric malonylated flavonoid glyconjugates in plant extracts with UPLC-ESI/MS/MS. *Phytochem. Anal.* **2010**, *19*, 444–452. [CrossRef]
47. Huang, D.; Zhou, X.; Si, J.; Gong, X.; Wang, S. Studies on cellulase-ultrasonic assisted extraction technology for flavonoids from *Illicium verum* residues. *Chem. Cent. J.* **2016**, *10*, 56. [CrossRef] [PubMed]

Article

Multifunctional Ingredient from Aqueous Flavonoidic Extract of Yellow Onion Skins with Cytocompatibility and Cell Proliferation Properties

Ștefania Adelina Milea [1], Oana Crăciunescu [2], Gabriela Râpeanu [1], Anca Oancea [2], Elena Enachi [1], Gabriela Elena Bahrim [1] and Nicoleta Stănciuc [1,*]

[1] Faculty of Food Science and Engineering, Dunărea de Jos University of Galati, 111 Domnească Street, 8002021 Galati, Romania; adelina.milea@ugal.ro (Ș.A.M.); gabriela.rapeanu@ugal.ro (G.R.); elena.ionita@ugal.ro (E.E.); gabriela.bahrim@ugal.ro (G.E.B.)

[2] National Institute of Research and Development for Biological Sciences, 296 Splaiul Independentei, 060031 Bucharest, Romania; oana_craciunescu2009@yahoo.com (O.C.); oancea.anca@gmail.com (A.O.)

* Correspondence: nsava@ugal.ro

Abstract: Significant quantities of onion are cultivated annually, such that industrial processing leads to an appreciable amount of by-products, estimated at around 500,000 tons. Onion skins are considered an important source of naturally occurring antioxidant compounds, particularly flavonoid compounds. Our study follows the development of a sustainable solution in order to manage the by-products of yellow onion skins by designing ingredients with multifunctional activities. A green solvent aqueous extraction of flavonoids was applied to obtain a safe, flavonoid-enriched extract, yielding a total flavonoid content of 50.21 ± 0.09 mg quercetin equivalent (QE)/g dry weight (DW), and an antioxidant activity of 250.81 ± 6.76 mM Trolox/g DW. Complex biopolymeric matrices consisting of whey protein isolates, whey protein hydrolysates, maltodextrin, and pectin were further dissolved in the flavonoid-enriched aqueous extract, followed by freeze-drying. Two powders were obtained, both showing satisfactory phytochemical content and good stability during storage. The application of confocal microscopy revealed that the microscopic structure of the powders have a distribution of the bioactive compounds within the biopolymeric matrices. The in vitro digestion suggested remarkable stability in the gastric tract and a flavonoid-controlled release in the intestinal phase. A significant compatibility range of up to 1 mg/mL for both powders was found, whereas concentrations between 10 and 250 µg/mL stimulated cell proliferation after 24 h of cultivation. The powders showed satisfactory thermal and pH stability, which favors their addition to different food matrices.

Keywords: yellow onion skins; flavonoids; extraction; antioxidant activity; multifunctional ingredients

Citation: Milea, Ș.A.; Crăciunescu, O.; Râpeanu, G.; Oancea, A.; Enachi, E.; Bahrim, G.E.; Stănciuc, N. Multifunctional Ingredient from Aqueous Flavonoidic Extract of Yellow Onion Skins with Cytocompatibility and Cell Proliferation Properties. *Appl. Sci.* **2021**, *11*, 7243. https://doi.org/10.3390/app11167243

Academic Editor: Emanuel Vamanu

Received: 25 June 2021
Accepted: 4 August 2021
Published: 6 August 2021

1. Introduction

Yellow onion skins (*Allium cepa* L.) are considered valuable raw material for the extraction of bioactives because they are rich, economical, sustainable, ubiquitous, and available worldwide; additionally, they are not considered a component of the global food supply. A significant amount of onion is cultivated annually, making it one of the most grown vegetables around the world, with an estimated total production of around 66–85.7 million tons per year [1]. Industrial processing leads to a significant amount of onion by-products, estimated to be around 500,000 tons [2], which is considered an important source of polyphenols. Furthermore, onions also represent a considerable amount of dietary phytochemicals such as organosulfur compounds, phenolic acids, flavonoids, thiosulfinates, and anthocyanins [3].

Flavonoids are well-known antioxidant compounds owing to their high redox potential, which allows them to act as reducing agents and singlet oxygen quenchers [4]. It has been reported that flavonoid levels in the edible portions of onions vary from 0.03 to 1 g/kg, whereas onion skins contain significantly higher levels of flavonoids at 2–10 g/kg [5].

Therefore, onion skins represent a valuable resource for flavonoid extraction. In most cases, solid–liquid conventional and assisted extraction methods are the most exploited technologies for the extraction of flavonoids from vegetables, but these techniques employ high volumes of solvent and energy [6]. However, high extraction yield, with its energetic feasibility and usage of environmentally friendly solvents, should be considered as an extraction technology. Hence, new solvents are being studied in order to carry out more selective extractions of flavonoid compounds from plants [7,8], while considering the reduction of extraction time, solvent, and energy.

When extracted from their natural matrices, polyphenols are highly unstable such that they undergo various chemical reactions because of the presence of unsaturated bonds in their molecular structure and are affected by oxidants, heat, light, and enzymatic reactions during storage [9]. Therefore, protective technologies should be applied to improve polyphenol stability and safeguard them against chemical damage before their industrial application. This can be achieved by applying suitable techniques such as microencapsulation [10].

It has been suggested that the use of microencapsulated flavonoids instead of the free compounds can overcome the drawbacks of their instability, improve their bioavailability as well as their shelf life [11], thus contributing to the widening of their industrial applications in food and in related fields such as the biopharmaceutical and cosmeceuticals industries [12].

Therefore, the objective of the present study was to test the use of a green, low-cost solvent to extract flavonoids from yellow onion skins. In order to provide a more stable, multifunctional, and safer powder—with multiple uses such as food ingredients, in biopharmaceuticals or in cosmeceuticals—a unique combination of biopolymeric matrices such as whey protein isolates (WPI), whey protein hydrolysates (WPH), maltodextrin (MD), and pectin (P) were dissolved in the flavonoid-enriched aqueous extract, followed by freeze-drying. The powders were evaluated for their functional properties, in terms of the flavonoids' entrapping efficiency, total flavonoid and polyphenol contents, antioxidant activity, their microstructural pattern by confocal microscopy, in vitro digestion as well as the thermal and pH stability of the phytochemicals. The cytocompatibility of the powders was tested on mouse fibroblasts to evaluate the release of possible toxic compounds and their proliferative effect. The results obtained in this study could bring certain benefits in terms of exploiting the bioactive potential of yellow onion skin flavonoids or for the development of ingredients with health benefits, thus upgrading their functional characteristics.

2. Materials and Methods

2.1. Chemicals

Whey protein isolates (protein content of 95%) and whey protein hydrolysates were purchased from Fonterra (Auckland, New Zealand). Quercetin, 2,2-diphenyl-1-picrylhydrazyl (DPPH), apple pectin and maltodextrin (DE dextrose equivalent 16.5–19.5), 6-Hydroxy-2,5,7,8-tetramethylchromane-2-carboxylic acid (Trolox), ethanol, sodium hydroxide, Folin-Ciocalteu reagent, and gallic acid were obtained from Sigma Aldrich, Steinheim, Germany.

2.2. Flavonoid Extraction from Yellow Onion Skins

Yellow onions (*Allium cepa*) were purchased from the local market (Galați, Romania) in May 2020. As a first step, onion skins were washed and manually detached, further washed with distilled water, and then blotted on paper towels. Skins were then dried for 2 h at 40 °C and further stored at 4 °C until extraction. Before extraction, the onion skins were ground to obtain a homogeneous batch with a particle size smaller than 0.5 × 0.5 mm. Furthermore, in order to obtain the flavonoid extract, about 40 g of milled onion skin was weighed and mixed with 1000 mL of hot water (70 °C). The extraction was performed at 500 rpm and 70 ± 2 °C for 2 h, followed by centrifugation at 6000 rpm for 15 min and 4 °C.

The supernatant was collected, and the extraction was repeated 5 times. The supernatant was further used for the characterization and experiments in powder development.

2.3. Powder Formula

The materials were dissolved in an aqueous extract, with a volume of 180 mL, at a ratio of WPI:WHP:MD:P (2:0.5:1:0.5) (Variant 1), and WPI:MD:P (2:1:0.5) (Variant 2). The two variants were allowed to hydrate by maintaining them overnight on a magnetic stirrer. The samples were further frozen at $-70\,°C$, and the ice crystals were then removed by freeze-drying (CHRIST Alpha 1–4 LD plus, Germany) at $-42\,°C$ under a pressure of 0.10 mBar for 48 h. Finally, the powders were collected and packed into metalized bags and stored in the freezer at $4\,°C$ for further analysis. Each experiment was duplicated.

2.4. Extract and Powder Characterization

The methods described in a previous study by Milea et al. [13] were further performed to determine the entrapping efficiency of flavonoids, and to quantify the phytochemical contents and antioxidant activity of the powders. Briefly, the extract and the powders were characterized in terms of total flavonoids, total polyphenols, and antioxidant activity by using the aluminum chloride method, the Folin-Ciocâlteu method, and the DPPH method, respectively. The colorimetric method, based on the aluminum chloride capacity of forming stable acid complexes with the flavones or flavonols, was used to measure the total flavonoids. In order to determine antioxidant activity, the protocol for measuring antiradical activity on DPPH was used. Discolorations were measured after incubation for 90 min at $25\,°C$ in the dark. The entrapping efficiency was determined by assessing the surface flavonoid content (SFC) and total flavonoid content (TFC) of the powders, as described by Milea et al. [13]. The entrapping efficiency (EE, %) was calculated using the following equation:

$$EE\ (\%) = \frac{(TFC - SFC)}{TFC} \times 100 \tag{1}$$

2.5. In Vitro Digestion

To test the protective effect of biopolymeric matrices upon flavonoids, a static digestion model was applied using the method described prior by Oancea et al. [14], a method that reproduces the digestion process in gastrointestinal conditions. Briefly, for the in vitro digestion experiments, the powders were first mixed with a Tris-HCl buffer. The simulated gastric juice, consisting of pepsin and HCl, was added to the mixture and the pH was adjusted to 2.0. After two hours of incubation, the previous mixture was mixed with the simulated intestinal juice consisting of pancreatin and sodium bicarbonate, and the pH was adjusted to 7.8. During the entire experiment, the samples were incubated for 4 h at 150 rpm and $37\,°C$, while the total flavonoid content of the samples was measured every 30 min.

2.6. Thermal and pH Stability of the Extracted Phytochemicals

In order to test the thermal and pH stability of the microencapsulated bioactives, considering their introduction into different types of food, which involves ensuring their preservability by heat treatment at high temperatures and/or the use of different pH values, the behavior of flavonoids were studied at increasing temperatures and pH levels. The powders were dissolved in various ratios ranging from 0.1–0.6% in distillated water, and 10 mL of the resulting solutions were treated at different temperatures ranging from 25 to $100\,°C$ for 15 min. Additionally, an amount of 100 mg was dissolved in 10 mL buffer, at a pH varying from 2.0 to 8.0, in order to test the stability of the phytochemicals in terms of total flavonoids, total polyphenols, and antioxidant activity.

2.7. Cell Culture and Treatment

Minimum Essential Medium (MEM) supplemented with 10% (v/v) fetal calf serum (FCS), 2 mM L-glutamine, and 1% (v/v) antibiotic mixture (penicillin–streptomycin–neomycin) was used for the culture and treatment of the mouse fibroblasts from the NCTC clone L929 cell line (ECACC, Sigma-Aldrich, Darmstadt, Germany). The culture medium was humidified with 5% CO_2 at 37 °C, until subconfluence. The stock solutions were prepared by adding the powders to the culture medium as stock solutions, at a concentration of 1 mg/mL, followed by incubation at 37 °C for 24 h, then filtered through 0.22 μm membrane (MilliporeSigma, Merck, Darmstadt, Germany). The preparation of the cell suspension was performed using trypsin over the cells in order to detach them. The cells were seeded in 96-well microplates at a density of 4×10^4 cells/mL and incubated for 24 h at 37 °C in a humidified atmosphere with 5% CO_2. Afterwards, the culture medium was replaced with fresh medium containing various concentrations of powders ranging from 10 to 1000 μg/mL. The samples were again incubated for 24 and 48 h, respectively. A negative control culture without any sample and a positive control using H_2O_2 were considered.

2.8. Cell Viability

The cell viability was evaluated using a spectrophotometric method described by Gaspar et al. [15], with neutral red (NR) as an Indicator. After the incubation time, the culture medium was removed from the wells and the cells were incubated in a 50 μg/mL NR solution, at 37 °C for 3 h. The cells were washed, and the colorant was released by gentle shaking for 15 min into 1% (v/v) acetic acid solution with 50% (v/v) ethanol. The amount of the uptaken colorant was directly proportional to the number of viable cells. A Sunrise microplate reader (Tecan, Austria) was used for measuring the optical density of the cells. A control culture with 100% viability was considered and the obtained results were reported as the percentage relative to it. The release of possible toxic compounds from the powders and the cell viability were evaluated in the non-tumorigenic L929 fibroblast cell culture by NR assay.

2.9. Cell Morphology

After 48 h of incubation, the morphology of the cells in the presence of the powders was analyzed using light microscopy. Firstly, the cells were washed and fixed with methanol and the identification was made using a Giemsa solution. The micrographs were acquired at an Axio Observer D1 optical microscope equipped with a digital camera (Carl Zeiss, Oberkochen, Germany).

2.10. Confocal Laser Microscope Spectroscopy

The confocal laser scanning microscopy (CLSM) analysis was used to determine the structural features and characteristics of the two powders. The CLSM analysis was employed by capturing the images with a type of Zeiss confocal laser scanning system (LSM710), equipped with various types of lasers such as the DPSS laser (diode pumped solid state—561 nm), diode laser (405 nm), Ar-laser (458,488,514 nm), and HeNe-laser (633 nm). The structural and morphological pattern of the powders was detected with a $20\times$ apochromatic objective, at 0.6 zoom; the samples were subsequently dyed using Red Congo as a fluorescent marker. The obtained 3D images were rendered and analyzed using ZEN 2012 SP1 software (Black Edition, Jena, Germany).

2.11. Statistical Analysis

Unless otherwise stated, the data reported in this study represent the average of triplicate analyses and were reported as mean $\pm$ standard deviation (SD). After running the normality and homoscedasticity test, the data were analyzed using a one-way analysis of variance (ANOVA). The Tukey method with a 95% confidence interval was employed for the post hoc analysis; $p < 0.05$ was statistically significant. The statistical analysis was carried out using the Minitab 18 software. The statistical analysis of the data was performed using the one-tailed paired Student's *t*-test for the cytocompatibility tests. Significant differences were considered at $p < 0.05$.

3. Results

3.1. Yellow Onion Skin Extract and Powder Characterization

By using hot water as a solvent for the extraction of polyphenolic compounds from yellow onion skins, the extract showed a total flavonoid content of 50.21 ± 0.09 mg QE/g DW, a total polyphenol content of 21.68 ± 0.69 mg GAE/g DW, and an antioxidant activity of 250.81 ± 6.76 mM Trolox/g DW. Piechowiak et al. [16] demonstrated that the total yield of the antioxidant compounds from onion skin extraction, using methanol as solvent, varied depending on temperature and process time as well as on the ratio of onion skin mass to the methanol volume. The obtained onion skin extract was characterized by a high antioxidant activity (ranging from 441.4 to 593.9 mg/g), due to a high concentration of flavonoid compounds such as quercetin (315.6 mg/g), quercetin-3-glucoside (40.3 mg/g), isorhamnetin (14.8 mg/g), and kaempferol (10.9 mg/g).

Although freeze-drying involves higher cost, lower rate, and industrial flexibility, the technique brings certain advantages when considering the preservation of high quality products with minimal thermal and oxidative degradation and higher entrapment efficiency (Ballesteros et al., 2017; Pasrija et al., 2015). In our study, the resulting powders obtained by freeze-drying displayed a different phytochemical profile, as shown in Table 1.

Table 1. Global phytochemical profile of the powders.

Variants	Entrapping Efficiency, %	Total Flavonoids, mg QE/g DW	Total Polyphenols, mg GAE/g DW	Antioxidant Activity, mM Trolox/g DW
1	76.55 ± 2.62 *,b	26.91 ± 2.93 a	21.24 ± 0.57 a	102.76 ± 0.82 b
2	83.51 ± 1.81 a	20.75 ± 0.78 b	18.57 ± 0.32 b	109.28 ± 0.12 a

Values are represented as mean $\pm$ standard errors (*). Superscript values in the same column with distinct letters (a and b) are significantly different at $p < 0.05$, based on the Tukey method.

From the data presented in Table 1, it can be observed that using WPI, MD, and P of the yellow-skin flavonoid-enriched extract by freeze-drying allowed for a higher entrapping efficiency of flavonoids at approximately 84% when compared with the biopolymeric coating, including WPI, MD, WPH, and P. The variant 1 biopolymeric combination allowed for a better retention of polyphenols, whereas variant 2 yielded a higher concentration of flavonoids. Based on the results showed in Table 1, it can be concluded that when considering the effect of freeze-drying on the phytochemical profile of the powders, several factors should be examined such as the core-to-coating ratio, chemical properties, and interaction effects. Although higher total flavonoid content was found in variant 1, the availability of surface flavonoid content was higher when compared with variant 2, thus favoring degradation. In contrast, the surface flavonoid content in variant 2 contributed to the total increment of entrapping efficiency, in spite of the lower phenolic content. The freeze-dried ingredients are expected to have a favored phytochemical profile and antioxidant activity as well as a protective effect, especially during the simulated digestion, thus facilitating their addition to different food matrices.

The matrices were selected in terms of developing a more stable, multifunctional, and safer powder. A brief analysis of previous work highlighted the preponderant use of gum arabic, MD, or whey protein, in single form or in combination, for the protection and stabilization of polyphenols. The selection of WPI was based on several selected characteristics such as good physicochemical properties, thus contributing to the intermolecular cross-links with other biopolymers as well as to its high nutritional value, biodegradability, biocompatibility, and pepsin-resistant ability. Moreover, due to its high surface activity, this protein acts as film-forming and emulsifying agent. MD is often used as coating material due to its good water solubility, low viscosity, and low sugar content. MD with a DE value of 16.5–19.5 is easily digestible, either moderately sweet or almost flavorless. Solubility, hydroscopicity, osmolality, and its effectiveness in reducing the freezing point increases with increasing DE. Additionally, WPH were used as a source of bioactive peptides for functional foods, with antibacterial and antivirus activity and anti-inflammatory properties.

The results showed that variant 2 displayed lower flavonoid and polyphenol content and a higher antioxidant activity. This could be a result of the influence of the bioactive peptides, which might display a potentially higher antioxidant activity, and also to the possible entrapping of some other polyphenol classes that might likewise be responsible for the higher antioxidant activity. The powders were tested for phytochemical stability after 30 days of storage at 4 °C. The results are shown in Table 2.

Table 2. Phytochemical stability in the powders during storage.

Variants	Total Flavonoids, mg QE/g DW	Total Polyphenols, mg GAE/g DW	Antioxidant Activity, mM Trolox/g DW
1	22.35 ± 0.60 [*,a]	19.56 ± 0.05 [a]	82.23 ± 0.92 [b]
2	22.74 ± 0.18 [a]	18.91 ± 0.18 [b]	86.99 ± 1.15 [a]

Values are represented as mean ± standard errors (*). Superscript values in the same column with distinct letters (a and b) are significantly different at $p < 0.05$, based on the Tukey method.

From Table 2, it can be observed that variant 2 showed a higher phytochemical stability, leading to an increase in flavonoids by approximately 10%, whereas in variant 1, a decrease in flavonoids by 17% was found. Polyphenols were found to be stable in variant 2, while they decreased by approximately 8% in variant 1. The antioxidant activity decreased in both variants by 20%.

Hamid et al. [17] studied the effect of maltodextrin concentrations on the microencapsulation of wild pomegranate flavedo (peel) phenolics through freeze-drying based on various functional properties and structural morphology. These authors reported that microencapsulation allowed for the retention of higher phenolics (78.47 ± 0.39 GAE/g) but lower flavonoids (4.18 ± 0.03 QE/g) as compared to our study. In another study, Horincar et al. [18] used WPI and different combinations of polymers, including chitosan, MD, and P to microencapsulate yellow onion skin flavonoids by freeze-drying. These authors also suggested a significantly different phytochemical profile for the powders as a function of the coating materials with a lower flavonoid content at 5.84 ± 0.23 mg QE/g DW in the WPI-chitosan combination, and a higher one at 104.97 ± 5.02 mg QE/g DW in the WPI-MD-P combinations. When compared with the data reported in this study, Horincar et al. [18] showed a significantly higher antioxidant activity for the powders, at 175.93 ± 1.50 mM Trolox/g DW and 269.20 ± 3.59 mM Trolox/g DW, respectively, particularly due to the different flavonoid content of the initial extract.

In a previous study, Milea et al. [12] encapsulated flavonoids from yellow onion skins, using MD, P, and WHP in different ratios as coating materials. These authors reported that the obtained powder showed comparable levels of flavonoids, varying from 98.12 ± 0.55 to 103.75 ± 0.57 mg QE/g DW, whereas significantly higher polyphenol content, varying from 53.53 ± 1.71 to 69.26 ± 1.03 mg GAE/g DW, and antioxidant activity from 280.61 ± 3.08 to 337.57 ± 0.89 mM Trolox/g DW, were reported for different variants.

3.2. The Microstructural Design of the Powders by Confocal Microscopy

The CLSM analysis of the two variants of powders displayed important characteristics in regard to their morphology and structural pattern. The images were obtained by point-by-point scanning, at a high resolution within a wide field, with the help of digital focusing; the images were captured after the staining of the samples (Figure 1a,b).

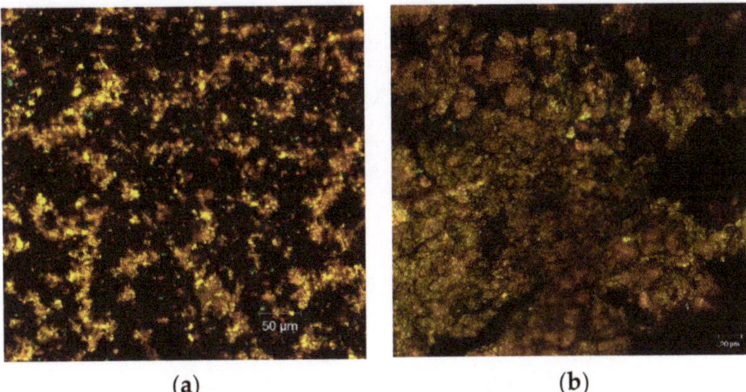

Figure 1. Confocal laser scanning microscopy images of the two powder samples obtained with the ZEN 2012 SP1 software (Black Edition) by encapsulating the flavonoids extracted from the yellow onion skins within different materials: Variant 1—WPI:WHP:MD:P—2:0.5:1:0.5 (**a**), Variant 2—WPI:MD:P—2:1:0.5 (**b**).

Flavonoids represent a valuable large group of polyphenol compounds with a high antioxidant capacity; they are of plant origin with over 4500 different compounds, delivering a variable absorption-emission spectra, which depends on the biochemical profile of the extraction source. Through confocal microscopy analysis, the microstructural design of the obtained powders was assessed. By fluorescent labeling of the two fluorescently stained powders, a well-individualized matrix formed of very small microspherosomes filled with flavonoids and polyphenols could be observed, together with the biopolymeric matrix. Between the two obtained powders, the finest powder was variant 1 (Figure 1a), which displayed a rather uniform microscopic appearance in the form of small spherosomes, with diameters smaller than 2 μm, emitting predominantly around 550 nm. Thus, it can be stated that adding WPH in variant 1 modulated the microscopic appearance and behavior of the powder. For variant 2, the observed spherosomes displayed the tendency to assemble into bigger clusters. From the images, it can be observed that the addition of WPH in variant 1 imprinted slightly different characteristics in terms of the spherosomes' tendency to assemble, which is well-correlated with the results of the entrapping efficiency. Therefore, it can be stated that this tendency of biopolymeric matrices to assemble is responsible for the higher efficiency of entrapping flavonoids. The agglutination process of the polyphenolic compounds and flavonoids in the form of coacervates of larger dimensions observed for the variant 2 powder could be correlated with the more complex biopolymeric coating material. Furthermore, the complexity of the emission spectra, ranging from the green (starting from 500 nm) to the red domain (until 650 nm), revealed the presence of polyphenolic compounds and flavonoids inside the coating material. This microscopic detail supports the results of the biochemical analysis, which showed a significantly high content of flavonoids and polyphenols associated with a remarkable antioxidant activity. Therefore, the confocal microscopy analysis highlighted the fact that the applied technique was efficient and generated multifunctional ingredients with a high bioactive content, which in turn produces high antioxidant activity.

3.3. In Vitro Digestion of Flavonoids

The in vitro behavior of the flavonoids under simulated gastric (SGS) and intestinal (SIF) conditions was observed. Significant flavonoid stability was found in SGS (Figure 2a) for both variants, with a slight increase in flavonoids, up to approximately 1% in variant 2. Therefore, it can be stated that the flavonoids were successfully protected by the selected coating materials.

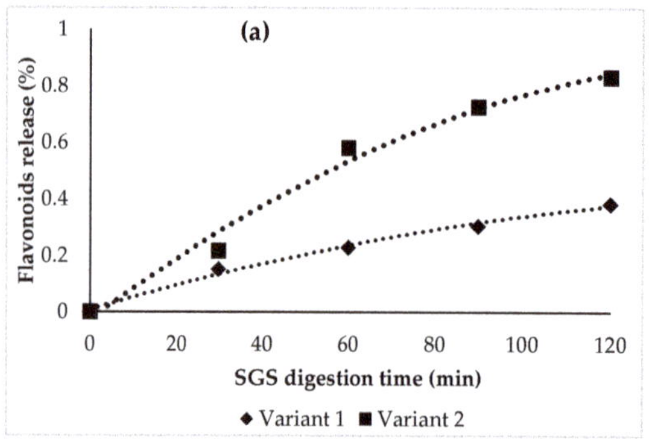
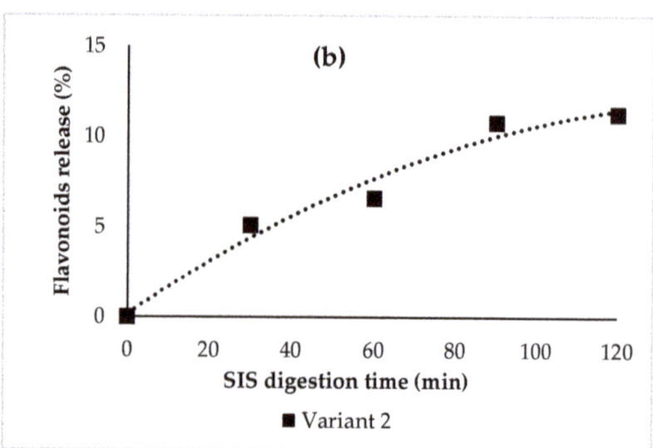

Figure 2. Flavonoid release (%) after in vitro digestion in the simulated gastric (**a**) and intestinal juices (**b**).

No flavonoids were detected in variant 1 in SIS, suggesting a possible maximum absorption capacity. In variant 2, the results indicated that the concentration of the total flavonoids was released during the 120 min period under the simulated intestinal conditions, with a maximum value recorded at 11%.

3.4. In Vitro Cytotoxicity of the Microencapsulated Powders

The possible presence and release of any toxic compounds from the powders and the cell viability were analyzed in the L929 fibroblast cell culture by NR assay. The results showed that the powders did not present a cytotoxic effect in the range of the tested concentrations after 24 h and 48 h of cultivation (>80% cell viability) (Figure 3).

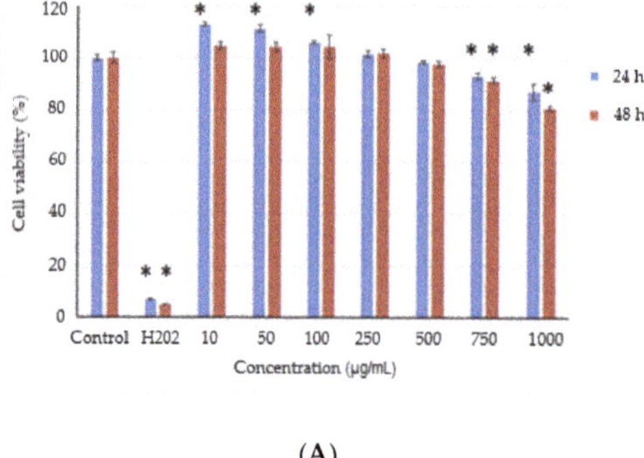

(**A**)

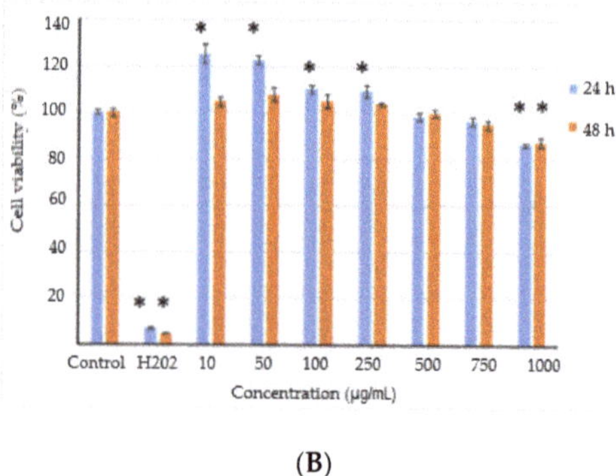

(**B**)

Figure 3. Cell viability of L929 fibroblasts cultivated in the presence of the powders variant 1 (**A**) and variant 2 (**B**) for 24 h and 48 h, respectively, determined by NR assay. The results were expressed as a percentage relative to the control culture (untreated), considered 100% viable. The values represent mean ± SD ($n = 3$). * $p < 0.05$, compared to the control.

The values of the cell viability were ranked between 81 and 113% for variant 1 and between 87 and 125% for variant 2, and then decreased by increasing the concentration. These data indicate that the cytocompatibility of the variants are up to 1 mg/mL. Moreover, concentrations between 10 and 100 µg/mL for variant 1 and between 10 and 250 µg/mL for variant 2 stimulated cell proliferation after 24 h of cultivation, in contrast to the control culture, with the highest values for variant 2. A slight decrease in the cell viability was recorded for higher concentrations of 750–1000 µg/mL for variant 1 and 1000 µg/mL for variant 2, until 80% and 87%, respectively.

Cell morphology examinations were correlated with the NR quantitative data (Figure 4). The images showed that the cells treated with the variants maintained their normal fusiform phenotype, a characteristic of fibroblast cells, similar to the untreated culture.

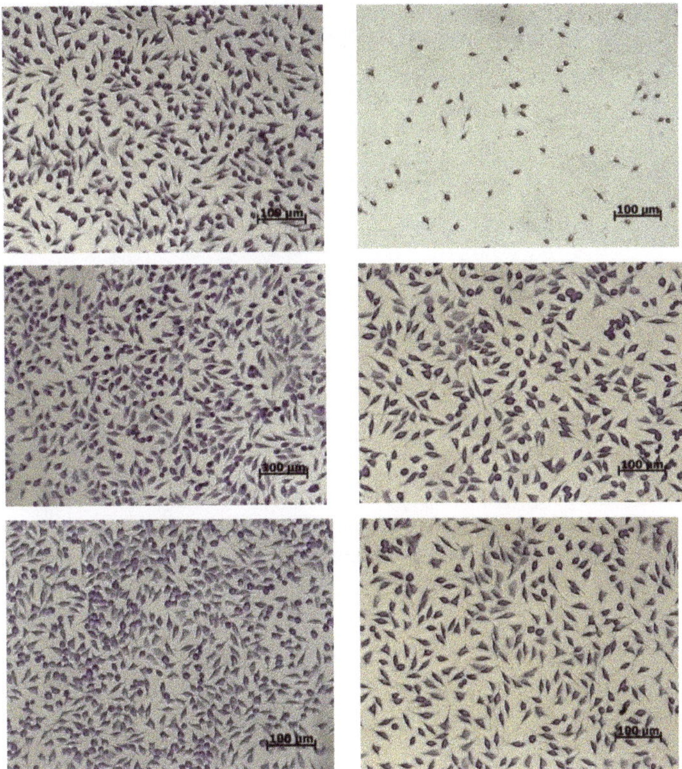

Figure 4. Light micrographs of L929 cells treated with microencapsulated variants of yellow onion skin extracts for 48 h.

The cell distribution on the culture plate was uniform, whereas the density was comparable to that of the untreated culture. At concentrations between 10 and 100 µg/mL, a cell density higher than that of the control was observed. When higher concentrations of powders were added, the cell density decreased in the culture plates.

Similar studies have shown that flavonoid-rich extracts from onion skins [19] did not exhibit toxicity in a normal cell culture within the range of the tested concentrations of 50–200 µg/mL. Our results indicate that the freeze-drying of the extract into different biopolymeric matrices could lead to an increase of the cytocompatibility domain. Thus, the cell viability was higher than 80% for concentrations of up to 1 mg/mL, especially in variant 2.

3.5. Thermal and pH Stability

In order to use the powders as functional ingredients, it is imperative to investigate the thermal and pH behavior of the bioactive nutrients in a simulated system, due to the fact that many functional foods containing bioactive compounds susceptible to degradation are usually subjected to thermal treatment and pH variation. Therefore, the selected concentration of powders, varying from 0.1 to 0.6%, were dissolved in distilled water and subjected to various temperatures (from 25 °C to 100 °C) for 15 min. The flavonoid content was measured after the thermal treatment (Figure 5). Heating caused an increase in flavonoid content, probably due to the thermal degradation of the matrices and the release of the bioactives.

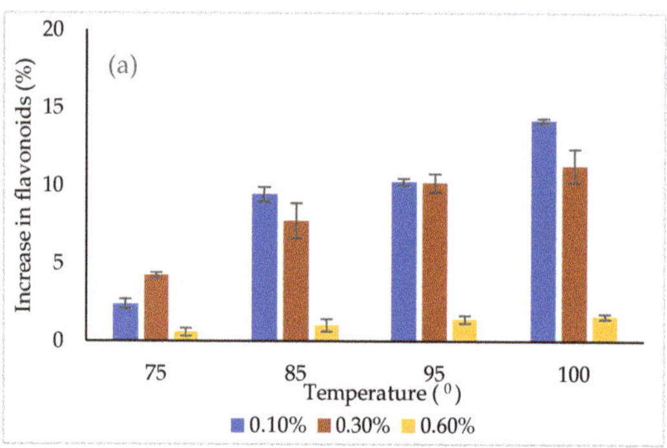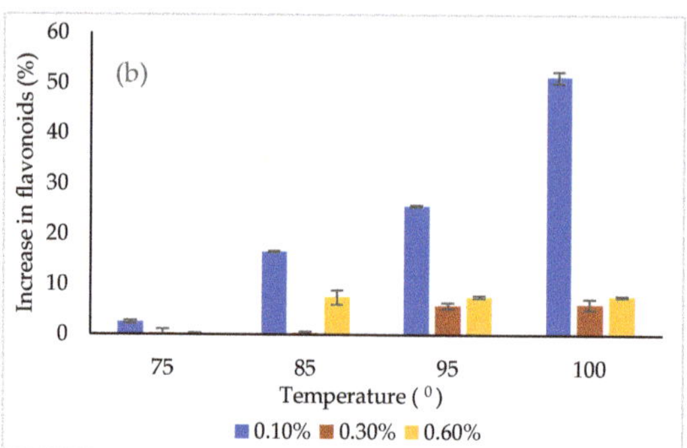

Figure 5. The effect of thermal treatment on the flavonoids released from the powders (**a**) variant 1, (**b**) variant 2.

The flavonoids' release from the powders was found to be temperature and concentration dependent. In variant 1, the highest release of approximately 14% was found at 0.1% and 100 °C, whereas in variant 2, a significantly higher release of 51% was highlighted under similar conditions. Therefore, it could be understood that variant 2 was less protective when a thermal treatment was applied.

Regarding the stability at different pH values, the results are shown in Figure 6.

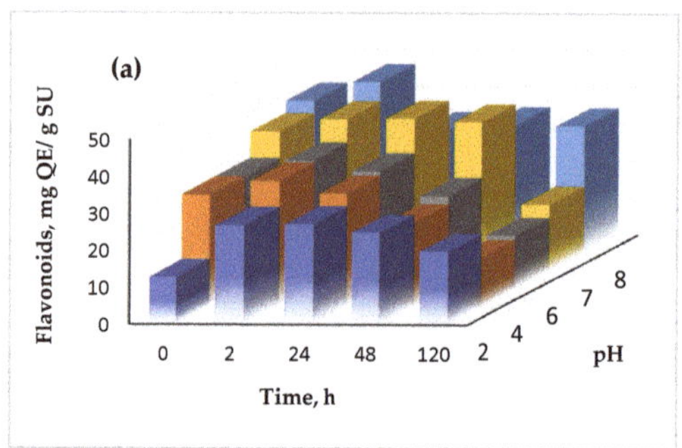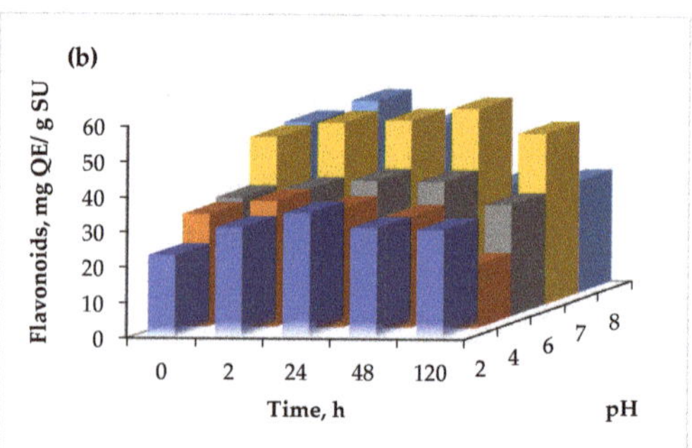

Figure 6. The effect of pH on the flavonoid content of the powders (**a**) variant 1, (**b**) variant (2).

The pH stability was checked at a constant value for variable periods, ranging from 0 to 120 h. From Figure 6, it can be seen that in general, the flavonoid content increased with time, up to 24 h, and started to slowly decrease at a more prolonged time with every test, regardless of the variant. The highest increase of the flavonoid content was found at pH 2.0, after 2 h, when a two-fold increase of the flavonoid content was found in variant 1.

4. Conclusions

The extraction of the flavonoid compounds from yellow onion skins was successfully completed using hot-water extraction, thus allowing for the obtainment of a safe, flavonoid-enriched extract. The achieved results permitted us to conclude that a hot aqueous extraction of the flavonoids from onion skins can be used as an alternative solvent for the selective extraction of flavonoid compounds. The aqueous extract was successfully valorized into biopolymeric matrices based on proteins, peptides, and complex carbohydrates, with an entrapping efficiency higher than 76%. The freeze-dried powders showed a satisfactory phytochemical content and antioxidant activity, whereas the selected matrices provided a protective effect during the simulated digestion. The microstructural analysis showed that the microscopic structure of the powders was evenly distributed and presented a rather uniform structural pattern. A significant cytocompatibility range was found, up to 1 mg/mL, whereas at lower concentrations, the powders stimulated cell proliferation. The powders showed an acceptable thermal and pH stability, which favors their addition to different food matrices. The obtained results demonstrate that simple, green strategies for an efficient valorization of onion skins may be applied in the design of multifunctional ingredients for food, biopharmaceutical, or nutraceutical applications.

Author Contributions: Conceptualization, N.S.; methodology, N.S.; validation, Ș.A.M. and N.S.; formal analysis, Ș.A.M., E.E. and O.C.; investigation, Ș.A.M., E.E. and O.C.; resources, G.R., A.O. and G.E.B.; writing—original draft preparation, Ș.A.M. and N.S.; writing—review and editing, N.S.; visualization, N.S.; supervision, G.R., A.O. and G.E.B.; project administration, N.S.; funding acquisition, G.R., A.O., and G.E.B. All authors have read and agreed to the published version of the manuscript.

Funding: This work was funded by a grant of the Romanian Ministry of Research and Innovation, CCCDI-UEFISCDI, project number PN-III-P1-1.2-PCCDI-2017-0569-PRO-SPER (10PCCI), within the PNCDI program.

Acknowledgments: The Integrated Center for Research, Expertise, and Technological Transfer in Food Industry and The Romanian Center for Modeling of Recirculating Aquaculture Systems (MORAS) are acknowledged for providing technical support.

Conflicts of Interest: The authors declare no conflict of interest.

References

1. Ren, F.; Niana, Y.; Perussello, C.A. Effect of storage, food processing and novel extraction technologies on onions flavonoid content: A review. *Food Res. Int.* **2020**, *132*, 108953. [CrossRef] [PubMed]
2. Choi, I.; Cho, E.; Moon, J.; Bae, H. Onion skin waste as a valorization resource for the by-product's quercetin and biosugar. *Food Chem.* **2015**, *188*, 537–542. [CrossRef] [PubMed]
3. Zeng, Y.; Li, Y.; Yang, Z.; Pu, X.; Du, J.; Yang, X.; Yang, T.; Yang, S. Therapeutic Role of Functional Components in Alliums for Preventive Chronic Disease in Human Being. *Evid. Based Complement. Alternat. Med.* **2017**, *2017*, 9402849. [CrossRef] [PubMed]
4. Tsao, R.; Yang, R. Optimization of a new mobile phase to know the complex and real polyphenolic composition: Towards a total phenolic index using high-performance liquid chromatography. *J. Chromat. A* **2003**, *1018*, 29–40. [CrossRef] [PubMed]
5. Akdeniz, N.; Sumnu, G.; Sahin, S. Microencapsulation of phenolic compounds extracted from onion (*Allium cepa*) skin. *J. Food Process. Preserv.* **2018**, *42*, e13648. [CrossRef]
6. Panja, P. Green extraction methods of food polyphenols from vegetable materials. *Curr. Opin. Food Sci.* **2018**, *23*, 173–182. [CrossRef]
7. Cui, Q.; Liu, J.-Z.; Wang, L.-T.; Kang, Y.-F.; Meng, Y.; Jiao, J.; Fu, Y.-J. Sustainable deep eutectic solvents preparation and their efficiency in extraction and enrichment of main bioactive flavonoids from sea buckthorn leaves. *J. Clean. Prod.* **2018**, *184*, 826–835. [CrossRef]
8. Meng, Z.; Zhao, J.; Duan, H.; Guan, Y.; Zhao, L. Green and efficient extraction of four bioactive flavonoids from Pollen Typhae by ultrasound-assisted deep eutectic solvents extraction. *J. Pharmaceut. Biomed. Anal.* **2018**, *161*, 246–253. [CrossRef] [PubMed]
9. Cheynier, V. Polyphenols in foods are more complex than often thought. *Am. J. Clin. Nutr.* **2005**, *81*, 223S–229S. [CrossRef] [PubMed]
10. Saiki, S.; Kumar, N.; Charu, M.; Mahanta, L. Optimisation of phenolic extraction from Averrhoa carambola pomace by response surface methodology and its microencapsulation by spray and freeze-drying. *Food Chem.* **2015**, *171*, 144–152. [CrossRef] [PubMed]
11. Çam, M.; Cihatİçyer, N.; Erdoğan, F. Pomegranate peel phenolics: Microencapsulation, storage stability and potential ingredient for functional food development. *LWT* **2014**, *55*, 117–123. [CrossRef]
12. Milea, A.S.; Aprodu, I.; Vasile, A.M.; Barbu, V.; Râpeanu, G.; Bahrim, G.E.; Stănciuc, N. Widen the functionality of flavonoids from yellow onion skins through extraction and microencapsulation in whey proteins hydrolysates and different polymers. *J. Food Eng.* **2019**, *251*, 29–35. [CrossRef]
13. Milea, Ș.A.; Vasile, M.A.; Crăciunescu, O.; Prelipcean, A.M.; Bahrim, G.E.; Râpeanu, G.; Oancea, A.; Stănciuc, N. Co-microencapsulation of flavonoids from yellow onion skins and lactic acid bacteria lead to a multifunctional ingredient for foods and pharmaceutics applications. *Pharmaceutics* **2020**, *12*, 1053. [CrossRef] [PubMed]
14. Oancea, A.M.; Hasan, M.; Vasile, A.M.; Barbu, V.; Enachi, E.; Bahrim., G.; Rapeanu, G.; Silvi, S.; Stănciuc, N. Functional evaluation of microencapsulated anthocyanins from sour cherries skins extract in whey proteins isolate. *LWT* **2018**, *95*, 129–134. [CrossRef]
15. Craciunescu, O.; Gaspar, A.; Toma, L.; Utoiu, E.; Moldovan, L. Evaluation of antioxidant and cytoprotective activities of Arnica montana L. and Artemisia absinthium L. ethanolic extracts. *Chem. Central J.* **2012**, *6*, 97–107. [CrossRef] [PubMed]
16. Piechowiak, T.; Grzelak-Błaszczyk, K.; Bonikowski, R.; Balawejder, M. Optimization of extraction process of antioxidant compounds from yellow onionskin and their use in functional bread production. *LWT* **2020**, *117*, 108614. [CrossRef]

17. Thakur, H.N.S.; Thakur, A. Microencapsulation of wild pomegranate flavedo phenolics by lyophilization: Effect of maltodextrin concentration, structural morphology, functional properties, elemental composition, and ingredient for development of functional beverage. *LWT* **2020**, *133*, 110077. [CrossRef]
18. Horincar, G.; Aprodu, I.; Barbu, V.; Râpeanu, G.; Bahrim, G.E.; Stănciuc, N. Interactions of flavonoids from yellow onion skins with whey proteins: Mechanisms of binding and microencapsulation with different combinations of polymers. *Spectrochim. Acta A Mol. Biomol. Spectrosc.* **2019**, *215*, 158–167. [CrossRef] [PubMed]
19. Shi, G.Q.; Yang, J.; Liu, J.; Liu, S.N.; Song, H.X.; Zhao, W.E.; Liu, Y.Q. Isolation of flavonoids from onion skin and their effects on K562 cell viability. *Bangladesh J. Pharmacol.* **2016**, *11*, S18–S25. [CrossRef]

Communication

In Vitro Coliform Resistance to Bioactive Compounds in Urinary Infection, Assessed in a Lab Catheterization Model

Emanuel Vamanu [1,*], Laura Dorina Dinu [1], Cristina Mihaela Luntraru [2] and Alexandru Suciu [2]

[1] Faculty of Biotechnology, University of Agronomic Sciences and Veterinary Medicine of Bucharest, 59 Marasti Street, 011464 Bucharest, Romania; lauradorina@yahoo.com

[2] Hofigal Export Import SA, Intrarea Serelor No. 2, 4 District, 042124 Bucharest, Romania; cristina_m_baicea@yahoo.com (C.M.L.); alexandru.suciu287@gmail.com (A.S.)

* Correspondence: email@emanuelvamanu.ro; Tel.: +40-742218240

Abstract: Bioactive compounds and phenolic compounds are viable alternatives to antibiotics in recurrent urinary tract infections. This study aimed to use a natural functional product, based on the bioactive compounds' composition, to inhibit the uropathogenic strains of *Escherichia coli*. *E. coli* ATCC 25922 was used to characterize the IVCM (new in vitro catheterization model). As support for reducing bacterial proliferation, the cytotoxicity against a strain of *Candida albicans* was also determined (over 75% at 1 mg/mL). The results were correlated with the analysis of the distribution of biologically active compounds (trans-ferulic acid-268.44 ± 0.001 mg/100 g extract and an equal quantity of Trans-p-coumaric acid and rosmarinic acid). A pronounced inhibitory effect against the uropathogenic strain *E. coli* 317 (4 log copy no./mL after 72 h) was determined. The results showed a targeted response to the product for tested bacterial strains. The importance of research resulted from the easy and fast characterization of the functional product with antimicrobial effect against uropathogenic strains of *E. coli*. This study demonstrated that the proposed in vitro model was a valuable tool for assessing urinary tract infections with *E. coli*.

Keywords: urinary infection; in vitro; *Escherichia coli*; antimicrobial

Citation: Vamanu, E.; Dinu, L.D.; Luntraru, C.M.; Suciu, A. In Vitro Coliform Resistance to Bioactive Compounds in Urinary Infection, Assessed in a Lab Catheterization Model. *Appl. Sci.* **2021**, *11*, 4315. https://doi.org/10.3390/app11094315

Academic Editor: Francisco Arrebola

Received: 15 April 2021
Accepted: 6 May 2021
Published: 10 May 2021

Publisher's Note: MDPI stays neutral with regard to jurisdictional claims in published maps and institutional affiliations.

1. Introduction

Catheter-associated urinary tract infections (UTIs) are the most common type of nosocomial infections and are mainly significant because of the severe sequelae that affect the patient's health. *E. coli* is a causative agent in 80% of catheter-associated UTIs, and it has specific factors that attach to the catheter surface and then to the uroepithelium [1]. The bladder is an organ where bacterial multiplication occurs, and most symptoms result from physiological manifestations at this level. Natural control of *E. coli* infection is a current method based on the antimicrobial capacity of some components isolated from raw materials obtained from plants. They act by blocking microbial multiplication, reducing oxidative processes, and lowering the pressure exerted by regulating the physiological processes [2]. Functional extracts reduce biofilm formation in the urinary tract and are also an essential aspect of managing the antibiotic resistance [3].

These problems occur when, in patients with a catheter, a particular strain becomes predominant. In this situation, disturbing the microbial balance makes traditional methods inefficient. The novelties of the study in the in vitro model are the possibility of modulating simulation parameters and its adaptation to the situations in clinical practice [4]. The laboratory study provides an image of the microbiota dynamics in the urinary tract, where recurrent infections occur. New possibilities for reducing antibiotic dependence can be established and tested, which is a current trend. The modular structure allows permanent adaptation to in vivo conditions and the possibility of conducting interdisciplinary in vitro studies [5].

139

In vitro models provide a better representation of the physiological environment in the urinary tract, and they can be a valuable preclinical approach to compare and select new product formulation or to test technological innovations for the prevention of UTI. Although in vitro tests do not truly represent the complex environment in vivo, they offer many advantages, including relatively low cost, high throughput, flexibility and timeframes for testing, and no ethical concerns associated with in vivo tests [6].

Recent data indicate that herbal therapy is not inferior to antibiotic treatment in the acute phase of lower uncomplicated UTI [7,8]. This feature and the increase in interest in the use of products with low toxicity have led to the prophylactic use of extracts from plant substrates. A vital part is a combination of the antimicrobial effect with the antioxidant and anti-inflammatory effects. The correlation of these characteristics is based on the heterogeneous composition of functional extracts, which can exert several properties against *E. coli* infections [8]. The administration of extracts is based on the knowledge of traditional medicine. Using in vitro simulators contributes to the characterization and understanding of the intrinsic mechanisms that reduce *E. coli* at the urogenital level [9]. The study to assess a functional product with antibacterial properties against uropathogenic *E. coli* strains using the in vitro lab catheterization model. The research objectives of the study were the bioactive compounds determination by chromatographic analysis, antimicrobial effect of the product by qPCR sustained by evaluation of antimicrobial resistance and antioxidant activity. This data will provide a validation of a new in vitro catheterization model-IVCM.

2. Materials and Methods

2.1. Biological Materials

Two strains of *E. coli*, one ATCC 10536, from the Faculty of Biology, University of Bucharest, Romania, *E. coli* 317 (uropathogenic clinical isolate) and *C. albicans* from the Faculty of Biotechnology, University of Agronomic Sciences and Veterinary Medicine of Bucharest, Romania collection, kept at $-80\ ^\circ$C in glycerol, were used. The *E. coli* strains were revitalized by using the Lysogeny broth (LB) medium [10], and *C. albicans*, by using the Yeast-extract Peptone Dextrose (YPD) medium [11]. The strains were cultured in the mentioned medium at 30 $^\circ$C, for 24 h, under 120 rpm. After the incubation period, the medium was removed, and the cells were washed once with phosphate-buffered saline and resuspended in the same buffer (sterile) until they were used [12].

2.2. Solvent Extraction

An amount of 20 g of the natural product for the treatment of urinary tract infections cranberry fruit (*Vaccinium vitis idaea* L.), St John's-wort aerial parts (*Hypericum perforatum*), thyme aerial parts (*Thymus vulgaris* L.), propolis tincture with 27% dry mass, thyme essential oil, rosemary essential oil, and inactive ingredients) was mixed with 100 mL of 50% ethanol. It was stored in a Duran bottle for 48 h at 25 $^\circ$C, under 150 rpm. At the end of the period, the mixture was filtered under a vacuum and filled up to the original volume with 50% ethanol to obtain an extract in a simple tincture form. The extract was kept in brown bottles in the refrigerator and used after 48 h [13]. The product formula (Supplementary Figure S1) was realized to be active against urinary bladder infections both as an adjuvant in antibiotic treatment and mild forms of the infection. It can be used to prevent recurrent infections. Different biological materials were chosen because the fruit of mountain cranberry contains polyphenols and flavonoids with anti-inflammatory and diuretic action, stimulating the excretion of urine procyanidins with strong antioxidant activity and bacterial antiadhesion activity [14]. Thyme powder and volatile oil have diuretic, antiseptic, and antispasmodic properties given by polyphenols, tannins, and volatile compounds [15]. Saint John's wort contains polyphenols, flavonoids and is used to alleviate inflammation, infections, and pain in urinary interstitial cystitis [16]. Propolis tincture contains polyphenols and flavonoids in high amounts, active compounds that ensure its high antioxidant activity, and contains antibacterial, anti-inflammatory, anesthetic, analgesic, and antibiotic properties [17]. Rosemary volatile oil has antioxidant, antiseptic, and antispasmodic properties [18].

2.3. Determination of Extract Cytotoxicity

Candida albicans was used as an in vivo instrument. The strain came from the Faculty of Biology, University of Bucharest, and was revitalized using the YPG environment. The protocol used the Bioscreen C MBR computer system. Different concentrations of the extract were used to determine the critical concentration, 0.062–1 mg/mL. As a control, ethanol was used (solvent 50%). The reaction mixture (YPD medium supplemented with the sample) had a maximum volume of 300 µL/well, and the inoculation ratio was 10% culture 10^8 CFU/mL. The plate was introduced at 32 °C for 24 h, with stirring and Optical Density (O.D.) was measured at 600 nm, every 15 min. The YPD environment was used as a positive control. The following formula was used to assess the degree of toxicity:

$$\text{Cytotoxicity (\%)} = (1 - (\frac{O.D._{sample}}{O.D._{control}})) \times 100 \ [19].$$

2.4. In Vitro Catheterization Model

An in vitro model was used to evaluate the efficacy of the tested products. The steps of the method and the structure of the in vitro model were (Figure 1):

1. Duran vessel (minimum volume 1 L) provided with two orifices containing sterilized artificial urine [20];
2. Peristaltic pump for regular administration of urine from the vessel of point 1, flow rate 0.4 mL/min, controlled by a timed outlet;
3. Inoculation point with the tested strain-silicone septum (a stock culture with a minimum viability 1×10^5 CFU/mL, in 0.9% sterile NaCl);
4. Theft made of silicone finished in a Millipore filter, diameter 0.45 µm;
5. In vitro bladder simulation vessel (minimum volume 250 mL);
6. Reaction mixture (simulated urine, the tested product, microbial strain);
7. Temperature maintenance system—37 °C;
8. Urinary catheter;
9. Sterile sample collection vessel—Duran vessel provided with two orifices.

The heating system was based on the recirculation of water through a circuit surrounding the main simulation vessel. The model included a heating plate, a temperature sensor, a peristaltic pump that ensured a constant flow of one mL/min, and a filling vessel with a lid with three holes [21].

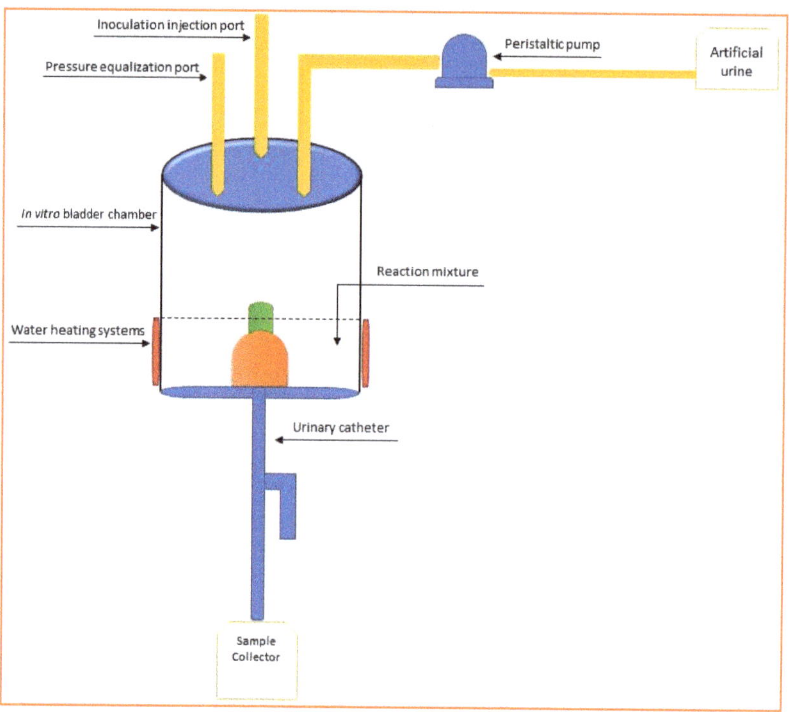

Figure 1. Schematic representation of the in vitro catheterization model.

2.5. DNA Extraction and Quantitative Detection of E. coli Strains by qPCR

According to the manufacturer's instructions for bacteria, genomic DNA from the bacterial cultures and samples (1 mL) was extracted using the Quick-DNA Miniprep Plus kit (Zymo Research, Irvine, USA). All quantitative polymerase chain reactions (qPCR) were performed using the Rotor-Gene Q (Qiagen, West Sussex, UK) instrument and software to generate the standard curve and microbial quantification. The final volume after all reactions was 20 µL, among them 1 µL of template DNA, 4 µL of 5× Hot FirePol EvaGreen qPCR Mix Plus (ROX) (Solis BioDyne, Tartu, Estonia), and 0.5 µL of each universal primer: forward primer 27F (5′-AGAGTTTGATCCTGGCTCAG-3′) and reverse primer 939R (5′-CTTGTGCGGGCCCCCGTCAATTC-3′) were used for the amplification of the 16S rDNA gene fragment. The cycling parameters were: 12 min at 95 °C, followed by 40 cycles for 30 s at 95 °C, 40 s at 60 °C, and 60 s at 72 °C. Reactions were carried out in triplicate. In all cases, negative control of amplification was included using 1 µL of water instead of a DNA template. A fluorescence threshold was chosen to determine the first cycle (CT) where the signals were above the threshold value. Standard curves were generated using 10-fold serial dilutions of DNA extracted from *E. coli* ATCC 10536 and strain 317, covering a range from 10^{-4} to 10^{-10}. The coefficient of correlation (R2) obtained for the standard curve of *E. coli* ATCC 10536 and *E. coli* 317 was 0.9986 and 0.9909, respectively, with the efficiency of the slope E = -3.32 ± 0.11 [22–24].

Evaluation of antibiotic susceptibility using antibiotics prescribed for treatment of recurrent urinary infections was performed by disk diffusion method on Muller-Hinton agar [25].

2.6. Determination of Bioactive Compounds

2.6.1. Total Polyphenolic Content

A colorimetric method was used, which was based on the Folin-Ciocalteu reagent. The calibration curve was obtained with standard solutions of gallic acid concentrations ranging from 1 to 5 µg/mL. The reaction mixture of 1 mL extract, 5 mL Folin-Ciocalteu reagent, and 4 mL sodium carbonate solution was incubated for one hour. The absorbance was measured (UV-VIS spectrophotometer, Jasco V-530, Tokyo, Japan) at 765 nm against water [26].

2.6.2. Procyanidins Content

A colorimetric method was used based on the following reagents: ethanol 70% (*v*/*v*), hydrochloric acid 250 g/L solution, and butanol. The reaction mixture was 50.0 mL of the obtained extract, which was evaporated in a round-bottomed flask to about 3 mL and transferred to a separating funnel. The round-bottomed flask was rinsed sequentially with 10 mL and 5 mL of water and transferred to a separating funnel. The combined solution was shaken with three quantities, each of 15 mL, of butanol. The organic layers were combined and diluted to 100.0 mL with butanol. The solution's absorbance was measured at 555 nm (UV-VIS spectrophotometer, Jasco V-530, Tokyo, Japan) [27].

2.6.3. HPLC Assay

An aliquot of 3 mL of the sample was weighed and diluted with 3 mL of methanol. The solution was filtered through a 0.2 µm polypropylene filter before high pressure liquid chromatography (HPLC) injection. Approximately 0.8 g of extract was ultrasonicated for 30 min in 20 mL methanol, and the solution was filtered through a 0.2 µm polypropylene filter before the HPLC injection. A Hitachi Chromaster HPLC system was used, equipped with a 5160 pump, 5310 column oven, 5260 thermostated autosamplers, and a 5430 DAD detector. The separation was done on a ZORBAX SB-C18 4,6 × 150 mm, 3.5 µm column. An adapted, reversed-phase HPLC (RP-HPLC) method was developed. The mobile phase had a mixture of acetonitrile, methanol, and water, and the elution was gradient. The analytical wavelength was set at 320, 369, 285, and 267 nm. The calibration curves were made by diluting stock solutions of each reference standard to five concentration levels. R2 values were between 0.996 and 0.999 [28].

2.7. In Vitro Antioxidant Activity-CUPRAC (CUPric Reducing Antioxidant Capacity) Assay

The method used a calibration curve with standard solutions of Trolox (6-hydroxy-2,5,7,8-tetramethylchroman-2-carboxylic acid), with concentrations ranging from 10 to 60 µg/mL. A 1 g sample was extracted with 50 mL of ethanol, 50% (*v*/*v*), under a reflux condenser for 30 min. The reagents used were: Copper sulfate pentahydrate solution 10^{-2} M, neocuproine ethanolic solution 7.5×10^{-3} M, ammonium acetate buffer solution 1 M, and pH 7. The reaction mixture contained 1 mL of each reagent (copper sulfate pentahydrate solution, neocuproine ethanolic solution, and ammonium acetate buffer solution), 0.1 mL sample, and 1.0 mL water and was incubated for 30 min. The absorbance was measured (UV-VIS spectrophotometer, Jasco V-530, Tokyo, Japan) under 450 nm, against a blank prepared with 0.1 mL water, instead of 0.1 mL sample [29].

2.8. Statistical Analysis

All the parameters investigated were evaluated in triplicate, with the results expressed as the mean ± standard deviation (SD). The IBM SPSS Statistics software package (IBM Corporation, Armonk, NY, USA) was used to calculate the mean and SD values. The significance level was: significant at $p \leq 0.05$; very significant at $p \leq 0.01$; and highly significant at $p \leq 0.001$, using the normal distribution of the variables. The differences were assessed by analysis of variance (ANOVA) followed by a Tukey post-hoc analysis. The experimental data analysis and correlation were performed using the IBM SPSS Statistics software package (IBM Corporation, Armonk, NY, USA) [30,31].

3. Results and Discussion

3.1. Determination of Bioactive Compounds

The total polyphenol content from the natural product was 146 mg/g gallic acid equivalents. An amount of 14 mg/g pyrogallol, equivalent to tannin content, and 1.5 mg/g procyanidins, expressed as cyanidin chloride were also identified.

Chromatographic analysis of the polyfunctional extract showed a heterogeneous composition, and after the fractionation phase, the polyphenolcarboxylic acids were predominant. Trans-p-coumaric acid (203.86 ± 0.009 mg/100 g extract), rosmarinic acid (197.41 ± 0.006 mg/100 g extract), and trans-ferulic acid (268.44 ± 0.001 mg/100 g extract) showed high values compared to caffeic acid (30.93 ± 0.004 mg/100 g extract).

The flavonoid composition showed the presence of rutin (30.93 ± 0.005 mg/100 g extract) and quercetin (26.03 ± 0.001 mg/100 g extract) as the major components. In addition, a much smaller amount of pinocembrin (14.67 ± 0.008 mg/100 g extract) and chrysin (16.74 ± 0.004 mg/100 g extract) were calculated.

3.2. The Antibacterial Effect of the Extracts in the In Vitro System

Our study tested the synergistic activity of thyme and rosemary essential oils and other ingredients used to treat urinary infections, contained in a natural product, using the in vitro catheterization model. It allowed a real-time evaluation of the microbial profile, with no infections from other strains. The product's effect on the in vitro model resulted in the inhibition of both strains, when tested after 72 h of simulation. The tested product (Figure 2) was shown to have a more pronounced inhibitory effect against the clinical/uropathogenic strain isolated from *E. coli* 317, with viability of approximately 4 log CFU/mL ($p \leq 0.001$) at 72 h. The control strain, *E. coli* ATCC 10536 had superior resistance to the presence of inhibitory compounds present in the samples, 6 log CFU/mL ($p \leq 0.05$) at 72 h. These results provided an image of the role that the administration of a functional compound has on some target strains. The decrease at 72 h against the uropathogenic strain was at least 50% compared to the control sample.

These results provided an image of the role that the administration of a functional compound has on some target strains. The decrease at 72 h against the uropathogenic strain was at least 50% compared to the control sample.

Antimicrobial resistance has also been shown to be widely variable between uropathogenic E. coli clinical isolates and the standard laboratory strains [32]. Therefore, in this study we used a clinical isolate with resistance to the third-generation cephalosporin (ceftriaxone-30 µg/mL and ciprofloxacin-5 µg/mL). Compared to the sensitive control, an additional explanation could be the result of the specific motility of each strain that could influence the exposure time of the strain to bioactive compounds [33].

Essential oils are complex mixtures containing up to 60 components, with 1–3 major ingredients at high concentrations. These make up a significant part of the antimicrobial activity of the essential oils. The antibacterial activity of *Rosmarinus officinalis* essential oil on a standard strain of *E. coli* ATCC 25922 and other 60 clinical strains of *E coli*, including the urinary tract and hospital equipment, was proved by Sienkiewicz et al. [34]. Other works compared the antibacterial activity of rosemary essential oil against different uropathogenic isolates. They noticed that the highest effect was obtained using a concentration of 10,000 ppm on *E. coli*, *Klebsiella* sp., *Staphylococcus aureus*, and *Candida* isolates [35]. Moreover, the combination of thyme and tea tree essential oils increased the guideline-recommended antibiotic activity to treat uncomplicated lower UTIs—fosfomycin and pivmecillinam, but not nitrofurantoin—against uropathogenic *E. coli* isolates [36]. Linalool, the major component of thyme essential oil, was shown to act as a bactericidal, fungicidal and had anti-inflammatory effects [37].

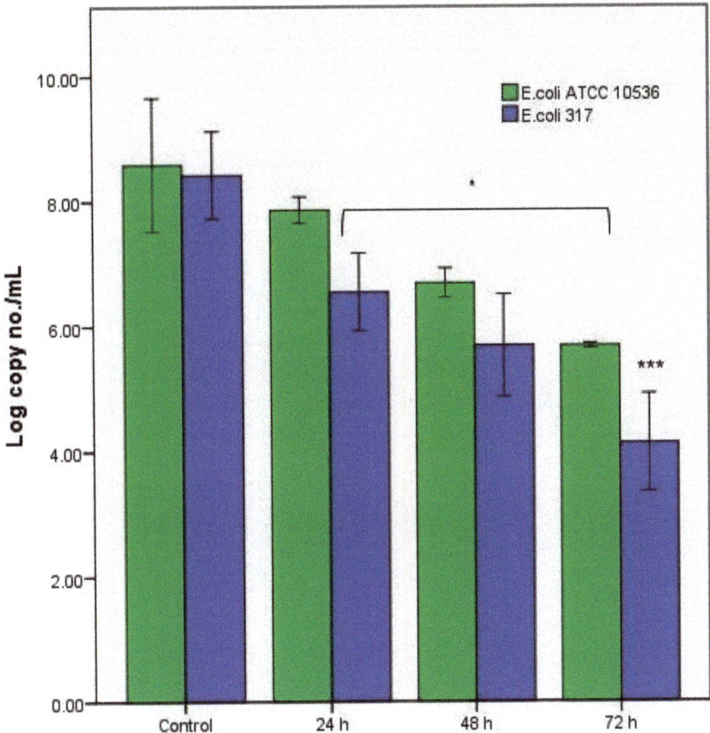

Figure 2. The antibacterial effect of the product in the in vitro catheterization system. Different asterisks represent significant statistical differences (control vs. samples; $p \leq 0.05$), n = 3.

It has been proved that the presence of rosmarinic acid, corresponded to the antimicrobial effect's correlation with the antioxidant one [38]. The tested product showed the antioxidant activity of 326.5 mg/g Trolox equivalent, which decreased to a quarter of its initial value at the end of the in vitro process. This result demonstrated that the bioactive load was released gradually, which gradual protection against endogenous factors. It is possible that other molecules were present în the extract (polysaccharides), influenced the bioactivity in vitro, and released the total bioactive compounds quantity [39].

3.3. The Cytotoxicity Tests against C. Albicans

The cytotoxicity results against *C. albicans* (Figure 3) demonstrated a gradual increase in the extract concentration. An average value of $47 \pm 1.63\%$ ($p \leq 0.01$) was calculated for an extract of 0.062 mg/mL. These results were interpreted as a functional property of the extract to modulate this strain's proliferation resistance, a relevant fact in the formation of the biofilm and the preservation of urinary tract homeostasis. The antifungal response differences were due to the variable presence of functional compounds that could be correlated with the type of bioactive molecules and biotransformations due to the environment, in vitro [40]. This feature could be considered as a limiting factor of in vitro simulators. This study demonstrated that the proposed in vitro model was a useful tool for managing UTI with *E. coli*. Such a laboratory model can bring the results as close as possible to those obtained in in vivo clinical trials [41].

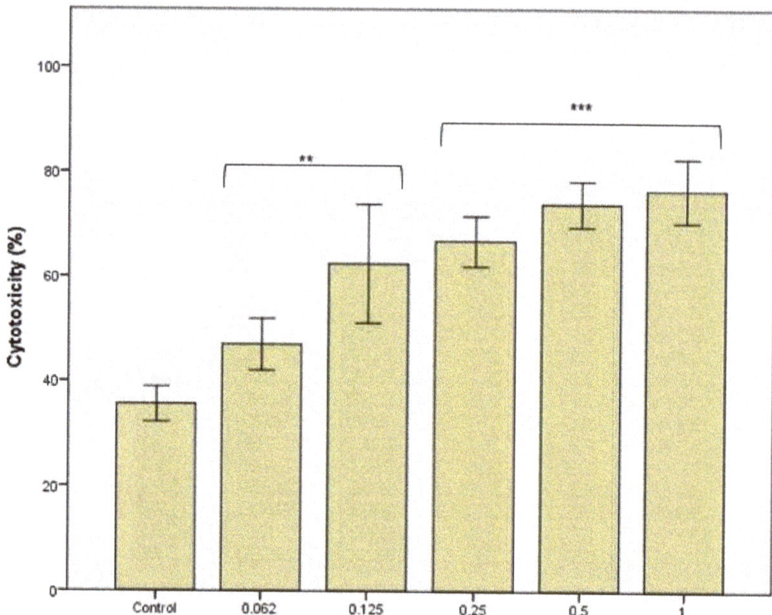

Figure 3. The cytotoxicity tests against *Candida albicans*. Different asterisks represent significant statistical differences (control vs. samples; $p \leq 0.05$), n = 3.

These results were supported by the presence of quercetin, which was reported to have urinary tract protection effects. Quercetin is used as an adjuvant to conventional treatments [42]. These data support previous studies that have shown a reduction in biofilm formation to about 65% and an inhibition in the proliferation of the *Candida* genus strains [43]. In addition, these compounds have prevented the formation (synthesis) of the self-produced extracellular matrix that helps build the biofilm and the recurrence of inflammatory processes. These data may have been associated with the action of other compounds, such as myricetin or catechin [44].

Another compound that expressed an antibacterial effect against *E. coli* was rosmarinic acid. Its action was synergistic in UTIs because an antioxidant effect was also determined. The expression of these effects is a specific one, with the antimicrobial effect being due to an action on the cell membrane [38,45]. The same mechanism has been demonstrated for ferulic acid [46] and in reducing the microbial biofilm [47] by decreasing the adhesion capacity of bacterial cells. The product's effectiveness was also due to the presence of coumaric acid, which is known to have similar effects in reducing bacterial cells as ferulic acid, the majority component [48]. The presence of these compounds also supported the cytotoxic effect, being correlated with the gradual decrease of the viability of the *Candida* strain [49].

Thus, the purpose of the study was achieved, given the inhibitory capacity of the tested extract. Static in vitro tests targeting the current model did not consider urinary epithelial interactions that may influence the bioavailability of specific compounds. This aspect can be viewed as a negative point of this type of study.

As a practical application, the realization of this model was useful for in vitro research studies that will be able to simulate certain recurrent infections. Testing the formulation of functional products and drugs with antimicrobial effects associated with inhibiting biofilm formation is another research perspective with this model in vitro. Other uropathogenic strains that can be identified with the biofilm associated with the urinary catheter can also be tested. New quantification and microbial analysis methods will meet practical needs regarding the presence of bacteria in the urinary tract [50].

4. Conclusions

This study addressed a target strain and was integrated research on the assessment of multifunctional product used in recurrent microbial infections. The study of the causes of microbial biofilm development at the urinary level will determine a clear understanding of the appearance of antibiotic resistance [51]. The in vitro model can also be adapted to other pathogenic bacterial strains and allows a complex evaluation of other functional products. Personalized analysis of the urinary microbiota response also gives the possibility of studying new therapeutic targets based on natural compounds [52].

Supplementary Materials: The following are available online at https://www.mdpi.com/article/10.3390/app11094315/s1, Figure S1: The aspect of functional product for UTIs tested in IVCM.

Author Contributions: E.V. designed the experiments, analyzed the data, and wrote the paper; A.S. and C.M.L. performed the chromatographic analysis; L.D.D. contributed with biological analyses of the samples. The authors discussed and made comments on the results. All authors have read and agreed to the published version of the manuscript.

Funding: This research received no external funding.

Institutional Review Board Statement: Not applicable.

Informed Consent Statement: Not applicable.

Data Availability Statement: Not applicable.

Conflicts of Interest: The authors declare no conflict of interest.

References

1. Flores-Mireles, A.L.; Walker, J.N.; Caparon, M.; Hultgren, S.J. Urinary tract infections: Epidemiology, mechanisms of infection and treatment options. *Nat. Rev. Microbiol.* **2015**, *13*, 269–284. [CrossRef]
2. van Seventer, J.M.; Hochberg, N.S. Principles of Infectious Diseases: Transmission, Diagnosis, Prevention, and Control. *Int. Encyclop. Public Health* **2017**, 22–39. [CrossRef]
3. Kwiecińska-Piróg, J.; Skowron, K.; Śniegowska, A.; Przekwas, J.; Balcerek, M.; Załuski, D.; Gospodarek-Komkowska, E. The impact of ethanol extract of propolis on biofilm forming by Proteus mirabilis strains isolated from chronic wounds infections. *Nat. Prod. Res.* **2019**, *33*, 3293–3297. [CrossRef]
4. Azevedo, A.S.; Almeida, C.; Gomes, L.C.; Ferreira, C.; Mergulhão, F.J.; Melo, L.F.; Azevedo, N.F. An in vitro model of catheter-associated urinary tract infections to investigate the role of uncommon bacteria on the Escherichia coli microbial consortium. *Biochem. Eng. J.* **2017**, *118*, 64–69. [CrossRef]
5. Ceprnja, M.; Oros, D.; Melvan, E.; Svetlicic, E.; Skrlin, J.; Barisic, K.; Starcevic, L.; Zucko, J.; Starcevic, A. Modeling of Urinary Microbiota Associated With Cystitis. *Front. Cell. Infect. Microbiol.* **2021**, *11*, 643638. [CrossRef] [PubMed]
6. Cortese, Y.J.; Wagner, V.E.; Tierney, M.; Devine, D.; Fogarty, A. Review of Catheter-Associated Urinary Tract Infections and In Vitro Urinary Tract Models. *J. Health. Eng.* **2018**, *2018*, 2986742. [CrossRef]
7. Wagenlehner, F.M.; Abramov-Sommariva, D.; Höller, M.; Steindl, H.; Naber, K.G. Non-Antibiotic Herbal Therapy (BNO 1045) versus Antibiotic Therapy (Fosfomycin Trometamol) for the Treatment of Acute Lower Uncomplicated Urinary Tract Infections in Women: A Double-Blind, Parallel-Group, Randomized, Multicentre, Non-Inferiority Phase III Trial. *Urol. Int.* **2018**, *101*, 327–336. [PubMed]
8. Vamanu, E.; Pelinescu, D.; Sarbu, I. Comparative Fingerprinting of the Human Microbiota in Diabetes and Cardiovascular Disease. *J. Med. Food* **2016**, *19*, 1188–1195. [CrossRef] [PubMed]
9. Leme, D.E.M.; Rodrigues, A.B.; de Almeida-Apolonio, A.A.; Gomes da Silva Dantas, F.; Norman Negri, M.F.; Svidzinski, T.I.E.; da Silva Mota, J.; Lima Cardoso, C.A.; Pires de Oliveira, K.M. In Vitro Control of Uropathogenic Microorganisms with the Ethanolic Extract from the Leaves of Cochlospermum regium (Schrank) Pilger. *Evid. Based Complement. Alt. Med.* **2017**, *2017*, 4687154. [CrossRef] [PubMed]
10. Vamanu, E.; Pelinescu, D.; Avram, I.; Nita, S. An in vitro evaluation of antioxidant and colonic microbial profile levels following mushroom consumption. *Biomed Res. Int.* **2013**, *2013*, 289821. [CrossRef] [PubMed]
11. Forsyth, V.S.; Armbruster, C.E.; Smith, S.N.; Pirani, A.; Springman, A.C.; Walters, M.S.; Nielubowicz, G.R.; Himpsl, S.D.; Snitkin, E.S.; Mobley, H.L.T. Rapid Growth of Uropathogenic Escherichia coli during Human Urinary Tract Infection. *mBIO* **2018**, *9*, e00186-18. [CrossRef]

12. Chauhan, N.; Kruppa, M.D. Standard Growth Media and Common Techniques for Use with *Candida albicans*. In *Candida albicans. Methods in Molecular Biology*; Cihlar, R.L., Calderone, R.A., Eds.; Humana Press: Totowa, NJ, USA, 2009; Volume 499. [CrossRef]

13. Dabulici, C.M.; Sârbu, I.; Vamanu, E. The Bioactive Potential of Functional Products and Bioavailability of Phenolic Compounds. *Foods* **2020**, *9*, 953. [CrossRef] [PubMed]

14. Jurikova, T.; Skrovankova, S.; Mlcek, J.; Balla, S.; Snopek, L. Bioactive Compounds, Antioxidant Activity, and Biological Effects of European Cranberry (*Vaccinium oxycoccos*). *Molecules* **2018**, *24*, 24. [CrossRef]

15. Uritu, C.M.; Mihai, C.T.; Stanciu, G.D.; Dodi, G.; Alexa-Stratulat, T.; Luca, A.; Leon-Constantin, M.M.; Stefanescu, R.; Bild, V.; Melnic, S.; et al. Medicinal Plants of the Family Lamiaceae in Pain Therapy: A Review. *Pain Res. Manag.* **2018**, *2018*, 7801543. [CrossRef] [PubMed]

16. Olajide, O.A. Inhibitory effects of St. John's Wort on inflammation: Ignored potential of a popular herb. *J. Diet. Suppl.* **2009**, *6*, 28–32. [CrossRef] [PubMed]

17. Ahangari, Z.; Naseri, M.; Vatandoost, F. Propolis: Chemical Composition and Its Applications in Endodontics. *Iran. Endod. J.* **2018**, *13*, 285–292. [CrossRef] [PubMed]

18. Parham, S.; Kharazi, A.Z.; Bakhsheshi-Rad, H.R.; Nur, H.; Ismail, A.F.; Sharif, S.; RamaKrishna, S.; Berto, F. Antioxidant, Antimicrobial and Antiviral Properties of Herbal Materials. *Antioxidants* **2020**, *9*, 1309. [CrossRef] [PubMed]

19. Hudz, N.; Makowicz, E.; Shanaida, M.; Białoń, M.; Jasicka-Misiak, I.; Yezerska, O.; Svydenko, L.; Wieczorek, P.P. Phytochemical Evaluation of Tinctures and Essential Oil Obtained from *Satureja montana* Herb. *Molecules* **2020**, *25*, 4763. [CrossRef]

20. Kabir, M.A.; Hussain, M.A.; Ahmad, Z. Candida albicans: A Model Organism for Studying Fungal Pathogens. *ISRN Microbiol.* **2012**, *2012*, 538694. [CrossRef]

21. Wang, R.; Neoh, K.G.; Kang, E.T.; Tambyah, P.A.; Chiong, E. Antifouling coating with controllable and sustained silver release for long-term inhibition of infection and encrustation in urinary catheters. *J. Biomed. Mater. Res. B* **2015**, *103*, 519–528. [CrossRef]

22. Vamanu, E. In Vitro System for Testing for Urinary Tract Infection with Escherichia coli and Test Method. Patent Application OSIM 00387/2020, 8 July 2020.

23. Dinu, L.D.; Bach, S. Detection of viable but non-culturable *Escherichia coli* O157: H7 from vegetable samples using quantitative PCR with propidium monoazide and immunological assays. *Food Control* **2013**, *31*, 268–273. [CrossRef]

24. Rusthen, S.; Kristoffersen, A.K.; Young, A.; Galtung, H.K.; Petrovski, B.É.; Palm, Ø.; Enersen, M.; Jensen, J.L. Dysbiotic salivary microbiota in dry mouth and primary Sjögren's syndrome patients. *PLoS ONE* **2019**, *14*, e0218319. [CrossRef]

25. Meroni, G.; Soares Filipe, J.F.; Martino, P.A. In Vitro Antibacterial Activity of Biological-Derived Silver Nanoparticles: Preliminary Data. *Vet. Sci.* **2020**, *7*, 12. [CrossRef] [PubMed]

26. Xi, Z.Q.; Xiao, F.; Yuan, J.; Wang, X.F.; Wang, L.; Quan, F.Y.; Liu, G.W. Gene expression analysis on anterior temporal neocortex of patients with intractable epilepsy. *Synapse* **2009**, *63*, 1017–1028. [CrossRef] [PubMed]

27. Hatami, T.; Emami, S.A.; Miraghaee, S.S.; Mojarrab, M. Total Phenolic Contents and Antioxidant Activities of Different Extracts and Fractions from the Aerial Parts of Artemisia biennis Willd. *Iran. J. Pharm. Res.* **2014**, *13*, 551–559. Available online: https://www.ncbi.nlm.nih.gov/pmc/articles/PMC4157030/ (accessed on 23 March 2021).

28. Cheng, A.; Chen, X.; Wang, W.; Gong, Z.; Liu, L. Contents of Extractable and Non-extractable Polyphenols in the Leaves of Blueberry. *Czech J. Food Sci.* **2013**, *31*, 275–282. Available online: https://www.agriculturejournals.cz/publicFiles/92405.pdf (accessed on 23 March 2021). [CrossRef]

29. Roman, M.C. Determination of catechins and caffeine in camillia sinensis raw materials, extracts, and dietary supplements by HPLC-uv: Single-laboratory validation. *J. AOAC Int.* **2013**, *96*, 933–941. [CrossRef]

30. Vamanu, E.; Gatea, F.; Pelinescu, D.R. Bioavailability and Bioactivities of Polyphenols Eco Extracts from Coffee Grounds after In Vitro Digestion. *Foods* **2020**, *9*, 1281. [CrossRef] [PubMed]

31. Al-Rimawi, F.; Rishmawi, S.; Ariqat, S.H.; Khalid, M.F.; Warad, I.; Salah, Z. Anticancer Activity, Antioxidant Activity, and Phenolic and Flavonoids Content of Wild *Tragopogon porrifolius* Plant Extracts. *Evid. Based Complement. Alt. Med.* **2016**, *2016*, 9612490. [CrossRef]

32. Ramírez-Castillo, F.Y.; Moreno-Flores, A.C.; Avelar-González, F.J.; Márquez-Díaz, F.; Harel, J.; Guerrero-Barrera, A.L. An evaluation of multidrug-resistant Escherichia coli isolates in urinary tract infections from Aguascalientes, Mexico: Cross-sectional study. *Ann. Clin. Microbiol. Antimicrob.* **2018**, *17*, 34. [CrossRef]

33. Dusane, D.H.; Hosseinidoust, Z.; Asadishad, B.; Tufenkji, N. Alkaloids Modulate Motility, Biofilm Formation and Antibiotic Susceptibility of Uropathogenic *Escherichia coli*. *PLoS ONE* **2014**, *9*, e112093. [CrossRef]

34. Sienkiewicz, M.; Łysakowska, M.; Pastuszka, M.; Bienias, W.; Kowalczyk, E. The potential of use basil and rosemary essential oils as effective antibacterial agents. *Molecules* **2013**, *18*, 9334–9351. [CrossRef] [PubMed]

35. Khamael, L.S.; Hassan, M.R.; Entissar, F.A. Study of rosemary essential oil antibacterial effect on bacteria isolated from urinary tract infection in some hospital of Baghdad. *Curr. Res. Microb. Biotech.* **2018**, *6*, 1490–1495. Available online: http://crmb.aizeonpublishers.net/content/2018/1/crmb1490-1495.pdf (accessed on 23 March 2021).

36. Loose, M.; Pilger, E.; Wagenlehner, F. Anti-Bacterial Effects of Essential Oils against Uropathogenic Bacteria. *Antibiotics* **2020**, *9*, 358. [CrossRef]

37. Huo, M.; Cui, X.; Xue, J.; Chi, G.; Gao, R.; Deng, X.; Guan, S.; Wei, J.; Soromou, L.W.; Feng, H. Anti-inflammatory effects of linalool in RAW 264.7 macrophages and lipopolysaccharide-induced lung injury model. *J. Surg. Res.* **2013**, *180*, e47–e54. [CrossRef]

38. Nieto, G.; Ros, G.; Castillo, J. Antioxidant and Antimicrobial Properties of Rosemary (*Rosmarinus officinalis*, L.): A Review. *Medicines* **2018**, *5*, 98. [CrossRef] [PubMed]
39. Lyu, F.; Xu, X.; Zhang, L. Natural polysaccharides with different conformations: Extraction, structure and anti-tumor activity. *J. Mater. Chem. B* **2020**, *8*, 9652–9667. [CrossRef] [PubMed]
40. Lone, S.A.; Wani, M.Y.; Fru, P.; Ahmad, A. Cellular apoptosis and necrosis as therapeutic targets for novel Eugenol Tosylate Congeners against *Candida albicans*. *Sci. Rep.* **2020**, *10*, 1191. [CrossRef]
41. Chua, R.Y.R.; Lim, K.; Leong, S.S.J.; Tambyah, P.A.; Ho, B. An in-vitro urinary catheterization model that approximates clinical conditions for evaluation of innovations to prevent catheter-associated urinary tract infections. *J. Hosp. Inf.* **2017**, *97*, 66e73. [CrossRef]
42. Terlizzi, M.E.; Gribaudo, G.; Maffei, M.E. UroPathogenic *Escherichia coli* (UPEC) Infections: Virulence Factors, Bladder Responses, Antibiotic, and Non-antibiotic Antimicrobial Strategies. *Front. Microbiol.* **2017**, *8*, 1566. [CrossRef]
43. Roy, R.; Tiwari, M.; Donelli, G.; Tiwari, V. Strategies for combating bacterial biofilms: A focus on anti-biofilm agents and their mechanisms of action. *Virulence* **2018**, *9*, 522–554. [CrossRef]
44. Arita-Morioka, K.; Yamanaka, K.; Mizunoe, Y.; Tanaka, Y.; Ogura, T.; Sugimoto, S. Inhibitory effects of Myricetin derivatives on curli-dependent biofilm formation in *Escherichia coli*. *Sci. Rep.* **2018**, *8*, 8452. [CrossRef]
45. Nicu, I.; Pîrvu, L.; Vamanu, A. Antibacterial activity of ethanolic extracts from Agrimonia eupatoria L. and Epilobium hirsutum L. HERBA. *Sci. Bull. Ser. F. Biotechnol.* **2017**, *XXI*, 127–132. Available online: http://biotechnologyjournal.usamv.ro/pdf/2017/Art22.pdf (accessed on 6 May 2021).
46. Borges, A.; Ferreira, C.; Saavedra, M.J.; Simões, M. Antibacterial activity and mode of action of ferulic and gallic acids against pathogenic bacteria. *Microb. Drug Resist.* **2013**, *19*, 256–265. [CrossRef] [PubMed]
47. Borges, A.; Saavedra, M.J.; Simões, M. The activity of ferulic and gallic acids in biofilm prevention and control of pathogenic bacteria. *Biofouling* **2012**, *28*, 755–767. [CrossRef] [PubMed]
48. Liu, J.; Du, C.; Beaman, H.T.; Monroe, M.B.B. Characterization of Phenolic Acid Antimicrobial and Antioxidant Structure–Property Relationships. *Pharmaceutics* **2020**, *12*, 419. [CrossRef]
49. Choi, J.H.; Park, J.K.; Kim, K.M.; Lee, H.J.; Kim, S. In vitro and in vivo antithrombotic and cytotoxicity effects of ferulic acid. *J. Biochem. Mol. Toxicol.* **2018**, *32*, 1. [CrossRef] [PubMed]
50. Nzakizwanayo, J.; Pelling, H.; Milo, S.; Jones, B.V. An In Vitro Bladder Model for Studying Catheter-Associated Urinary Tract Infection and Associated Analysis of Biofilms. *Methods Mol. Biol.* **2019**, *2021*, 139–158. [CrossRef]
51. Aslam, B.; Wang, W.; Arshad, M.I.; Khurshid, M.; Muzammil, S.; Rasool, M.H.; Nisar, M.A.; Alvi, R.F.; Aslam, M.A.; Qamar, M.U.; et al. Antibiotic resistance: A rundown of a global crisis. *Infect. Drug Resist.* **2018**, *11*, 1645–1658. [CrossRef] [PubMed]
52. Davenport, M.; Mach, K.E.; Shortliffe, L.M.D.; Banaei, N.; Wang, T.H.; Liao, J.C. New and developing diagnostic technologies for urinary tract infections. *Nat. Rev. Urol.* **2017**, *14*, 296–310. [CrossRef]

Article

Low Doses of Gamma Irradiation Stimulate Synthesis of Bioactive Compounds with Antioxidant Activity in *Fomes fomentarius* Living Mycelium

Cristina Florentina Pelcaru [1,*], Mihaela Ene [1,*], Alina-Maria Petrache [2] and Daniel Constantin Neguț [1]

[1] "Horia Hulubei" National Institute for Physics and Nuclear Engineering (IFIN-HH), Reactorului Street 30, 077125 Măgurele, Romania; dnegut@nipne.ro

[2] Department of Proteomics and Metabolomics, Research Center for Advanced Medicine—MedFUTURE, "Iuliu Hațieganu" University of Medicine and Pharmacy Cluj-Napoca, Louis Pasteur Street 6, 400349 Cluj-Napoca, Romania; alina.mornea@gmail.com

* Correspondence: cristina.pelcaru@drd.unibuc.ro (C.F.P.); mene@nipne.ro (M.E.)

Abstract: Environmental changes generate free radicals and reactive oxygen species (ROS), resulting in abiotic stress in plants and fungi. Gamma ionizing radiation generates a significant amount of free radicals and ROS, thereby simulating natural environmental stressors. We used a ^{60}Co source of radiation to experimentally induce oxidative stress in living mycelium mass of the medicinal fungus *Fomes fomentarius*, in order to obtain a late response of stress tolerance by means of bioactive compounds synthesis. We measured the response at 24, 48, and 72 h after the irradiation. The highest improvement was found 24 h after exposure for antioxidant activity and for total phenolic compounds of methanolic extract, with a 1.89- and 1.64-fold increase, respectively. The total flavonoids in methanolic extract increased 1.68 times after 48 h from treatment and presented a more stable raising in the assessed time-lapse. For the three analyzed parameters, 300 Gy was the optimum absorbed dose to trigger a beneficial response, with potentially applications in pharmaceutics and nutraceutics. Gamma irradiation can be used as a biotechnological tool to direct the secondary metabolites synthesis upregulation in medicinal mushroom living mycelium.

Keywords: mycelium; antioxidant activity; gamma irradiation; *Fomes fomentarius*; mushroom; DPPH; flavonoids; polyphenols

Citation: Pelcaru, C.F.; Ene, M.; Petrache, A.-M.; Neguț, D.C. Low Doses of Gamma Irradiation Stimulate Synthesis of Bioactive Compounds with Antioxidant Activity in *Fomes fomentarius* Living Mycelium. *Appl. Sci.* **2021**, *11*, 4236. https://doi.org/10.3390/app11094236

Academic Editor: Claudio Medana

Received: 23 April 2021
Accepted: 1 May 2021
Published: 7 May 2021

Publisher's Note: MDPI stays neutral with regard to jurisdictional claims in published maps and institutional affiliations.

1. Introduction

Environmental changes generate free radicals and reactive oxygen species (ROS), resulting in abiotic stress in plants and fungi, also acting as natural regulators. This causes alterations in many metabolic pathways, especially those related to oxidative stress management [1].

Both polyphenols and flavonoids are chemical compounds, synthesized by the body, involved in antioxidant defense. They can block reactive oxygen species that lead to the onset of oxidative stress, which disorganizes the cell membrane and cell organs. Increased oxidative stress can lead to degradation of DNA molecules and oxidation of histones [2,3].

Gamma irradiation is itself considered to be a physically induced stress on living organisms or cells. Radiation treatment can be a much faster way to quantitatively improve the chemical synthesis of antioxidant compounds that may play a role in defending irradiated tissue [4,5].

Exposure to solar ultraviolet radiation and to atmospheric ozone activates both enzymatic and non-enzymatic antioxidant defense systems in response. Some flavonoids and phenolic compounds help in protecting the plants and their photosynthetic tissues from UV-B induced damages [1].

Gamma ionizing radiation generates significant amount of free radicals and ROS [6], thereby simulating natural environmental stressors.

Fomes fomentarius (tinder fungus) is a medicinal lignivorous tough polypore, producing a white-mottled rot mainly on beeches and birches. This wood-decaying macrofungus is well-known for its potential use in a wide range of biotechnological applications [7]. Its perennial woody fruiting body is shown to have cytotoxic activity against murine sarcoma S180 in vitro and to inhibit in vivo tumor growth activity of the same line cell [8]. Its ethanol extract exerts inhibition of cell growth and motility induction of apoptosis via targeting AKT in human breast cancer MDA-MB-231 cells and also decreases cell viability in six cancer cell lines; the fungus is also known to have anticancer, anti-inflammatory, and anti-diabetes effects [9,10].

According to Zhang et al., (2021) [11], the main chemical constituents of this fungus were suggested to be triterpenoids and ergosterols. The authors isolated new structures: two pentacyclic lupane-type triterpenes, 3-formyloxybetulin and 3-formyloxybetulinic acid; two rare carbon-degraded ergosterol derivatives, pyropolincisterols (A and B); along with ten known triterpenids and four erigosterols.

The most significant phenolics in *F. fomentarius* methanolic extract are benzoic acid, rutin, quercetin and protocatechuic acid; p-hydroxy benzoic acid, catechin, syringic acid, p-coumaric acid, benzoic acid, cinnamic acid, p-hydroxy benzoic acid, and vanillic acid were also found [12].

Gamma irradiation is an established sterilization method for medical devices and one which is particularly suited to plastics. A more recent application has been in the sterilization of single-use disposable components [13]. Due to DNA sensitivity, gamma irradiation was also used in plant mutation breeding, to generate crop varieties with improved treats [14,15] by selecting for useful permanent mutation in the genetic material. This latter application was however gradually abandoned in favor of targeted mutagenesis. Nowadays, gamma irradiation treatment for the improvement of various biological-active properties begins to develop as a result of its accessibility, due to the spread of industrial irradiators.

We used a ^{60}Co source of radiation to experimentally induce oxidative stress in living mycelium of *Fomes fomentarius* fungus, in order to obtain a late response of oxidative stress tolerance after treatment. We measured the response by mean of free radical scavenging capacity (RSC) on DPPH, total phenols, and total flavonoids in methanolic extract, at 24, 48, and 72 h after the irradiation exposure of living mycelium. Our aim was to find out in what range sublethal doses of gamma irradiation can be used as a biotechnological tool to elevate the level of some classes of compounds with antioxidant capacity in *Fomes fomentarius* mycelium.

2. Materials and Methods

2.1. Chemicals and Equipment

All chemicals used were purchased from Sigma—Aldrich Co. LLC St. Louis, MO, USA. Deionized water was used for the steps where it was needed. Absorbances were measured, using SpectraMax i3x Multimode Detection Platform (Molecular Devices LLC, San Jose, CA, USA). The 10 L bioreactor BIOSTAT® B (Sartorius Stedim Biotech, Göttingen, Germany) was used to obtain biomass.

2.2. Fungal Identification

The strain used in this work was isolated as pure culture dikaryotic mycelium from an active growing area of a young basidiocarp. It was harvested in a hardwood of a hilly area of Romania and kept under in vitro culture by periodic transfer on fresh culture media. The isolate was designated as FPM and was identified morphologically and confirmed by 18S ribosomal RNA gene sequencing (98% homology in NCBI database) as *Fomes fomentarius*.

2.3. Mycelial Biomass Generation

Fomes fomentarius mycelium was used as a pure strain. Stock cultures were maintained on Petri dishes on Potato Dextrose Agar (PDA), incubated at 23.5 °C.

Fresh mycelial biomass for irradiation was obtained by submerged fermentation in bioreactor in a total volume of 10 L PDB medium (Potato Dextrose Broth).

To generate the inoculum for the bioreactor culture, mycelium of *F. fomentarius* from a liquid culture was transfer to fresh PDB in Duran bottles, with Teflon membranes screw cap (pores of 0.45 μm—for sterile air passage). The experiment was performed under the following conditions: temperature of 23.5 °C, shaking speed of 100 rpm, and initial pH = 5.6 [16].

After approximately one month of incubation, the mycelium occupied the entire culture volume (400 mL). The bioreactor inoculum was prepared by triturating the biomass with sterile glass beads (Ø = 3 mm). The resulted mycelial suspension was aseptically inoculated into the bioreactor glass vessel. The experiment was performed under aeration 2 L/min, continuously.

After 7 days of growth, the bioreactor culture was ended, and the mycelium was drained through sieves with pores of 300 μm and 125 μm, then washed with deionized water to remove culture medium, and fairly drained again by settling. No vacuum pressure was applied, so the biomass was fully hydrated.

The drained biomass mycelium was divided into 4 replicative aliquots for each dose (20—500 Gy) and 8 replicative aliquots for control (0 Gy) (~ 50 g each), distributed in sterile 90 mm Petri dishes for irradiation.

2.4. Irradiation of the Mycelium of Fomes fomentarius

The irradiation was performed by using a ^{60}Co research irradiator GC-5000 (B.R.I.T.—Mumbai, India) located in IRASM Radiation Processing Department of "Horia Hulubei" National Institute of Physics and Nuclear Engineering (Romania).

The dose interval was chosen on the base of previous experiments with fungi, trying to cover from non-effective to saturating effect. Mycelial samples were acutely exposed to gamma rays at following average doses: 0 (control), 20, 50, 70, 100, 200, 300, 400, and 500 Gy, respectively, at a dose rate of 0.8 Gy/s. An alanine dosimetry system was used for dose evaluation. The reference material for the doses is water. Irradiation temperature, as measured inside the irradiation chamber, was in the range of 27–28 °C.

After irradiation, mycelial biomass aliquots were incubated at 23.5 °C for 24, 48, and 72 h, to allow metabolic recovery and response, then lyophilized.

2.5. Preparation of Samples and Obtaining the Extract

From each irradiated aliquot, three technical replicates of lyophilized mycelium were established. Each replicate of was separately ground and extracted twice in 2 mL of 80% methanol, with an incubation of 2 h, at −20 °C, each. The second extraction was preceded by sonication for 15 min. Both supernatants were combined and washed with n-hexane (*v:v*) to remove lipids and waxes.

After a final centrifugation to remove possible debris, the supernatant was evaporated to dryness and redissolved in 80% methanol. The obtained extract was stored at 4 °C until the assays were performed.

2.6. Evaluation of the Antioxidant Capacity of the Extract
2.6.1. Total Polyphenols Content (TPC)

The total content of polyphenols in the methanolic extracts of submerged cultivated mycelium after irradiation treatment was determined by using the Folin–Ciocalteu method [17]. This is based on the reduction of the Folin–Ciocalteu reagent by phenolic compounds, which will form chromogens that can be detected spectrophotometrically.

The polyphenols in *F. fomentarius* extract or gallic acid (3,4,5-trihydroxybenzoic acid) reacted with the Folin–Ciocalteu reagent, forming a blue complex which was quantified by measuring the absorbance at $\lambda = 765$ nm. Gallic acid has been used as a standard antioxidant compound for spectrophotometric determination of antioxidant activity. Its standard curve comprised 6 dilutions: 0.05, 0.1, 0.15, 0.2, 0.25, and 0.5 mg/mL, starting from a 1 mg/mL stock solution.

The total content of polyphenols in the 20 µL sample/reference substance was determined by adding 120 µL of Folin–Ciocalteu solution was diluted 1/10, followed by 1 min of incubation at room temperature. Subsequently, 120 µL of 7.5% Na_2CO_3 solution was added, and the final mixture was incubated for 60 min, at 25 °C. Finally, O.D. was measured at 765 nm, using 80% methanol as a blank [18,19].

The total polyphenol content of the methanolic extract was calculated based on the regression equation y = 0.0057x + 0.0955 and expressed in micrograms of gallic acid equivalents per mL extract (µg GAE/mL methanolic extract) [19].

Both the samples and the reference substance were worked in triplicate, and their average absorbance value was calculated.

2.6.2. Total Flavonoids Content (TFC)

The flavonoid content of methanolic extracts obtained from mycelium of *F. fomentarius* was quantified according to the colorimetric method of aluminum chloride, described by Lamaison and Carnet (1990) [20]. To achieve the reaction, 2.5% $AlCl_3$ solution was added over the sample/standard quercetin solution (in a 1:1 volumetric ratio), and the mixture was incubated at room temperature for 20 min. The standard curve was made of quercetin and comprised 5 dilutions (0.01, 0.02, 0.03, 0.04, and 0.05 mg/mL), starting from a stock solution of 1 mg/mL in 80% methanol. The absorbance was read at 420 nm, and the flavonoid content was calculated based on the regression equation y = 31.538x − 0.5262 and expressed in µg of quercetin equivalents (QE) per mL of extract [21].

2.6.3. The DPPH Method

The evaluation of the antioxidant properties of the methanolic extract obtained from the submerged mycelium of *F. fomentarius* was determined by using the radical 2,2-diphenyl-1-picrilhydrazyl (DPPH), according to the method of Burits [22].

The sample was reacted with the 0.25 mM DPPH solution in 80% methanol, in a ratio of 1/10 = *v/v*. The optical density was measured at 523 nm after incubation of the plates, in the dark, for 40 min, at room temperature.

DPPH reduction occurs when it reacts with an antioxidant compound that can donate hydrogen. Thus, the color varies with the reduction of DPPH (from purple to pale yellow), and the color intensity decreases depending on the concentration of antioxidant in the sample.

The free radical scavenging activity of DPPH (expressed as a percentage) was calculated by using the following formula [23]:

$$\% \text{ inhibition} = [(A_{blank} - A_{sample})/A_{blank}] \times 100$$

where A_{blank} = absorbance blank, and A_{sample} = absorbance sample.

2.7. Statistical Analysis

All extracts were performed in three technical replicates for each dose and each response period (time-lapse after irradiation), and each replicate was read three times ($n = 120$). Mean values were used to calculate antioxidant activity and standard deviation (SD) values for each experimental group. Comparisons between groups (control and mycelium subjected to gamma irradiation) were evaluated by one-way ANOVA. A value of $p < 0.05$ was considered statistically significant. The graphics were plotted by using OriginLab 9.0.

3. Results

3.1. Statistics

Both gallic acid and quercetin calibration curve rendered a very good correlation coefficient ($R^2 = 0.9986$ and $R^2 = 0.9997$, respectively).

A vast majority of results in the range of 200 to 300 Gy differ significantly from the non-irradiated control; statistical difference between response times can also be noticed, as per Table 1.

Table 1. A one-way ANOVA.

Assay	Compared Variables (Irradiation Doses)	24 h	48 h	72 h
		Statistical Significance of the Difference (p-Value)		
DPPH	0 Gy/200 Gy	0.0001	0.0014	0.0532
	0 Gy/300 Gy	0.0000	0.0000	0.0003
	0 Gy/400 Gy	0.0000	0.0002	0.0007
	0 Gy/500 Gy	0.0001	0.0003	0.0027
TPC	0 Gy/200 Gy	0.0000	0.0036	0.7250
	0 Gy/300 Gy	0.0000	0.0001	0.5286
	0 Gy/400 Gy	0.0000	0.0086	0.0382
	0 Gy/500 Gy	0.0000	0.0073	0.0006
TFC	0 Gy/200 Gy	0.0200	0.0006	0.6781
	0 Gy/300 Gy	0.0086	0.0254	0.0000
	0 Gy/400 Gy	0.0010	0.0433	0.0204
	0 Gy/500 Gy	0.1100	0.0879	0.0000

Although differences in results between treated samples at very low doses (20–100 Gy) and untreated control turned partially statistically insignificant (Table 2), a general trend for all tested parameters can be observed, namely that doses under 100 Gy proved inefficient in upregulating synthesis of active metabolites.

Table 2. A one-way ANOVA.

Assay	Compared Variables (Irradiation Doses)	24 h	48 h	72 h
		Statistical Significance of the Difference (p-Value)		
DPPH	0 Gy/20 Gy	0.0234	0.0012	0.0186
	0 Gy/50 Gy	0.1220	0.0003	0.0719
	0 Gy/70 Gy	0.6468	0.0001	0.0045
	0 Gy/100 Gy	0.0720	0.0000	0.0004
TPC	0 Gy/20 Gy	0.0250	0.0929	0.8858
	0 Gy/50 Gy	0.1468	0.0183	0.2233
	0 Gy/70 Gy	0.0002	0.6792	0.1678
	0 Gy/100 Gy	0.0006	0.0001	0.0002
TFC	0 Gy/20 Gy	0.2837	0.1741	0.5092
	0 Gy/50 Gy	0.0447	0.0162	0.0146
	0 Gy/70 Gy	0.0304	0.0026	0.4435
	0 Gy/100 Gy	0.0007	0.0029	0.0012

3.2. DPPH Method

Our work shows that a delay after the irradiation treatment, prior to extraction, increases the extract's capacity of scavenging the DPPH free radical (Figure 1) from $18.5 \pm 1.88\%$ (untreated control) to a range of 28.83 ± 1.4–$34.99 \pm 1.14\%$ (irradiated) after 24 h, from 29.8 ± 1.45% (untreated control) to a range of 38.06 ± 1.6–45.48 ± 2.45% (irradiated) after 48 h, and from 51.07 ± 2.53% (untreated control) to a range of 54.90 ± 1.94–60.19 ± 0.6% (irradiated) after 72 h. Overall, this represents a maximum mean enhancement of 1.89 times of scavenging capacity, recorded after 24 h, at 300 Gy (Figure 1). The three tested response times shown the maximum increase at 300 Gy. Very low doses, in the range of 20 to 100 Gy, had negative or non-significant effect on the antioxidant activity (Figure 1).

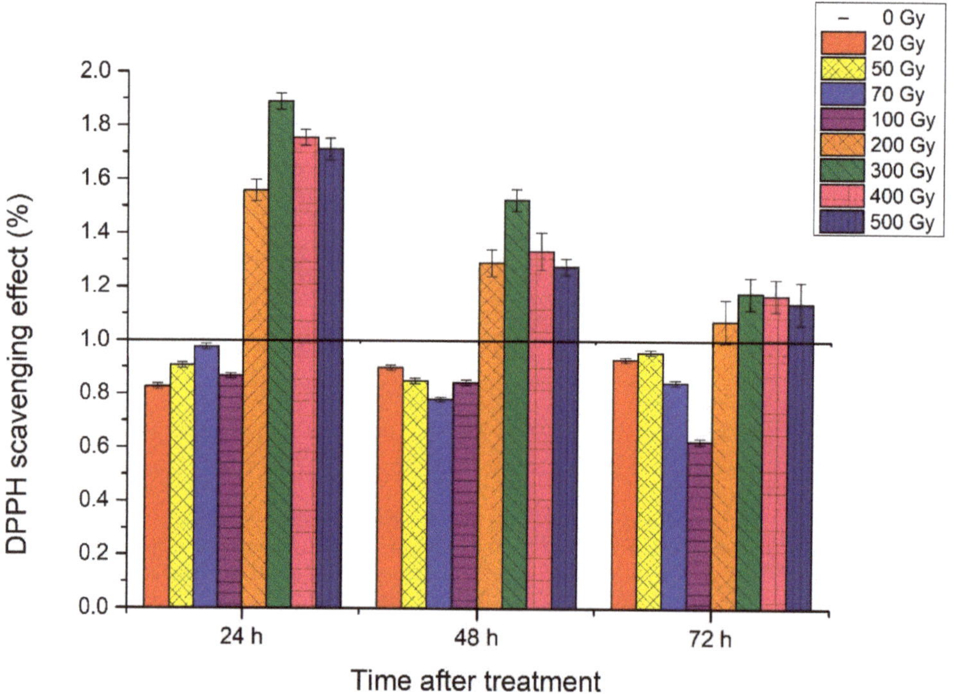

Figure 1. The effect of gamma irradiation and response-time on the antioxidant capacity in methanolic extract of *Fomes fomentarius* mycelium. Results are expressed as average irradiated/untreated control ratio (n = 120 per time). Bars are one standard deviation (SD).

3.3. Total Polyphenols Content (TPC)

In the range of 200 to 500 Gy, the total polyphenols content of methanolic extract increases from 241.14 ± 7.25 (untreated control) to a range of 312.95 ± 9.98–396.14 ± 20.83 μg GAE/mL (irradiated) after 24 h, from 375.64 ± 23.42 (untreated control) to a range of 431.43 ± 15.48–500.22 ± 14.48 μg GAE/mL (irradiated) after 48 h, and from 551.18 ± 26.29 (untreated control) to 560.64 ± 10.42 μg GAE/mL (in samples irradiated at 300 Gy) after 72 h. Overall, this represent a maximum mean enhancement of 1.64 times of total polyphenolic content, recorded after 24 h, at 300 Gy (Figure 2). Again, the three tested response times shown the maximum increase at 300 Gy. Very low doses (20–100 Gy) produced irrelevant variations in total polyphenols (Figure 2).

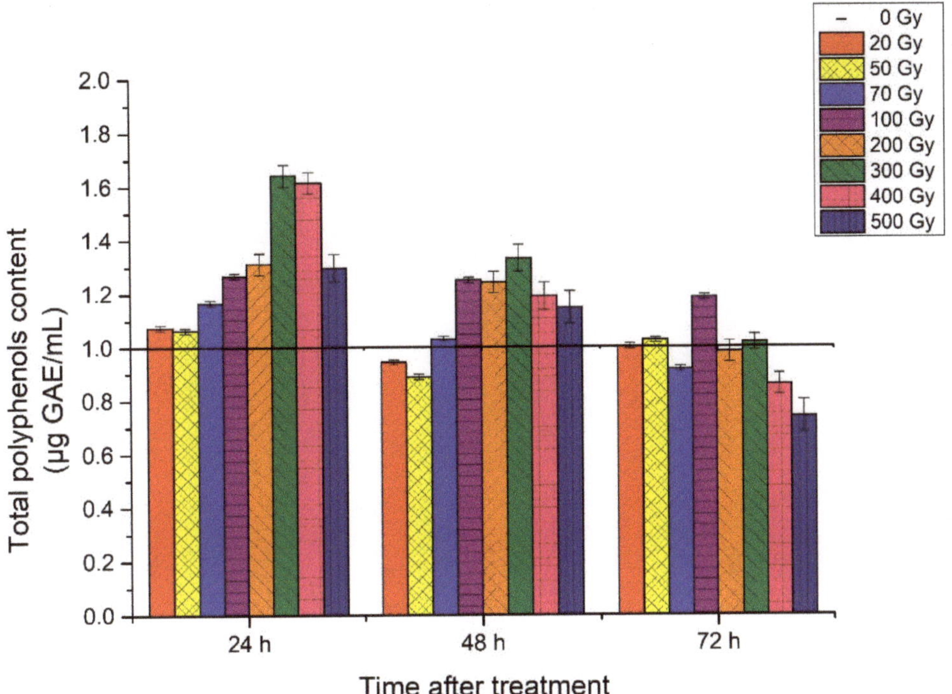

Figure 2. The effect of gamma irradiation and response-time after treatment on the total content of polyphenols in methanolic extract of *Fomes fomentarius* mycelium. Results are expressed in gallic acid equivalent as average irradiated/untreated control ratio (*n* = 120 per time). Bars are one standard deviation (SD).

3.4. Total Flavonoids Content (TFC)

For the total flavonoids content of methanolic extract, we encountered improvement in all doses. The significant differences over control were obtained in the range of 200 to 500 Gy, as follows: from 12.98 ± 0.72 (untreated control) to a range of 15.09 ± 1.22–21.63 ± 1.99 µg quercetin/mL (irradiated) after 24 h, from 14.84 ± 0.74 (untreated control) to a range of 18.6 ± 0.75–25.06 ± 0.67 µg quercetin/mL (irradiated) after 48 h, and from 10.44 ± 0.75 (untreated control) to a range of 11.24 ± 0.23–16.72 ± 1.79 µg quercetin/mL (irradiated) after 72 h. Overall, this represent a mean maximum enhancement of 1.68 times of total flavonoids, recorded after 48 h, at 300 Gy (Figure 3). As with previous measurements, the three tested response times shown the maximum increase at 300 Gy. Furthermore, a significant increase in flavonoids content was also generated by 100 Gy: 1.61 irradiated/control ratio after 24 h, 1.56 after 48 h, and 1.36 after 72 h. Irrelevant upregulations were recorded at doses under 100 Gy.

3.5. Response Time after Irradiation

Since the maximum yield of Total Polyphenols Content and DPPH scavenging effect of methanolic extract was obtained after 24 h, it is plausible that a shorter delay may reveal an even greater increase of these parameters, provided that optimal living conditions are ensured. For those parameters, the maximum response slowly quenches in time, with a difference of 1.53 fold after 48 h and 1.18 fold after 72 h, above the untreated control for DPPH scavenging, and a difference of 1.33 fold above the untreated control after 48 h and no difference after 72 h for total polyphenols.

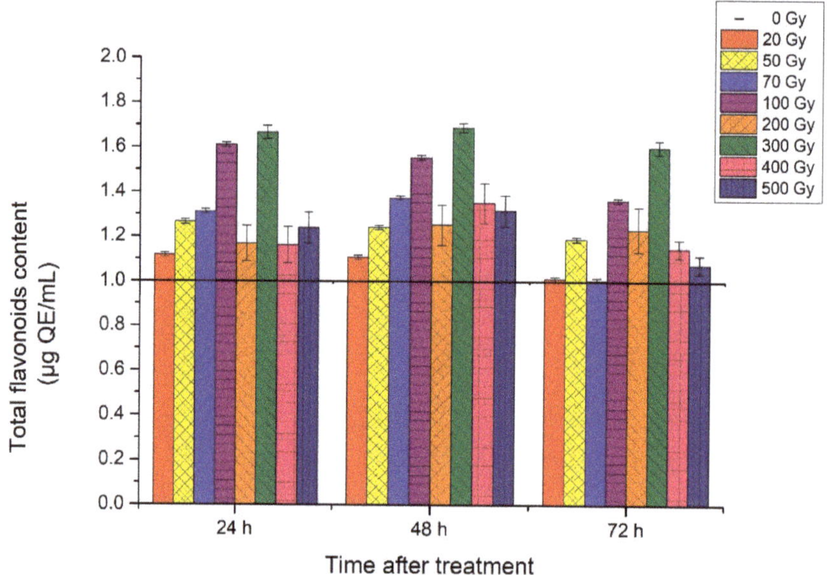

Figure 3. The effect of gamma irradiation and response-time after treatment on the total flavonoids content in methanolic extract of *Fomes fomentarius* mycelium. Results are expressed in quercetin equivalent as average irradiated/untreated control ratio (*n* = 120 per time). Bars are one standard deviation (SD).

Our results also show that the antioxidant capacity (DPPH free radical scavenging) and total polyphenols level increase over time, regardless of the irradiation dose, due to their extracellular nature. Instead of being diluted by the nutritive medium while in culture, secondary metabolites concentrate as a jelly matrix in the mycelial mass, outside the hyphae, partially consisting of exopolysaccharide (EPS), which were shown to present free radical scavenging activity [24]. Flavonoids instead, being accumulated in-cell, are consumed and in constant decline over time, regardless of the irradiation dose. The difference between doses however, maintains in flavonoids, in contrast to the total polyphenols and antioxidant activity by DPPH assay, which tends to fade after 72 h.

Response in terms of flavonoids synthesis showed more stable over time at all doses, with very small variation between maxima (namely increase of 1.66-fold after 24 h, 1.68- fold after 48 h, and 1.60-fold after 72 h, respectively, at 300 Gy). Flavonoids have great potential to counter the detrimental effects of the massive generation of reactive oxygen species (ROS). The notion that large alterations in the cell redox state trigger the biosynthesis of flavonoids [25,26], led Agati et al., (2020) [27] to hypothesize that flavonoids may complement the functions of primary antioxidants when their ROS detoxifying capacity declines in plants suffering from severe photo-oxidative stress. Barry Halliwell, who provided a series of definitions of "antioxidants", has always been critical of the effective ability of flavonoids to "delay, prevent or remove oxidative damage to a target molecule", a widely accepted definition of antioxidant given in the paper written by Gutteridge and Halliwell [27,28].

A general growth trend of all parameters with time was noticed, which was to be expected due to concentration of secreted secondary metabolites outside cells.

4. Discussion

Similar approaches of irradiating fresh living fungi or plants, in order to enhance their nutritional value, showed similar results, depending on the experimental design and the assessed compounds. Working with *Cordyceps militaris* medicinal fungus and using UV-B as source of radiation, Huang et al. [29] showed that the flavonoid content increased by

30.6 to 56% in living mycelium (depending on the preparation method) and by 10.4% in living fruiting bodies. UV-B irradiation also considerably increased total phenol content in mycelia, but had insignificant effect on mushrooms. However, we must notice that time after irradiation, up to the freezing step, is not mentioned and, according to our estimation, the absorbed energy (in Joules) was ~11 times higher in UV-B irradiated *Cordyceps* than in our experiment (gamma irradiated *Fomes fomentarius* mycelium).

Taheri et al. (2014) [30] found that acute gamma irradiation enriches the content of some polyphenols and flavonoids in *Curcuma alismatifolia* leaves of plants whose rhizomes have been previously irradiated. All analyzed polyphenols, namely salicylic acid, caffeic acid, catechin, epicatechin, cinnamic acid, ellagic acid, resorcinol and flavonoids, namely rutin, naringin, apigenin, quercetin, myricetin, showed an increase in a range similar to our work. Their experiment was performed at doses of about an order of magnitude smaller than ours (tens of grays), since plants are less radioresistant than fungi and their highest dose (20 Gy) presented the highest increase in both polyphenols and flavonoids.

Liu et al. [31] also found that irradiation with UV-B light after harvest enhanced the amount of flavonoids, phenols and total phenols in tomato fruits, while Du et al., [32] reported that low-dose UV-B light treatment to freshly cut carrot products could increase the total soluble phenolic content and enhance antioxidant activity. The penetration power of the two radiation sources is, however, different: while ultraviolet radiation acts only on the surface, gamma photons are highly penetrating, making results only partially comparable.

Köhler et al. [33] showed that the enzymatic (superoxide dismutase, SOD and total peroxidases, POD) and non-enzymatic antioxidant activity (total phenolic) increased significantly in *Deschampsia antarctica* plants in hydroponic system, in response to UV-B treatment. A peak in total phenolic content of *D. antarctica* was found at 3 h of exposure, which related to percentage of consumed DPPH and FRAP values, suggesting that these phenolics function as soluble antioxidant metabolites. Various polyphenols have antioxidant and anti-inflammatory properties that could have preventive and/or therapeutic effects for cardiovascular disease, neurodegenerative disorders, cancer, and obesity [34].

Submerged fermentation is a method at hand for producing fungal biomass of rare and/or slow growing species. The process is easily controllable, regardless of the season, and the total and specific compound yield are predictable. Mycelia and/or fermentation broth could be harvest in couple of days to few weeks, depending on the desired metabolite. Freeze-drying of live mycelium is a good method to both conserve the properties of compounds and to catch the differences between treatments and response time, since metabolic activity is completely interrupted.

Our experimental approach reveals gamma irradiation as a novel biotechnological tool to obtain medicinal mycelium with improved antioxidant properties. Such a natural resource is highly desirable for pharmaceutical, as well as for nutraceutical products. Due to very small doses that are necessary, industrial gamma irradiation becomes a cost-effective large-mass-applicable instrument to generate added value to raw biological material for extracts or new dietary supplements and functional foods. By density and water content, submerged cultivated mycelium (after draining) is very similar to fresh food. Doses are also in the same range, so the cost of mycelium irradiation for stimulation of bioactive compound synthesis is similar to fresh food irradiation, in industrial or semi-industrial plants; food irradiation for decontamination and shelf-life extension is a common practice, currently costing few cents per kilogram. Other oxidative stressors like exposure to ozone or hydrogen peroxide (H_2O_2) are a chemical approach with a dose-effect ratio that needs to be investigated on a case-by-case basis. Cellular interaction with heavy metal ions, such as iron, copper, cadmium, mercury, nickel, lead, and arsenic, also generates free radicals and oxidative stress, but fungi are particularly eager to absorb heavy metals potentially causing intoxications.

Stimulating fresh living mycelium with gamma irradiation is a completely different approach from irradiating either dry mycelium or even final extract of it. Apart from significantly higher doses, in the range of kGy to tens of kGy, in the latter strategy, gamma irradiation acts by different mechanism, probably by degrading or decomposing some compounds into antioxidant components [35]. In fact, Adamo et al. [36] proposed that the destructive processes of oxidation and γ-irradiation were capable of breaking the chemical bonds of polyphenols, thereby releasing soluble phenols of low molecular weight. This may change the content and composition of antioxidant components, thereby affecting the antioxidant properties. Basically, it is not a real enrichment, since no new active bio-synthesis is made, but rather an improved image of its bioactivity. In contrast, applying low oxidative stress triggers a cellular response, implying conversion of different stored precursors into metabolites with radioprotective and scavenging abilities on free radicals, meant to counteract and prevent future damages. In a certain window, this response could be used in the benefit of human welfare.

5. Conclusions

So far, our results point out that 300 Gy is the optimum gamma irradiation dose for stimulation of antioxidant response in living mycelium for all the investigated parameters (TPC, TFC, and DPPH). If an interval is to be considered for the delivered dose, to cover for inherited errors, then 400 Gy is the end of the interval. The dose may appear surprisingly high for stimulation purposes, but we should take into account that fungi, especially melanized ones, are highly radioresistant and resilient species. Anyway, for stimulation approach, irradiation dose must be chosen in the radiation resistance window of the species, in order for cellular damages to be recoverable and to allow for proper response; this dose may differ from one species to another and must be confirmed experimentally.

Based on our results, a gamma photons exposure at 300 Gy, followed by a time response, could markedly enrich the living mycelium of the medicinal fungus *Fomes fomentarius* in total polyphenols and flavonoids and could potentiate its antioxidant activity. Doses below 100 Gy proved invariably inefficient in generating any relevant effect.

The response time after treatment depends on the targeted antioxidant compound, provided that the mycelium is in good state and optimum conditions (temperature and air access) are offered.

To our knowledge, this is the first record of the usage of gamma irradiation, at small sublethal doses, as a biotechnological tool to increase the antioxidant capacity of *Fomes fomentarius* living mycelium, as proven by methanolic extract.

Author Contributions: Conceptualization methodology and formal analysis, M.E.; statistics, software, and graphs, C.F.P.; strain isolation and identification, M.E.; extractions, assays, and data acquisition, C.F.P.; irradiation treatments, D.C.N.; extraction protocol optimization, A.-M.P.; writing—original draft preparation, C.F.P.; writing—review and editing, M.E.; project administration, M.E. All authors have read and agreed to the published version of the manuscript.

Funding: This work was supported by a grant of the Romanian Ministry of Research and Innovation, CCCDI—UEFISCDI, project number PN-III-P1-1.2-PCCDI-2017-0323/5PCCDI/2018, within PNCDI III.

Institutional Review Board Statement: Not applicable.

Informed Consent Statement: Not applicable.

Data Availability Statement: Not applicable.

Conflicts of Interest: The authors declare no conflict of interest. The funders had no role in the design of the study; in the collection, analyses, or interpretation of data; in the writing of the manuscript; or in the decision to publish the results.

References

1. Sharma, S.; Chatterjee, S.; Kataria, S.; Joshi, J.; Datta, S.; Vairale, M.; Veer, V. A Review on Responses of Plants to UV-B Radiation Related Stress. In *UV-B Radiation: From Environmental Stressor to Regulator of Plant Growth*, 1st ed.; Singh, V.P., Singh, S., Prasad, S.M., Parihar, P., Eds.; John Wiley & Sons Ltd.: Hoboken, NJ, USA, 2017; pp. 75–97. [CrossRef]

2. Reis, F.S.; Martins, A.; Barros, L.; Ferreira, I.C.F. Antioxidant Properties and Phenolic Profile of the Most Widely Appreciated Cultivated Mushrooms: A Comparative Study Between in Vivo and in Vitro Samples. *Food Chem. Toxicol.* **2012**, *50*, 1201–1207. [CrossRef] [PubMed]
3. Sarikurkcu, C.; Tepe, B.; Semiz, D.K.; Solak, M.H. Evaluation of Metal Concentration and Antioxidant Activity of Three Edible Mushrooms from Mugla, Turkey. *Food Chem. Toxicol.* **2010**, *48*, 1230–1233. [CrossRef]
4. Patil, A.S.; Suryavanshi, P.; Fulzele, D. Evaluation of Effect of Gamma Radiation on Total Phenolic Content, Flavonoid and Antioxidant Activity of in Vitro Callus Culture of *Artemisia Annua*. *Nat. Prod. Chem Res.* **2018**, *6*, 1–6. [CrossRef]
5. Charbaji, T.; Nabulsi, I. Effect of Low Doses of Gamma Irradiation on in Vitro Growth of Grapevine. *Plant. Cell Tissue Organ. Cult.* **1999**, *57*, 129–132. [CrossRef]
6. Pradhan, B.; Baral, S.; Patra, S.; Behera, C.; Nayak, R.; MubarakAli, D.; Jena, M. Delineation of Gamma Irradiation (^{60}Co) Induced Oxidative Stress by Decrypting Antioxidants and Biochemical Responses of Microalga, *Chlorella* sp. *Biocatal. Agric. Biotechnol.* **2020**, *25*, 101595. [CrossRef]
7. Gáper, J.; Gáperová, S.; Pristas, P.; Naplavova, K. Medicinal Value and Taxonomy of the Tinder Polypore, *Fomes Fomentarius* (Agaricomycetes): A Review. *Int. J. Med. Mushrooms* **2016**, *18*, 851–859. [CrossRef] [PubMed]
8. Zhang, Y.; Xiao, Y.; Wang, P.; Liu, Q. Compositions and Anti-Tumor Activity of *Pyropolyporus fomentarius* Petroleum Ether Fraction In Vitro and In Vivo. *PLoS ONE* **2014**, *9*, 109599. [CrossRef]
9. Lee, S.O.; Lee, M.H.; Lee, K.R.; Lee, E.O.; Lee, H.J. *Fomes Fomentarius* Ethanol Extract Exerts Inhibition of Cell Growth and Motility Induction of Apoptosis via Targeting AKT in Human Breast Cancer MDA-MB-231 Cells. *Int. J. Mol. Sci.* **2019**, *20*, 1147. [CrossRef] [PubMed]
10. Karimi, E.; Jaafar, H.Z.; Ahmad, S. Antifungal, Anti-Inflammatory and Cytotoxicity Activities of Three Varieties of *Labisia pumila* Benth: From Microwave Obtained Extracts. *BMC Complement. Altern Med.* **2013**, *13*, 1–10. [CrossRef]
11. Zhang, F.L.; Shi, C.; Sun, L.T.; Yang, H.X.; He, J.; Li, Z.H.; Feng, T.; Liu, J.K. Chemical Constituents and Their Biological Activities from the Mushroom *Pyropolyporus fomentarius*. *Phytochemistry* **2021**, *183*, 112625. [CrossRef] [PubMed]
12. Dogan, H.; Kars, M.; Özdemir, Ö.; Gunduz, U. *Fomes fomentarius* and *Tricholoma anatolicum* (Agaricomycetes) Extracts Exhibit Significant Multiple Drug Resistant Modulation Activity in Drug Resistant Breast Cancer Cells. *Int. J. Med. Mushrooms* **2020**, *22*, 105–114. [CrossRef]
13. Sandle, T. Gamma Radiation. In *Sterility, Sterilisation and Sterility Assurance for Pharmaceuticals*, 1st ed.; Sandle, T., Ed.; Woodhead Publishing: Cambridge, UK, 2013; pp. 55–68. ISBN 9781907568381.
14. Goldenberg, L.; Yaniv, Y.; Porat, R.; Carmi, N. Effects of Gamma-Irradiation Mutagenesis for Induction of Seedlessness, on the Quality of Mandarin Fruit. *Food Sci. Nutr.* **2014**, *5*, 943–952. [CrossRef]
15. Huang, Y.L.; Yuan, S.C.; Chang, K.W.; Chen, F.C. Gamma Irradiation Mutagenesis in *Monstera Deliciosa*. *Acta Hortic.* **2017**, *1167*, 213–216. [CrossRef]
16. Petre, M.; Petre, V.; Teodorescu, A.; Pătrulescu, F. Nutritive Mushroom Biomass Producing Through Submerged Fermentation of Agricultural Organic Wastes. *Studia UBB Ambient.* **2013**, *58*, 79–86.
17. Singleton, V.L.; Rossi, J.A. Colorimetry of Total Phenolics with Phosphomolybdic-Phosphotungstic Acid Reagents. *Am. J. Enol Vitic.* **1965**, *16*, 144–158.
18. Stan, M.S.; Voicu, S.N.; Caruntu, S.; Nica, I.C.; Olah, N.K.; Burtescu, R.; Balta, C.; Rosu, M.; Herman, H.; Hermenean, A.; et al. Antioxidant and Anti-Inflammatory Properties of a *Thuja occidentalis* Mother Tincture for the Treatment of Ulcerative Colitis. *Antioxidants* **2019**, *8*, 416. [CrossRef] [PubMed]
19. Korte, K.N.; Odamtten, G.T.; Obodai, M.; Appiah, V.; Akuamoa, F.; Adu-Bobi, A.K.; Annann, S.N.Y.; Armah, J.N.O.; Acquah, S.A. Evaluating the Effect of Gamma Radiation on the Total Phenolic Content, Flavonoids, and Antioxidant Activity of Dried *Pleurotus ostreatus* ((Jacq. ex. Fr) Kummer) Stored in Packaging Materials. *Adv. Pharm.* **2014**, 1–8. [CrossRef]
20. Lamaison, J.L.; Carnet, A. Teneurs en Principaux Flavonoids des Fleurs de *Crataegeus monogyna* Jacq et de *Crataegeus laevigata* (Poiret, D. C) en Fonction de la Vegetation. *Pharm. Acta. Helv.* **1990**, *65*, 315–320.
21. Chandra, S.; Khan, S.; Avula, B.; Lata, H.; Yang, M.H.; Elsohly, M.A.; Khan, I.A. Assessment of Total Phenolic and Flavonoid Content, Antioxidant Properties, and Yield of Aeroponically and Conventionally Grown Leafy Vegetables and Fruit Crops: A Comparative Study. *Evid Based Complement. Alternat Med.* **2014**, *2014*, 1–9. [CrossRef]
22. Burits, M.; Bucar, F. Antioxidant Activity of *Nigella sativa* Essential Oil. *Phytother Res.* **2000**, *14*, 323–328. [CrossRef]
23. Bakir, T.; Karadeniz, M.; Unal, S. Investigation of Antioxidant Activities of *Pleurotus ostreatus* Stored at Different Temperatures. *Food Sci. Nutr.* **2018**, *6*, 1040–1044. [CrossRef]
24. Zhang, Q.; Li, Y.; Zong, S.; Ye, M. Optimization of Fermentation of *Fomes fomentarius* Extracellular Polysaccharide and Antioxidation of Derivatized Polysaccharides. *Cell Mol. Biol.* **2020**, *66*, 56–65. [CrossRef]
25. Taylor, L.P.; Grotewold, E. Flavonoids as Developmental Regulators. *Curr. Opin. Plant. Biol.* **2005**, *8*, 317–323. [CrossRef] [PubMed]
26. Gayomba, S.R.; Muday, G.K. Flavonols Regulate Root Hair Development by Modulating Accumulation of Reactive Oxygen Species in the Root Epidermis. *Development* **2020**, *147*, dev185189. [CrossRef]
27. Agati, G.; Brunetti, C.; Fini, A.; Gori, A.; Guidi, L.; Landi, M.; Sebastiani, F.; Tattini, M. Are Flavonoids Effective Antioxidants in Plants? Twenty Years of Our Investigation. *Antioxidants* **2020**, *9*, 1098. [CrossRef]
28. Halliwell, B.; Gutteridge, J.M.C. *Free Radicals in Biology and Medicine*, 4th ed.; Oxford University Press: Oxford, UK, 2007.

29. Huang, S.J.; Lin, C.P.; Mau, J.L.; Li, Y.S.; Tsai, S.Y. Effect of UV-B Irradiation on Physiologically Active Substance Content and Antioxidant Properties of the Medicinal Caterpillar Fungus *Cordyceps militaris* (Ascomycetes). *Int. J. Med. Mushrooms.* **2015**, *17*, 241–253. [CrossRef] [PubMed]

30. Taheri, S.; Abdullah, T.L.; Karimi, E.; Oskoueian, E.; Ebrahimi, M. Antioxidant Capacities and Total Phenolic Contents Enhancement with Acute Gamma Irradiation in *Curcuma alismatifolia* (Zingiberaceae) Leaves. *Int. J. Mol. Sci.* **2014**, *15*, 13077–13090. [CrossRef] [PubMed]

31. Liu, C.; Han, X.; Cai, L.; Lu, X.; Ying, T.; Jiang, Z. Postharvest UV-B Irradiation Maintains Sensory Qualities and Enhances Antioxidant Capacity in Tomato Fruiting During Storage. *Postharvest Biol. Technol.* **2011**, *59*, 232–237. [CrossRef]

32. Du, W.X.; Avena-Bustillos, R.J.; Breksa, A.P., 3rd; McHugh, T.H. Effect of UV-B Light and Different Cutting Styles on Antioxidant Enhancement of Commercial Fresh-Cut Carrot Products. *Food Chem.* **2012**, *134*, 1862–1869. [CrossRef]

33. Köhler, H.; Contreras, R.A.; Pizarro, M.; Cortés-Antíquera, R.; Zúñiga, G.E. Antioxidant Responses Induced by UVB Radiation in *Deschampsia antarctica*. *Desv. Front. Plant. Sci.* **2017**, *8*, 921. [CrossRef] [PubMed]

34. Cory, H.; Passarelli, S.; Szeto, J.; Tamez, M.; Mattei, J. The Role of Polyphenols in Human Health and Food Systems: A Mini-Review. *Front. Nutr.* **2018**, *5*, 87. [CrossRef] [PubMed]

35. Huang, S.J.; Mau, J.L. Antioxidant Properties of Methanolic Extracts from *Antrodia camphorata* with Various Doses of γ-Irradiation. *Food Chem.* **2007**, *105*, 1702–1710. [CrossRef]

36. Adamo, M.; Capitani, D.; Mannina, L.; Cristinzio, M.; Ragni, P.; Tata, A.; Coppola, R. Truffles Decontamination Treatment by Ionizing Radiation. *Radiat. Phys. Chem.* **2004**, *71*, 167–170. [CrossRef]

applied sciences

Article

Submerged Cultivation of *Inonotus obliquus* Mycelium Using Statistical Design of Experiments and Mathematical Modeling to Increase Biomass Yield

Alexandru Petre [1,2,*], Mihaela Ene [1] and Emanuel Vamanu [2,*]

[1] Horia-Hulubei National Institute for Physics and Nuclear Engineering (IFIN-HH), 077125 Magurele, Romania; mene@nipne.ro
[2] Faculty of Biotechnology, University of Agricultural Sciences and Veterinary Medicine, 011464 Bucharest, Romania
* Correspondence: alexandru.petre@nipne.ro (A.P.); email@emanuelvamanu.ro (E.V.)

Abstract: Submerged culturing of mycelium is an efficient technique used to increase biomass yields, more so when employed with naturally slow-growing species of mushrooms. This paper is concerned with optimizing nutrient broth components used in *Inonotus obliquus* cultures for achieving high biomass yields. We modeled the effect of seven biotechnological parameters (six broth ingredients and the initial pH of nutritive broth) on mycelial biomass and predicted an optimum broth formula using response surface methodology. An analysis of variance showed that the elaborated model is significant (F-value of 2.76 and *p*-value of 0.0316). We used bioreactor cultures to confirm the model's optimum prediction and to compare these results with a general-purpose mycology medium, namely potato dextrose broth (PDB). The optimized bioreactor culture yielded 4.37 g/L (93.36% of the dry weight prediction), while the PDB bioreactor culture yielded 2.084 g/L, after 15 days of cultivation. The optimized formula was: 2.15299 g malt extract, 3.99296 g yeast extract, 11.0041 g fructose, 17.4 g soluble starch, 0.1 g $MgSO_4$, and 0.05 g $CaCl_2$ per liter of broth.

Keywords: Chaga; medicinal fungi; biomass yield; mathematical modeling; response surface methodology; bioreactor culture

Citation: Petre, A.; Ene, M.; Vamanu, E. Submerged Cultivation of *Inonotus obliquus* Mycelium Using Statistical Design of Experiments and Mathematical Modeling to Increase Biomass Yield. *Appl. Sci.* **2021**, *11*, 4104. https://doi.org/10.3390/app11094104

Academic Editor: Monica Gallo

Received: 6 April 2021
Accepted: 27 April 2021
Published: 30 April 2021

Publisher's Note: MDPI stays neutral with regard to jurisdictional claims in published maps and institutional affiliations.

1. Introduction

The therapeutic properties of many mushrooms, known to indigenous cultures across the globe, have been traced to their underlying chemical compounds with the advent of chemistry and molecular biology. Although the most well-documented are their use in cancer treatment due to beneficial glucan and proteoglycan synthesis [1], mushrooms have other therapeutic properties including antioxidant, antihypertensive, hepatoprotective, antifibrotic, anti-inflammatory, antidiabetic, antiviral, antimicrobial, and cholesterol-lowering properties [2].

Great improvements in biomass yield and bioactive compound production have been achieved with the help of biotechnological cultivation technics and conditions applied to mycology [3–6].

I. obliquus, also known as Chaga, is a mushroom that lives in certain parts of Europe and Asia at latitudes of 45–50° N as a parasite of birch trees [7]. In this environment, the Chaga mushroom grows very slowly and is not a reliable source of pharmaceutical compounds for industrial applications [8].

Recent studies have shown that the Chaga mushroom can produce polyphenols [9], flavonoids [10], melanins [11], and tannins [12]. It is also able to synthesize biologically active ergosterol peroxide, with a role in anticancer activity [13], as well as betulin, which has antiviral and anti-inflammatory properties [14].

Optimization of the various parameters or products of *I. obliquus* culture regularly employs a fractional factorial design, coupled with response surface methodology [15,16].

163

In this study, we attempted to increase the biomass of submerged cultures of Chaga mycelium by using a custom mix-process experimental design capable of predicting an optimum nutrient broth composition. Response surface methodology was used to evaluate the effects of different biotechnological parameters on biomass yield and played a role in the optimization process.

2. Materials and Methods

2.1. Inoculum Preparation

Inonotus obliquus (Ach. ex Pers.) Pilát (CBS 314.39) mycelium was purchased from Westerdijk Fungal Biodiversity Institute and grown on PDA agar (VWR Chemicals, Leuven, Belgium) for 7 days at 23.5 °C. Subsequently, 1×1 cm agar sectors were each transferred to 250 mL bottles with PTFE (polytetrafluoroethylene) membrane screw caps, which contained PDB seed culture medium (VWR Chemicals, Leuven, Belgium). The mycelia were incubated at the same temperature, with orbital shaking at 100 rpm until they occupied the entire volume of the culture vessels.

2.2. Optimization Design of the Nutrient Broth Components

Modern experimental designs fall into three main study types: factorial, mixture, or custom. The latter is used when the experiment requires adjustments that cannot be accommodated by a standard mixture design, such as when the differences between the high and low of all the mixture components are not the same, or when mixture and process variables are both used in the same design.

In order to optimize growth media in terms of biomass yield, the effect of six mixture components and one numeric factor was studied. Yeast extract (Merck KGaA, Darmstadt, Germany) and malt extract (Oxoid Ltd., Basingstoke, UK) were used as a protein source, while fructose (Schneekoppe GmbH, Buchholz, Germany) and soluble starch (Carl Roth GmbH, Karlsruhe, Germany) provided the carbohydrate source, and the latter also acted as a glucose substitute. We focused on fructose because it was reported that silver birch sap consists mostly of fructose (5.39% w/w) and, to a lesser extent, glucose (4.46% w/w) and sucrose (0.58% w/w) [17]. All growth media formulas were supplemented with equal amounts of magnesium sulfate heptahydrate (Merck KGaA, Darmstadt, Germany) and calcium chloride dihydrate (Honeywell GmbH, Seelze, Germany). Sterilization was performed at 121 °C for 15 min. The pH value of the growth media represented the single numeric factor, which was measured and entered into the experimental design after each media variant was sterilized, before being inoculated. The boundary values for each variable were chosen according to relevant literature data regarding submerged mushroom mycelium cultivation (Table 1).

Table 1. Design constraints.

Factor	Symbol	Low Value (g/L)	High Value (g/L)
Malt extract	A	1.10	2.30
Yeast extract	B	2.00	4.00
Fructose	C	11.00	19.00
Soluble starch	D	11.00	19.00
$MgSO_4$	E	0.10	0.10
$CaCl_2$	F	0.05	0.05
pH	G	4.97	5.67

The software Design-Expert version 11.1.0.1 (Stat-Ease Inc., Minneapolis, MN, USA) was used to achieve the mathematical modeling and statistical analysis of the experiment. By working with the type of design space described so far and inputting the above constraints, the software generates model points (runs) that are chosen algorithmically, though limited to as few runs as possible. These included five replicate points and five lack-of-fit points, supplementing the 20 required model points (Table 2).

Table 2. Experimental design.

Run	Malt Extract (g/L)	Yeast Extract (g/L)	Fructose (g/L)	Soluble Starch (g/L)	MgSO$_4$ (g/L)	CaCl$_2$ (g/L)	pH
1	1.14725	2	14.12359301	17.27915699	0.1	0.05	-
2	1.1	2	12.45	19	0.1	0.05	5.55
3	1.1	2.70875	19	11.74125	0.1	0.05	5.33
4	2.3	2.780591346	18.46940865	11	0.1	0.05	4.97
5	1.1	4	13.81857624	15.63142376	0.1	0.05	5.55
6	2.3	4	13.35966064	14.89033936	0.1	0.05	5.13
7	1.551558201	4	17.9984418	11	0.1	0.05	5.01
8	1.759389627	3.01120613	11.48678705	18.29261719	0.1	0.05	5.14
9	1.115150333	3.119596696	15.5597747	14.75547827	0.1	0.05	5.1
10	1.93376577	4	14.60469819	14.01153604	0.1	0.05	5.33
11	2.3	2	19	11.25	0.1	0.05	5.21
12	1.1	2	12.45	19	0.1	0.05	5.15
13	1.781034944	2	15.79942516	14.96953989	0.1	0.05	5.05
14	1.1	4	11	18.45	0.1	0.05	5.18
15	1.104378427	3.445621573	11	19	0.1	0.05	5.44
16	1.803890875	3.730662742	18.01544638	11	0.1	0.05	5.39
17	1.654403265	3.132022743	11.23077409	18.5327999	0.1	0.05	5.62
18	2.3	4	11	17.25	0.1	0.05	5.67
19	2.3	2	11.25	19	0.1	0.05	5.54
20	1.1	2	19	12.45	0.1	0.05	4.99
21	2.3	2.936180941	14.98784285	14.32597621	0.1	0.05	5.41
22	2.3	2.936180941	14.98784285	14.32597621	0.1	0.05	5.4
23	2.3	2	11.25	19	0.1	0.05	5.14
24	2.078933462	2	18.77101112	11.70005541	0.1	0.05	5.12
25	1.115150333	3.119596696	15.5597747	14.75547827	0.1	0.05	5.06
26	1.683913211	2.065251896	15.77432099	15.0265139	0.1	0.05	5.32
27	1.654403265	3.132022743	11.23077409	18.5327999	0.1	0.05	5.52
28	1.1	3.45	19	11	0.1	0.05	5.22
29	1.759389627	3.01120613	11.48678705	18.29261719	0.1	0.05	5.15
30	1.683913211	2.065251896	15.77432099	15.0265139	0.1	0.05	5.29

To determine the appropriateness of a design involving mixture components, prediction-based metrics such as fraction of design space (FDS) statistics are employed. The FDS graph is useful for calculating the volume of the design space with a prediction variance equal to or less than a specified value. This custom volume's ratio to the total volume is the fraction of the design space. The response surface was drawn by predicting the mean outcome as a function of inputs over the region of experimentation (Figure 1).

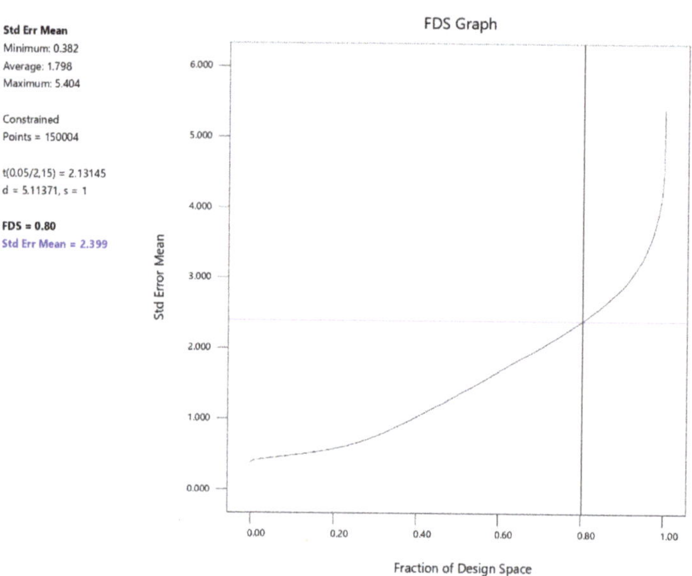

Figure 1. Fraction of design space graph for the evaluated model components and interactions.

The custom quadratic × linear model in which the terms A, B, C, and D, and interactions: AB, AC, AD, AG, BC, BD, BG, CD, CG, and DG, were evaluated by FDS statistics found that 80% of the design space was below a standard error mean of 2.399.

By plotting the standard error of design and the different model values for the A, B, and G terms, the latter's effect was shown to be the main source of the standard error mean value (Figure 2).

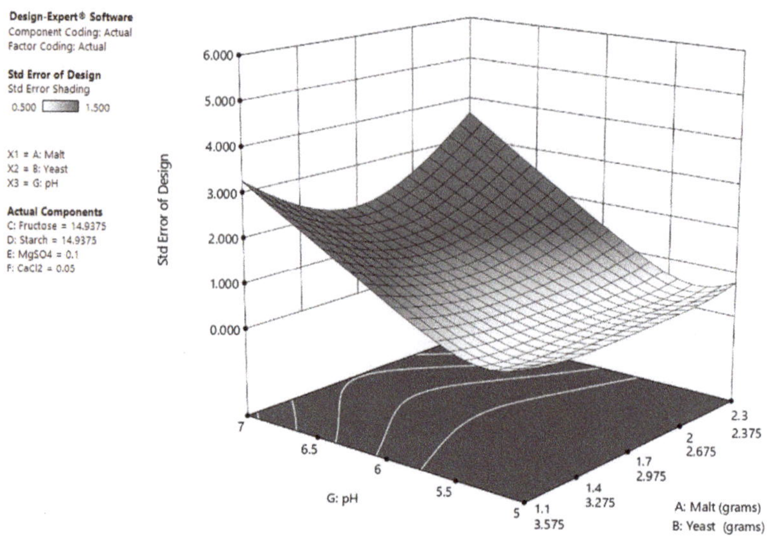

Figure 2. The 3D mix-process graph of the standard error of design.

If the interactions related to the G term were not evaluated, the standard error mean for the same fraction of design space became 0.489. However, this approach hinders the significance of the model's prediction power when evaluating the experimental response.

Because the values of the single numeric factor were not generated algorithmically but were inputted based on real data, their predictive power is low. As such, the optimization results relating to the effect of pH on the dry weight of mycelium were not considered.

2.3. Broth Variants Cultivation

The experimental broths were prepared at a volume of 500 mL in screw-cap bottles with air exchange provided by PTFE membranes (0.22 μm pores). Each variant was inoculated by transferring 20 mL (a 4% *v/v* inoculum) of triturated and homogenized mycelium suspension from the fully occupied culture vessels. Sterilized glass beads were added to the screw-cap bottles, and vigorous shaking of the mycelium with the beads was used to achieve trituration. The resulting inoculated broth variants were cultured at 26 °C and agitated at 115 rpm for 21 days.

2.4. Bioreactor Cultivation

A 10 L double wall (jacketed) Biostat B (Sartorius Stedim Biotech, Göttingen, Germany) bioreactor, equipped with a dissolved oxygen and pH sensor, temperature control, stirrer, and ring sparger, was used to ensure replicable, pilot scale cultures of *I. obliquus*. The submerged, aerobic cultivation lasted 15 days and employed a strategy of specific parameter variation, according to the age of the culture. The stirrer shaft speed was kept at 50 rpm until the 5th day of culture, when it was increased to a final value of 75 rpm. Aeration was provided through the ring sparger at a rate of 2 L/min by adjusting a flowmeter. A solenoid valve further limited the airflow to a chosen percentage of the time. At the moment of inoculation, this value was 1%. On the 8th day of culture, aeration was increased to 8% and then again on the 12th day to the final value of 10%. Airflow entry and exit points were passed through Millex-FG50 (Millipore Co., Billerica, MA, USA) filter units. The temperature was maintained at a constant 24 °C throughout the whole culture period.

All bioreactor batches were loaded with 8 L of nutrient broth, 6 L of which were sterilized in 2 L communicating vessels, while 2 L remained in the culture vessel. The communicating vessels were sterilized separately from the bioreactor, although for the same duration of time. This procedure allows for better sterilization efficiency, especially when dealing with custom nutrient broths with ingredients having a high microbial load. Sterilization time differed: for potato dextrose broth, we used 20 min at 121 °C; for the custom nutrient broth, we used 50 min at the same temperature.

Each batch was inoculated by transferring 250 mL (a 3.125% *v/v* inoculum) of triturated mycelium suspension from one of the fully occupied culture vessels. The dry weight of this inoculum was calculated to be 1.415 g by averaging the weight of three identical lyophilized cultures. Similarly to the shaken cultures, the bioreactor cultures were not pH corrected throughout the fermentation process. For the optimized broth, the pH value at the start of cultivation was 4.89, whereas for the PDB, this value was 5.4 at the moment of inoculation.

2.5. Mycelial Biomass Harvesting

All cultures were harvested by separating the mycelium from the broth through a 1 mm sieve, followed by further filtration of the broth through a 250 nm sieve. The mycelia were then washed three times with deionized water and frozen at −50 °C before being lyophilized (Biobase BK-FD12S vacuum freeze dryer). Mycelial biomass was estimated by dry weight and is expressed as g/500 mL for the optimization experiment and as g/L for the bioreactor cultures.

3. Results and Discussion

3.1. Analysis of the Dry Mycelium Response and Model Building

During the agitated incubation, one of the experimental variants was lost due to a mechanical impact within the incubator. Modeling of the dry weight response proceeded in the absence of this variant, without issue.

After the mycelia were lyophilized and weighed, each broth variant was assigned with the experimentally obtained weight values. Design-Expert software was used to carry out an analysis of the dry weight response and build a model of the interactions between the design factors and the response. This model was then able to generate predictions about the response for any given combination of factors within the design constraints. For the actual experimental runs, the model was also able to predict theoretical yields (Table 3).

Table 3. Statistical analysis of actual results and predicted values.

Run Order	Actual Value (g/500 mL)	Predicted Value (g/500 mL)	Residual	Leverage	Internally Studentized Residuals	Externally Studentized Residuals	Cook's Distance
2	0.8200	0.7275	0.0925	0.696	0.647	0.634	0.068
3	1.16	1.15	0.0052	0.577	0.031	0.030	0.000
4	0.7900	0.5846	0.2054	0.665	1.367	1.412	0.265
5	1.05	1.02	0.0309	0.736	0.232	0.225	0.011
6	0.1100	0.3745	−0.2645	0.711	−1.894	−2.098	0.630
7	0.7500	0.6654	0.0846	0.664	0.562	0.549	0.045
8	0.9300	0.7486	0.1814	0.309	0.841	0.832	0.023
9	0.3900	0.5534	−0.1634	0.299	−0.752	−0.741	0.017
10	0.8700	0.8672	0.0028	0.260	0.013	0.012	0.000
11	0.2800	0.6352	−0.3552	0.504	−1.944	−2.171	0.275
12	0.6200	0.6609	−0.0409	0.703	−0.289	−0.280	0.014
13	0.4300	0.6702	−0.2402	0.349	−1.147	−1.160	0.050
14	0.6100	0.3260	0.2840	0.706	2.017	2.282	0.696
15	0.6900	0.8875	−0.1975	0.329	−0.929	−0.924	0.030
16	1.19	1.27	−0.0845	0.595	−0.511	−0.498	0.027
17	1.21	1.34	−0.1346	0.354	−0.645	−0.632	0.016
18	1.68	1.63	0.0469	0.728	0.346	0.336	0.023
19	0.7700	0.9171	−0.1471	0.691	−1.020	−1.022	0.167
20	1.13	1.03	0.1040	0.605	0.637	0.624	0.044
21	0.6400	0.7846	−0.1446	0.331	−0.682	−0.669	0.016
22	1.28	0.7742	0.5058	0.323	2.368	2.892	0.191
23	0.8900	0.7551	0.1349	0.718	0.978	0.977	0.174
24	0.8600	0.7589	0.1011	0.311	0.469	0.457	0.007
25	0.3800	0.5139	−0.1339	0.338	−0.634	−0.621	0.015
26	0.7400	0.7358	0.0042	0.276	0.019	0.018	0.000
27	1.30	1.22	0.0826	0.269	0.372	0.361	0.004
28	1.12	1.14	−0.0187	0.400	−0.093	−0.090	0.000
29	0.5400	0.7604	−0.2204	0.300	−1.015	−1.016	0.032
30	1.01	0.7308	0.2792	0.254	1.246	1.271	0.038

To better illustrate the relationship between the first two terms of this table, their values were plotted together and color-coded by dry weight (Figure 3).

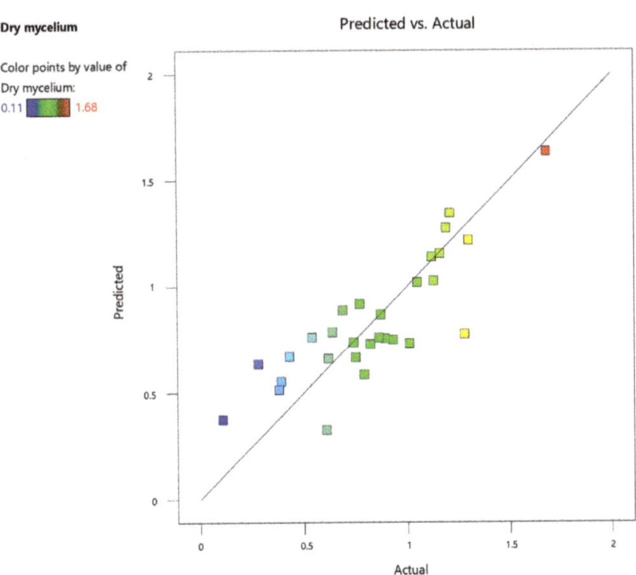

Figure 3. The predicted against actual results of dry mycelium weight (g/500 mL).

An analysis of variance between the predicted and actual dry mycelium weights was carried out for two-component interactions and for every factor (Table 4) to determine whether the response modeling was statistically significant and could reasonably predict an optimal broth composition.

Table 4. ANOVA for reduced quadratic × linear modeling of dry mycelium response.

Source	Sum of Squares	df	Mean Square	F-Value	*p*-Value
Model	2.42	13	0.1860	2.76	0.0316
[1] Linear Mixture	0.0752	3	0.0251	0.3721	0.7743
AB	0.1874	1	0.1874	2.78	0.1161
AC	0.0859	1	0.0859	1.28	0.2764
AD	0.1174	1	0.1174	1.74	0.2066
AG	0.0070	1	0.0070	0.1038	0.7518
BC	0.0356	1	0.0356	0.5291	0.4782
BD	0.0244	1	0.0244	0.3629	0.5559
BG	0.4574	1	0.4574	6.79	0.0199
CD	0.2314	1	0.2314	3.44	0.0836
CG	0.0031	1	0.0031	0.0456	0.8338
DG	0.0075	1	0.0075	0.1108	0.7439
Residual	1.01	15	0.0674		
Cor Total	3.43	28			

[1] Inference for linear mixtures uses Type I sums of squares.

The model F-value of 2.76 indicates the model is significant. There is only a 3.16% chance that an F-value this large can occur due to noise. The *p*-value of 0.0316 also indicates that the model is significant.

3.2. Predicted Optimum Nutrient Broth for I. obliquus Biomass Production

A mathematical analysis of the response, carried out using Design-Expert software, was able to generate a model of the broth component interactions and use it to extrapolate a theoretical response for a given combination of factors. The contour plot below (Figure 4), shows the combined effect of malt extract, yeast, and fructose on the dry weight response for a particular concentration of soluble starch.

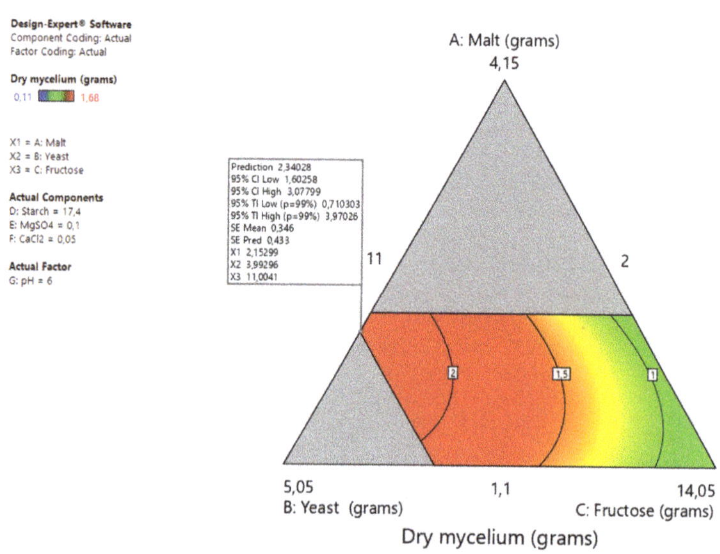

Figure 4. Contour plot of the modeled effect of broth components (g/500 mL) and interactions on *I. obliquus* dry mycelium weight.

By changing this last term, slightly higher or lower predicted responses fell within the threshold of the model. However, the modeled space remained largely proportional due to the orthogonal nature of the design and fixed constraints. A value of 17.4 g of soluble starch was found to elicit the best design space, from which a point with the highest dry weight prediction (2.34028 g/500 mL) was flagged to determine the values of other involved factors. The predicted optimum values (Table 5) were obtained easily due to the interdependence of the factors within the model.

Table 5. Optimized concentrations of broth components for dry mycelium yield.

Factor	Optimum Value (g/L)
Malt extract	2.15299
Yeast extract	3.99296
Fructose	11.0041
Soluble starch	17.40
$MgSO_4$	0.10
$CaCl_2$	0.05

3.3. Bioreactor Biomass Yield with Predicted Optimum and PDB

After culturing the submerged Chaga mycelium in the optimized broth, under strict aeration and agitation parameters guided by the Biostat B control tower according to the culture plan, we found the total recoverable dry biomass weight to be 34.96 g. This value corresponds to 2.185 g when scaled down to the 0.5 L of the optimization experiment, and surpasses its best result (1.68 g/500 mL).

A two-sided statistical test at a 95% confidence level, using the chosen optimum values for each factor and the previously established model fit, was carried out (Table 6) to determine the relationship between the data mean and the prediction intervals.

Table 6. Point prediction and confirmation of the dry mycelium response.

Analysis	Predicted Mean	Predicted Median	SD	n	SE Prediction	95% PI Low	Data Mean	95% PI High
Dry mycelium	2.34028	2.34028	0.259524	1	0.432599	1.41822	2.185	3.26234

Because the number of runs (n) was small, the spread of the low to high prediction interval is large (varying by ± 0.92206 g of the predicted mean) at the 95% confidence level. The data mean's value is 93.36% that of the predicted mean, so the difference between them is only equal to 6.64% of the predicted mean.

The culturing procedure was then repeated, under identical conditions, but using potato dextrose broth as a culture medium (Figure 5). Here we found the total dry mycelium to be 16.67 g. This value corresponds to approximately 1.042 g when scaled down to the 0.5 L of the optimization experiment.

Figure 5. Different macroscopic growth patterns of the 15 day aged bioreactor cultures of *I. obliquus* mycelium grown in potato dextrose broth (**left**), showing extracellular pigment secretion and dispersed globular morphology, and optimized broth (**right**), unpigmented occupation of the entire bioreactor.

4. Discussion

The submerged optimum broth bioreactor culture of *I. obliquus* mycelium yielded 93.36% of the dry weight prediction calculated by modeling 29 successful experimental cultures with different broth compositions. We studied the effect of six nutrient components and broth pH in a mixture-process type of experimental design and chose the highest yielding combination of factors using response surface methodology.

By using a stirred and aerated bioreactor, the cultivation period of the *I. obliquus* mycelium was reduced from 21 days (with static air exchange vessels in an incubator) to 15 days, all while the inoculum volume decreased from 4% to 3.125 % *v/v*. With these parameters, the mycelium grown in the optimized broth occupied the entire volume of the bioreactor, showing a 24.7-fold increase in dry weight at the end of the culture period.

We also grew *I. obliquus* mycelium in potato dextrose broth under the same bioreactor conditions to compare growth speed and final yield. Here we found that the culture grew more slowly, with an 11.78-fold increase in dry weight at the end of the cultivation period. We also found differences in the morphology of this mycelium, which accreted in dark-colored globules compared with the one grown in the optimized broth. Here, the color of the broth changed from a pale yellow at the beginning of cultivation to a reddish-brown at the end, indicating the possible release of soluble melanin pigments.

In a study of *I. obliquus* optimization, potato dextrose broth was also used in conjunction with submerged culture technics, yielding between 1.65 and 2.36 dry g/L for Difco PDB and fresh potato broth, respectively. The cultivation period lasted 12 days at 26 °C, with 100 mL flasks containing 25 mL of broth being used [18]. Our findings show similar yields (2.084 g/L) with 8 L of PDB inoculated in a stirred, aerated bioreactor culture.

In another study, shake flask cultures were performed with 250 mL flasks containing 50 mL of glucose, yeast extract, and $MgSO_4$-based broth. Cultivation lasted 8 days at 25 °C and 150 rpm, and we used an orthogonal design to vary the three broth components. Mycelium dry mass was found to be between 3.6 and 8.11 g/L. Furthermore, a 5 L stirred tank bioreactor was used (3.5 L working volume) at identical temperature and agitation speed for 10 days. This yielded a maximum of 8.24 g/L after 8 days [14]. Our results confirm high yields when using optimized liquid culture media containing yeast extract, $MgSO_4$, and a simple monosaccharide. However, we found that after 15 days of bioreactor culture with an optimized liquid medium, the entire 8 L volume was completely occupied by mycelium. The yields reported in the aforementioned study present a 1.88-fold increase in dry mycelium compared with a bioreactor culture that is already fully colonized.

For *I. obliquus*, we found that the maximum airflow of 0.2 L/min, combined with an agitation value of 75 rpm, was adequate for keeping the broth at an oxygen saturation of 50% for the final 3 days of culture.

In this paper, we provide a low-cost optimized broth formula for the rapid growth of *I. obliquus* mycelium and present a bioreactor method of culture based on incremental aeration and agitation.

The submerged cultivation of mycelium is well-suited for fast biomass production, and more so when paired with stirred tank bioreactor methods. Fast production of therapeutic metabolites such as betulin and ergosterol peroxide is possible by using this biotechnological approach.

A study of antioxidant constituents identified 22 extracellular and intracellular phenolic compounds in submerged cultures of *I. obliquus* grown in different culture media (control, H_2O_2, and arbutin-supplemented broths). The addition of hydrogen peroxide and arbutin increased the total levels of intracellular phenols and decreased these same compounds secreted extracellularly [8]. By this procedure, the mycelial biomass was enriched in antioxidant compounds. Such methods can easily be coupled with biomass accumulation strategies for a synergistic approach.

Author Contributions: A.P., E.V., and M.E. designed the experiments. M.E. was responsible for the administration of the project and resources. A.P. performed the experiments, analyzed the data and wrote the paper. The authors discussed and made comments on the results. All authors have read and agreed to the published version of the manuscript.

Funding: This work was supported by a grant of the Romanian Ministry of Research and Innovation, CCCDI—UEFISCDI, project number 171PED/2017 (PN-III-P2-2.1-PED-2016-0132) and 5PCCDI/2018 (PN-III-P1-1.2-PCCDI-2017-0323), within PNCDI III.

Institutional Review Board Statement: Not applicable.

Informed Consent Statement: Not applicable.

Data Availability Statement: Not applicable.

Conflicts of Interest: The authors declare no conflict of interest.

References

1. Kidd, P.M. The use of mushroom glucans and proteoglycans in cancer treatment. *Altern. Med. Rev.* **2000**, *5*, 4–27. [PubMed]
2. Jong, S.C.; Donovick, R. Antitumour and antiviral substances from fungi. *Adv. Appl. Microbiol.* **1989**, *34*, 183–262. [CrossRef] [PubMed]
3. Hou, X.; Chen, W. Optimization of extraction process of crude polysaccharides from wild edible BaChu mushroom by response surface methodology. *Carbohydr. Polym.* **2008**, *72*, 67–74. [CrossRef]
4. Hsieh, C.; Tsai, M.J.; Hsu, T.H.; Chang, D.M.; Lo, C.T. Medium optimization for polysaccharide production of *Cordyceps sinensis*. *Appl. Biochem. Biotechnol.* **2005**, *120*, 145–157. [CrossRef]
5. Jonathan, G.S.; Ayodele, S.; Damilola, A. Optimization of sub-merged culture conditions for biomass production in *Pleurotus florida* (Mont.) Singer, a Nigerian edible fungus. *Afr. J. Biotechnol.* **2006**, *5*, 1464–1469.
6. Wasser, S.P.; Elisashvili, V.; Tan, K.K. Effects of carbon and nitrogen sources in the medium on *Tremella mesenterica* Retz:Fr. (Heterobasidiomycetes) growth and polysaccharide production. *Int. J. Med. Mushrooms* **2003**, *5*, 49–56. [CrossRef]
7. Handa, N.; Yamada, T.; Tanaka, R. An unusual lanostane-type triterpenoid, spiroinonotsuoxodiol, and other triterpenoids from *Inonotus obliquus*. *Phytochemistry* **2010**, *71*, 1774–1779. [CrossRef] [PubMed]
8. Zheng, W.; Zhang, M.; Zhao, Y.; Wang, Y. Accumulation of antioxidant phenolic constituents in submerged cultures of *Inonotus obliquus*. *Bioresour. Technol.* **2009**, *100*, 1327–1335. [CrossRef] [PubMed]
9. Lee, I.K.; Yun, B.S. Highly oxygenated and unsaturated metabolites providing a diversity of hispidin class antioxidants in the medicinal mushrooms *Inonotus* and *Phellinus*. *Bioorg. Med. Chem.* **2007**, *15*, 3309–3314. [CrossRef] [PubMed]
10. Zheng, W.F.; Gu, Q.; Chen, C.F.; Yang, S.Z.; Wei, J.C.; Chu, C.C. Aminophenols and mold-water-extracts affect the accumulation of flavonoids and their antioxidant activity in cultured mycelia of *Inonotus obliquus*. *Mycosystema* **2007**, *26*, 414–426.
11. Babitskaya, V.G.; Shcherba, V.V.; Lkonnikova, N.V. Melanin complex of the fungus *Inonotus obliquus*. *Appl. Biochem. Microbiol.* **2000**, *36*, 377. [CrossRef]
12. Yang, S.Z.; Zheng, W.F. Factors affecting the accumulation of hydrolysable tannins in cultured mycelia of *Inonotus obliquus*. *Chin. Tradit. Herb. Drugs* **2007**, *38*, 798–803.
13. Kang, J.H.; Jang, J.E.; Mishra, S.K.; Lee, H.J.; Nho, C.W.; Shin, D.; Jin, M.; Kim, M.K.; Choi, C.; Oh, S.H. Ergosterol peroxide from Chaga mushroom (*Inonotus obliquus*) exhibits anti-cancer activity by down-regulation of the β-catenin pathway in colorectal cancer. *J. Ethnopharmacol.* **2015**, *173*, 303–312. [CrossRef] [PubMed]
14. Bai, Y.H.; Feng, Y.Q.; Mao, D.B.; Xu, C. Optimization for betulin production from mycelial culture of *Inonotus obliquus* by orthogonal design and evaluation of its antioxidant activity. *J. Taiwan Inst. Chem. Eng.* **2012**, *43*, 663–669. [CrossRef]
15. Chen, H.; Xu, X.; Zhu, Y. Optimization of hydroxyl radical scavenging activity of exo-polysaccharides from *Inonotus obliquus* in submerged fermentation using response surface methodology. *J. Microbiol. Biotechnol.* **2010**, *20*, 835–843. [PubMed]
16. Xiang, C.; Xiang, Q.X.; Ling, H.Z. Submerged culture optimisation for triterpenoids production by *Inonotus obliquus* using response surface methodology. *Adv. Mat. Res.* **2012**, *560–561*, 784–790. [CrossRef]
17. Kūka, M.; Čakste, I.; Geršebeka, E. Determination of bioactive compounds and mineral substances in Latvian birch and maple saps. *Proc. Latv. Acad. Sci. Sect. B Nat. Exact Appl. Sci.* **2013**, *67*, 437–441. [CrossRef]
18. Cho, N.S.; Shin, Y.S. Optimization of in vitro cultivation of *Inonotus obliquus*. *J. Korean Wood Sci. Technol.* **2005**, *33*, 92–98.

Article

Determination of the Qualitative Composition of Biologically Active Substances of Extracts of In Vitro Callus, Cell Suspension, and Root Cultures of the Medicinal Plant *Rhaponticum carthamoides*

Lyudmila Asyakina [1,2], Svetlana Ivanova [3,4,*], Alexander Prosekov [5], Lyubov Dyshlyuk [1,3], Evgeny Chupakhin [2], Elena Ulrikh [6], Olga Babich [2] and Stanislav Sukhikh [1,2]

1   Department of Bionanotechnology, Kemerovo State University, Krasnaya Street 6, Kemerovo 650043, Russia; alk_kem@mail.ru (L.A.); soldatovals1984@mail.ru (L.D.); stas-asp@mail.ru (S.S.)
2   Institute of Living Systems, Immanuel Kant Baltic Federal University, A. Nevskogo Street 14, Kaliningrad 236016, Russia; ECHupakhin@kantiana.ru (E.C.); oobabich@kantiana.ru (O.B.)
3   Natural Nutraceutical Biotesting Laboratory, Kemerovo State University, Krasnaya Street 6, Kemerovo 650043, Russia
4   Department of General Mathematics and Informatics, Kemerovo State University, Krasnaya Street 6, Kemerovo 650043, Russia
5   Laboratory of Biocatalysis, Kemerovo State University, Krasnaya Street 6, Kemerovo 650043, Russia; a.prosekov@inbox.ru
6   Kuzbass State Agricultural Academy, Markovtseva Street, 5, Kemerovo 650056, Russia; elen.ulrich@mail.ru
*   Correspondence: pavvm2000@mail.ru; Tel.: +7-384-239-6832

Citation: Asyakina, L.; Ivanova, S.; Prosekov, A.; Dyshlyuk, L.; Chupakhin, E.; Ulrikh, E.; Babich, O.; Sukhikh, S. Determination of the Qualitative Composition of Biologically Active Substances of Extracts of In Vitro Callus, Cell Suspension, and Root Cultures of the Medicinal Plant *Rhaponticum carthamoides*. *Appl. Sci.* **2021**, *11*, 2555. https://doi.org/10.3390/app11062555

Academic Editor: Emanuel Vamanu

Received: 14 February 2021
Accepted: 9 March 2021
Published: 12 March 2021

Publisher's Note: MDPI stays neutral with regard to jurisdictional claims in published maps and institutional affiliations.

Abstract: This work aims to study the qualitative composition of biologically active substance (BAS) extracts in vitro callus, cell suspension, and root cultures of the medicinal plant *Rhaponticum carthamoides*. The research methodology is based on high-performance liquid chromatography, and 1H nuclear magnetic resonance (NMR) spectra, to study the qualitative and quantitative analysis of BAS. The results of the qualitative composition analysis of the dried biomass extracts of in vitro callus, cell suspension and root cultures showed that the main biologically active substances in the medicinal plant *Rhaponticum carthamoides* are 2-deoxy-5,20,26-trihydroxyecdyson (7 mg, yield 0.12%), 5,20,26-trihydroxyecdyson 20,22-acetonide (15 mg, yield 0.25%), 2-deoxy-5,20,26-trihydroxyecdyson 20,22-acetonide (6 mg, yield 0.10%), 20,26-dihydroxyecdyson 20,22-acetonidecdyson 20,22-acetonide (5 mg, yield 0.09%), and ecdyson 20,22-acetonide (6 mg, yield 0.10%). In the future, it is planned to study the antimicrobial, antioxidant, and antitumor activity of BAS of extracts of in vitro callus, cell suspension, and root cultures of the medicinal plant *Rhaponticum carthamoides*, for the production of pharmaceuticals and dietary supplements with antitumor, antimicrobial and antioxidant effects.

Keywords: in vitro callus; cell suspension and root cultures; biologically active substances; ecdysteroids; HPLC; ^{1}H NMR spectra; *Rhaponticum carthamoides*

1. Introduction

Medicinal plants are sources of large amounts of biologically active substances (BAS), such as antioxidants, alkaloids, phenylpropanoids, terpenoids, and many others. Each of these compounds, to a greater or lesser extent, have medicinal properties and potential pharmaceutical applications [1–4]. Some compounds have several properties at once; for example, antioxidant, adaptogenic, anticancer, and even anti-aging properties are attributed to ginsenosides [5]. Ginseng is good at reducing fatigue and increasing resistance to cancer [5]. Tanshinones (diterpenoids) and salvianolic acid (phenylpropanoids) have antioxidant, anti-inflammatory, antibacterial, antitumor, and cytotoxic properties, and are actively used in the comprehensive treatment of cardiovascular diseases [6]. Sesquiterpene-artemisinin is effectively used in antimalarial therapy [7].

175

It is believed that plants have many natural compounds, such as terpenoids (isoprenoids, more than 50,000 different structures are distinguished), the production of which has two pathways (the cytosolic mevalonate or plastid methylerythritol phosphate) [8,9]. The cytosolic mevalonate pathway is considered to be the main pathway for producing sesquiterpenoids (in particular artemisinin) and triterpenoids (ginsenosides) [7,10–14], while the plastid methylerythritol phosphate pathway is considered to be the main pathway for diterpenoids (such as tanshinones and taxol) and monoterpenoids (such as limonene) [9,13,15]. Phenylpropanoids (anthocyanin, catechin, salvianolic acid, etc.) are derivatives of the shikimate pathway, synthesized from phenylalanine and tyrosine [6,16].

It is known that plants often synthesize biologically active compounds in small amounts [1], and only in some of the tissues [10,17]. It was established that compounds such as tanshinones, ginsenosides and flavones are formed and accumulate in the roots of *Salvia miltiorrhiza*, ginseng (*Panax*), and *Scutellaria baicalensis* [10,18,19]. It is known that catharantin accumulates in all tissues and organs, while vinblastine and vincristine only accumulate in the aerial parts of the *Catharanthus roseus* plant [20,21]. In *Artemisia annua*, artemisinin is formed in the glandular secretory trichomes of leaves [7]. Natural medicinal plants often require many years of growth before biologically active compounds are synthesized in their parts, for example, in the roots, usually in very small doses [22]. Moreover, the environmental and ecological problems lead to a deterioration in the quality characteristics of both plants and their derivatives [23,24]. The study and research of alternative methods of obtaining compounds with useful and unique characteristics from plant materials is relevant. The biosynthesis of in vitro transformed callus, cell suspension, and root cultures (hairy roots), from the point of view of the controlled production of biologically active substances with desired properties, has potential, and is of interest to both researchers and producers [25].

It is known that hairy roots grow faster than any roots of intact plants [26,27] and accumulate a higher content of valuable compounds in their cells [19,28,29]. The total content of tanshinone in the hairy roots of transgenic *S. miltiorrhiza* was several times higher than the content in the roots of field plants (15.4 and 1.7–9.7 mg/g dry weight, respectively) [19,28]. The total content of vilforin in the hairy roots of *Tripterygium wilfordii* Hook.f. was significantly higher than in the adventitious roots [29]. There is an opinion that various BAS with new characteristics can be produced in vitro using hairy root cultures. New compounds of cadaverine and natural triterpene saponins, which were absent in both the leaves and roots of intact plants, were found in hairy roots of *Brugmansia candida* and *Medicago truncatula* when varying the parameters of the biosynthesis process.

It was also found that in vitro hairy root cultures can be used as model systems, both for identifying new genes and for fast characterization of their functions. It is possible to control BAS biosynthesis using genetic engineering methods (genome editing), and obtain genetically modified hairy root cultures. Blocking via the RNA biotransformation of the initial precursor, with the simultaneous optimization of substrates with unique compositions, makes it possible to construct hairy root cultures and produce specific BAS artificially.

At present, it is of great interest to use in vitro callus, cell suspension, and root cultures of medicinal plants as biologically active additives to increase the economic efficiency of the pharmaceutical industry. One of such plants is *Rhaponticum carthamoides*. It contains the following amino acids: aspartic (3.5–3.9 g/100 g), glutamic (2.5–3.0 g/100 g), leucine (1.9–2.0 g/100 g), tyrosine (1.2–1.4 g/100 g), lysine (1.6 g/100 g), vitamin C (62.0–77.0 mg/%), vitamin E (6.2 mg/%), carotenoids (65.0–113.0 mg/%), PP (vitamin PP is nicotinic acid, 11.5 mg/%), ecdysteroids (418.0–2170.0 mg/kg), and flavonoids (3.0–7.2% of dry matter). Due to the presence of various biologically active substances, the study of the qualitative and quantitative composition of in vitro callus, cell suspension, and root cultures of *Rhaponticum carthamoides* is a relevant task [30].

This work aims to study the possibility of producing and using extracts of BAS from *Rhaponticum carthamoides* callus, cell suspension, and root cultures as pharmaceutical substances.

2. Materials and Methods

2.1. Research Objects

BAS complexes isolated from extracts of freeze-dried biomass of in vitro callus, cell suspension, and root cultures of the medicinal plant *Rhaponticum carthamoides* (family *Compositae*), collected in the Kemerovo region (Siberia, Russia) in 2020, were the objects of this research. To obtain the biomass of in vitro callus, cell suspension, and root cultures, the seeds of *Rhaponticum carthamoides* were pre-washed with a detergent (acetic acid 3%), then immersed for 1 min in a 75% ethanol solution, transferred to a laminar box, and sterilized for 15 min in a 20% sodium hypochlorite solution (5% of active chlorine). All chemical reagents were purchased from Akvilon (Moscow, Russia). After sterilization, the sterilizing substance was washed off; for this, the seeds were washed for 20 min in distilled sterile water three times. The explants were then placed in a sterile 100 mL flask with 30 mL of Murashige-Skoog culture medium containing 3% sucrose, 0.7% agar-agar (Laverna XXI vek, Moscow, Russia), without growth stimulants, and were illuminated by compact fluorescent lamps (Economy 11W/865 11W E27 3U 6500K 6y CDL Philips 871150031502110 (Philips, Eindhoven, The Netherlands)) while maintaining a temperature of 25 °C. Seedlings that were 1.5-month-old were used for transformation. Lyophilization of germinated biomass of in vitro callus, cell suspension, and root cultures was carried out using a Triad freeze-dryer by Labconco (Kansas City, Missouri, USA). Lyophilization conditions: vacuum 0.05 mbar and cooler temperature −80 °C. The extracts were obtained as follows: a weighed portion of the studied biomass sample was weighed on an analytical balance (Oxaus PX85, New York, NY, USA), and transferred into a polyethylene Falcon tube. An organic solvent (ethanol) was added in an amount of 1:5 according to the experimental procedure, and the extraction process was carried out. The duration and temperature of the experiment varied up to 360 min and from 25 °C to boiling, respectively. Furthermore, the filtration process was carried out, followed by centrifugation of the filtrate at a rotor speed of 3900 ± 100 rpm. The filtrate was centrifuged in a PE-6900 centrifuge (Ekros, Moscow, Russia) to remove suspended particles. The solvent was evaporated from the extract on an IKA RV 8 V rotary evaporator (IKA, Staufen, Germany), under reduced pressure from a 100 mL flask pre-weighed on a CAS CUW420H balance (CAS Corporation Ltd., Seoul, Korea). The flask was weighed, and the yield of the extract was determined [31].

2.2. Drying of the BAS Complex

The drying of the BAS complex, isolated from lyophilized biomass extracts of in vitro callus, cell suspension, and root cultures, was also carried out by lyophilization. Lyophilization was performed using a Triad freeze-dryer by Labconco (USA). Lyophilization conditions made it possible to optimize the temperature and drying time of the samples. The stable conditions for lyophilization were selected: vacuum 0.05 mbar and temperature of the cooler −80 °C. The temperature regime and duration of the lyophilization process were individually selected for each sample. The residual solvent content was the controlled parameter [3].

2.3. Separation and Identification of Individual BAS

The isolated BAS complexes, from the lyophilized biomass extracts of in vitro callus, cell suspension, and root cultures of *Rhaponticum carthamoides*, were additionally separated by preparative HPLC, using a Shimadzu chromatograph (Shimadzu, Kyoto, Japan), flow rate 10 mL/min, phase A–B gradient 1–90% in 15 min, phase A—0.1% trifluoroacetic acid, B—acetonitrile (Laverna XXI vek, Moscow, Russia) [32], and a column ZORBAX Eclipse XDB-C18 Semi-Preparative 250 mm × 9.4 mm × 5 μm. Each fraction was evaporated to dryness, weighed, the yield was determined, and the structure of the compounds was identified by ^{1}H NMR spectrometry.

To identify the BAS of in vitro callus, cell suspension, and root culture extracts of *Rhodiola rosea*, a mixed stock solution was prepared immediately before the experiment, containing 1 mg/mL of each biologically active substance in ethanol. Standard solutions, prepared by serial dilution to the final concentration (from 0.1 to 100.0 µg/mL) of the stock solution with ethanol, were used to construct the calibration curve with R^2 0.987.

The solutions were chromatographed and eluted. We used a H_2O:MeCN eluent system, with an acetonitrile gradient of 0–20% with a step of 2%. Trifluoroacetic acid (Lavrena XXI vek, Moscow, Russia), in an amount of 0.1%, was used as a modifier. The content of each BAS was calculated based on the calibration curves of the relationship between the peak regions and the concentrations of the standard solutions.

^{1}H NMR spectra were obtained using a Bruker Avance NMR spectrometer (Bruker, Leipzig, Germany), with an operating frequency of 500 MHz [33]; $CDCl_3$ (Chloroform-*d*) was used as a solvent for all compounds (Laverna XXI vek, Moscow, Russia).

2.4. Statistical Analysis

All experimentations were achieved in triplicates and results were given as a mean. The differences in the extracts were investigated by using student t-test ($p < 0.05$), and this test was performed in Statistica 10.0 (StatSoft Inc., 2007, Tulsa, OK, USA).

3. Results

Figure 1 shows a chromatogram after fractionation of the extract of in vitro callus, cell suspension, and root cultures of *Rhaponticum carthamoides* by HPLC.

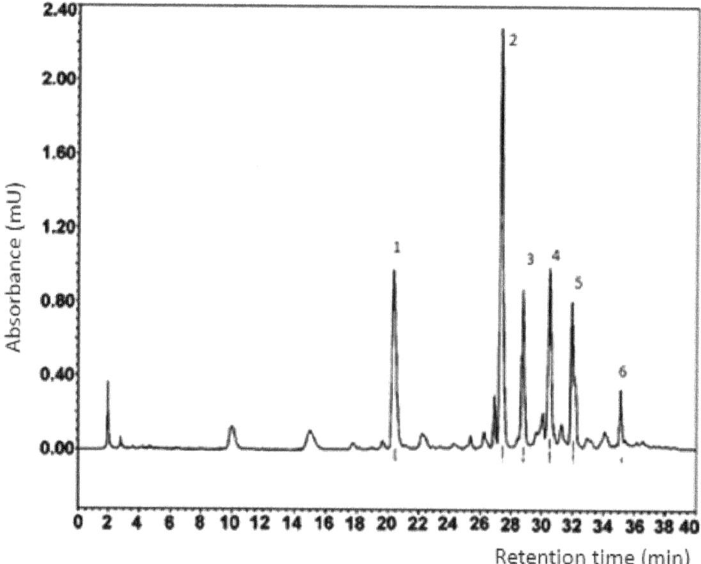

Figure 1. Results of preparative separation of the ecdysteroid fraction of in vitro callus, cell suspension, and root cultures of *Rhaponticum carthamoides*.

For each BAS compound of in vitro callus, cell suspension and root cultures of *Rhaponticum carthamoides*, which corresponds to one of the numbered peaks (for small peaks, the yield of compounds was less than 0.5 mg, which did not allow identifying their structure by NMR), the structure was established by ^{1}H NMR (see ESI with correlation of proton signals in NMR spectra and number, Figures S1–S5). The formulas of the identified substances are presented in Figure 2.

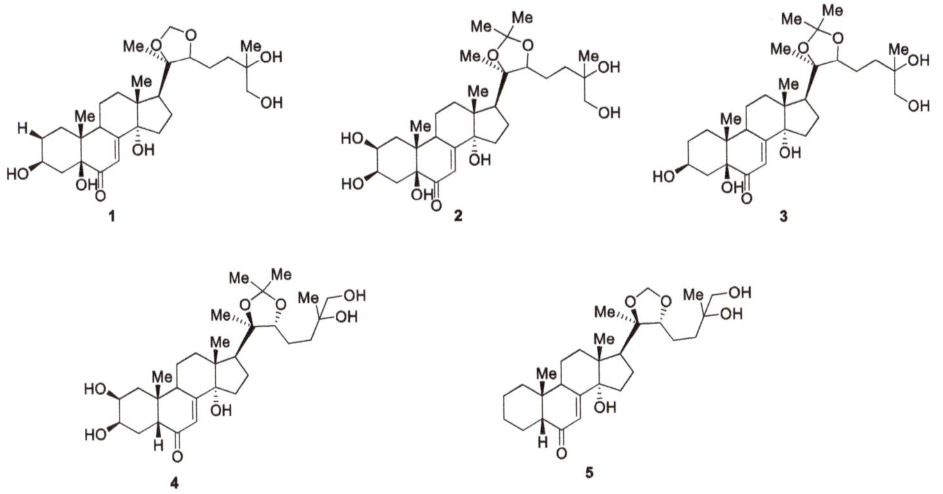

Figure 2. Formulas of compounds present in the extract of callus, cell suspension, and root cultures of *Rhaponticum carthamoides*, identified by NMR spectra (the formula number corresponds to the peak number).

Identification of compounds of in vitro callus, cell suspension, and root cultures of BAS complex extracts from *Rhaponticum carthamoides* dried biomass by HPLC (Figure 1) resulted in a determination of the ecdysteroid fractions, in particular, 2,3,5,14,20,22,25-Heptahydroxycholest-7-en-6-one, with a mobility factor of 0.5 units.

For compound No. 1 (2-deoxy-5,20,26-trihydroxyecdyson), 7 mg of the substance was obtained, and the yield was 0.12% (Appendix A). For compound No. 2 (5,20,26-trihydroxyecdyson 20,22-acetonide), 15 mg of the substance was obtained, and the yield was 0.25% (Appendix B). For compound No. 3 (2-deoxy-5,20,26-trihydroxyecdyson 20,22-acetonide), 6 mg was isolated, and the yield was 0.10% (Appendix C). For compound No. 4 (20,26-dihydroxyecdyson 20,22-acetonide), 5 mg was isolated, and the yield was 0.09% (Appendix D). For compound No. 5 (ecdyson 20,22-acetonide), 6 mg was isolated, and the yield was 0.10% (Appendix E). The identified compounds are presented in Appendices A–E

It was not possible to identify compound No. 6 of the chromatogram (Figure 1); the NMR and mass spectrum have complex sets of signals, indicating a mix of two substances.

4. Discussion

The results of the qualitative composition analysis of the dried biomass extracts of in vitro callus, cell suspension, and root cultures, showed that the main BAS of the medicinal plant *Rhaponticum carthamoides* are 2-deoxy-5,20,26-trihydroxyecdyson, 5,20,26-trihydroxyecdyson 20,22-acetonide, 2-deoxy-5,20,26-trihydroxyecdyson 20,22-acetonide, 20,26-dihydroxyecdyson 20,22-acetonidecdyson and 20,22-acetonide. Stereochemistry for corresponding isomers of ecdysone was determined by the value of the spin–spin constant; for compound 4 and 5, value J increased, and this fact and conclusion led to the conclusion about the diastereomeric nature of the obtained compounds. We also took into account the values of the spin–spin interaction constant of the methylene group and protons of the cyclopentanoperhydrophenanthrene system. In this case, a decrease in the value of the spin–spin interaction constant was observed, which indicates the trans-configuration of methylene groups. In the side chain, a close value of the spin–spin interaction and chemical shift constants, with already known compounds, was found. From this, an assumption was made about a similar character of stereoisomerism for the obtained compounds.

In the root culture of *Leuzea carthamoides*, Yu et al. [33] found three new ecdysteroids (polypodine B 20,22-acetonide, 20-hydroxyecdysone 2,3; 20,22-diacetonide and isovitexiron together with 20-hydroxyecdysone, 20-hydroxyecdysone 2,3-acetonide, 20-hydroxyecdysone 20,22-acetonide, ajugasterone C, macisterone A and polypodyne B).

The described data confirm the presence of ecdysteroids in the *Rhaponticum carthamoides* root cultures. The molecular structure of each compound was determined by NMR spectroscopy and HRMS analysis. A method for the isolation of an individual compound with high brightness and selectivity, and subsequent identification by NMR spectra, which characterize the magnitude of the chemical shift signal and the multiplet structure, is described

The BAS profile of callus, cell suspension, and root cultures of *Thevetia peruviana* (an ornamental shrub growing in many tropical regions of the world) is described [34]. The biologically active compounds of this plant have unique characteristics and can be used as pharmaceutical substances for drugs under development. The research objects were the 50% aqueous ethanol and ethyl acetate extracts. The work used thin-layer chromatography and standard chemical tests for phytochemical analysis. High-performance liquid chromatography was used to analyze the phenolic chemical profile. The total amount of phenolic and flavonoid, the total amount of cardiac glycosides, and the total antioxidant activity in callus, cell suspension, and root cultures, were determined. All the samples under study had common biological activity; antioxidants, amino acids, alkaloids, flavonoids, phenols, cardiac glycosides, leukoanthocyanidins, triterpenes, and sugars were found in their composition. Dihydroquercetin, which has anticancer activity, has been identified using high-performance liquid chromatography. The presented results demonstrate the usefulness of *T. peruviana* callus, cell suspension, and root cultures for the production of valuable pharmaceutical compounds. The data described in [34] confirm the high accumulation of flavonoids and ecdysteroids in in vitro callus, cell suspension, and root cultures of medicinal plants.

In [35], ginseng callus, cell suspension and root cultures, and their extracts (alcohol content 30–70%), were studied. Organic compounds were determined by thin-layer chromatography. For each plant, quercetin, magneferin, luteolin, rutin, quercetin-2-D-glucoside, malvidin, caffeic, cinnamic, ferulic, and sinapic acids were identified. The described results confirm the high accumulation of flavonoids and ecdysteroids in the extracts of in vitro callus, cell suspension, and root cultures of medicinal plants as well.

5. Conclusions

The qualitative composition analysis of extracts of callus, cell suspension, and root cultures showed that ecdysteroids and flavonoids are the most promising BAS, from the point of view of industrial and technological production. These compounds make the greatest contribution to the BAS complex of extracts of callus, cell suspension, and root cultures; their biological activity has been established, and their technological production is cost-effective, since it allows these compounds to be sold on the existing market, thereby reducing economic risk. Many people nowadays prefer natural dietary supplements to synthetic medicines. Therefore, the in vitro biotechnological production of callus, cell suspension, and root cultures, under controlled conditions, represents a cost-effective way for the commercial mass production of phytochemicals. The studied extracts of callus, cell suspension, and root cultures of the medicinal plant *Rhaponticum carthamoides* are readily available and are considered effective, with fewer side effects compared to modern drugs, in the treatment of various diseases.

Substances such as flavonoids, ecdysteroids and anthocyanins found in the in vitro callus, cell suspension, and root cultures of *Rhaponticum carthamoides* have antioxidant activity, the ability to trap free radicals, and cardioprotective, antidiabetic, anti-inflammatory and anti-allergic effects, while some other phytonoid and ecdysteroid compounds show potential antiviral activity. Recently, the flavonoids and ecdysteroids of *Rhaponticum carthamoides* were proved to be the most effective anticancer agents, due to apoptosis, which causes cell cycle arrest and the inhibition of key enzymes involved in tumor promotion. The use of flavonoids and ecdysteroids as potential drugs for the prevention of many chronic diseases is a topical trend in pharmacology [36].

In the future, it is planned to study the antimicrobial, antioxidant, and antitumor activity of the BAS of extracts of in vitro callus, cell suspension, and root cultures of the medicinal plant *Rhaponticum carthamoides* for the production of pharmaceuticals and dietary supplements with antitumor, antimicrobial and antioxidant effects.

Supplementary Materials: The following are available online at https://www.mdpi.com/2076-3417/11/6/2555/s1, Figure S1: ^{1}H NMR spectrum of identified compound 2-deoxy-5,20,26-trihydroxyecdyson; Figure S2: ^{1}H NMR spectrum of identified compound 5,20,26-trihydroxyecdyson 20,22-acetonide/; Figure S3: ^{1}H NMR spectrum of identified compound 2-deoxy-5,20,26-trihydroxyecdyson 20,22-acetonide; Figure S4: ^{1}H NMR spectrum of identified compound 20,26-dihydroxyecdyson 20,22-acetonide; Figure S5: ^{1}H NMR spectrum of identified compound ecdyson 20,22-acetonide.

Author Contributions: S.I., A.P. and O.B. conceived and designed the research; L.A., L.D. and S.S. analyzed and interpreted the data; L.D., E.U. and S.S. contributed reagents, materials, analysis tools or data; S.I., A.P., E.C., E.U. and O.B. wrote the paper. All authors have read and agreed to the published version of the manuscript.

Funding: This research was funded by RUSSIAN SCIENCE FOUNDATION, grant number 18-75-10066.

Institutional Review Board Statement: Not applicable.

Informed Consent Statement: Not applicable.

Data Availability Statement: Not applicable.

Conflicts of Interest: The authors declare no conflict of interest.

Appendix A. The Identified Compound No. 1 (1 2-deoxy-5,20,26-trihydroxyecdyson)

$[\alpha]_D^{25}$ +41 (c 0.05 MeOH); UV (MeOH) λmax (log ε) 243 nm (4.02) nm; ^{1}H NMR 1H NMR (500 MHz, Chloroform-d) δ 6.02 (*d*, *J* = 1.1 Hz, 1H), 4.81 (s, 1H), 4.74 (*d*, *J* = 12.4 Hz, 1H), 4.59–4.51 (m, 2H), 4.43 (*d*, *J* = 7.9 Hz, 1H), 4.27 (s, 1H), 4.10–3.99 (m, 2H), 3.73–3.66 (m, 1H), 3.58–3.47 (m, 2H), 2.20–2.08 (m, 2H), 2.08–2.01 (m, 2H), 1.92–1.50 (m, 15H), 1.50–1.40 (m, 1H), 1.33 (t, *J* = 1.5 Hz, 3H), 1.26 (s, 3H), 1.10 (*d*, *J* = 1.4 Hz, 3H), 0.86 (s, 3H).
HRESIMS m=z 497.3108 (calculated for C27H45O8, 497.3102).

Appendix B. The Identified Compound No. 2 (5,20,26-trihydroxyecdyson 20,22-acetonide)

$[R]_D^{25}$ +89 (c 0.05 MeOH); UV (MeOH) λmax (log ε) 242 (3.76) nm; ^{1}H NMR (500 MHz, Chloroform-d) δ 5.86 (1H, *d*, *J* 2.8 Hz, CH), 3.99 (1H, q, *J* 3.0 Hz, CH), 3.95 (1H, ddd, *J* 10.0, 7.4, 3.6 Hz, CH), 3.695 (1H, t, *J* 6.0 Hz, CH), 3.375 (1H, *d*, *J* 11.0 Hz, CH), 3.355 (1H, *d*, *J* 11.0 Hz, CH), 3.19 (1H, ddd, *J* 11.3, 7.0, 2.7 Hz, OH), 2.31 (1H, *dd*, *J* 9.4, 8.1 Hz, OH), 2.12 (1H, *td*, *J* 13.1, 5.0 Hz, CH), 2.075 (1H, *dd*, *J* 14.7, 3.0 Hz, CH), 2.03 (1H, m, CH), 1.96 (1H, *dd*, *J* 12.4, 6.5 Hz, CH), 1.87 (1H, m, CH), 1.86 (1H, m, CH), 1.81 (1H, m, CH), 1.77 (1H, *dd*, *J* 14.9, 3.0 Hz, CH), 1.74 (1H, m, CH), 1.73 (2H, m, CH), 1.71 (1H, m, CH), 1.61 (1H, m, CH), 1.55 (1H, m, CH), 1.53 (1H, m, CH), 1.52 (1H, m, CH), 1.39 (3H, s, CH3), 1.32 (3H, s, CH3), 1.18 (3H, s, CH3), 1.15 (3H, s, CH3), 0.915 (3H, s, CH3), 0.83 (3H, s, CH3); 13C NMR (CD3OD, 125 MHz) δ 202.5, 167.4, 120.7, 108.2, 86.0, 85.3, 83.6, 80.4, 73.6, 70.7, 70.4, 68.6, 50.5, 48.7, 45.5, 39.2, 37.2, 36.3, 34.3, 32.6, 31.8, 29.5, 27.3, 24.0, 23.9, 22.7, 22.65, 22.5, 17.8, 17.1; ESIMS m/z 575 [M + Na]+ (46), 553 [M + H]+ (100), 537 [M - CH3]+ (5), 535 [M + H - H2O]+ (2), 520 [M + H - H2O - CH3]+ (2), 495 [M + H - acetone]+ (59), 481 (1), 477 (3), 437 (3), 359 (3), 328 (14).
HRESIMS m=z 553.3366 (calculated for C30H49O9, 553.3363).

Appendix C. The Identified Compound No. 3 (2-deoxy-5,20,26-trihydroxyecdyson 20,22-acetonide)

$[R]_D^{25}$ +25 (c 0.05 MeOH); **UV (MeOH) λmax (log ε) 238 (4.08) nm;** ^{1}**H NMR (CD3OD, 500 MHz, Chloroform-d)** δ 5.86 (1H, s, br, OH), 4.08 (1H, s, br, CH), 3.70 (1H, m, OH), 3.37 (1H, *d*, *J* 11.0 Hz, CH), 3.36 (1H, *d*, *J* 11.0 Hz, CH), 3.28 (1H, m, CH), 2.32 (1H, *t*, *J* 8.7 Hz,

CH), 2.12 (1H, *td*, *J* 12.4, 5.7 Hz, CH), 2.04 (1H, m, CH), 2.035 (1H, m, CH), 1.97 (1H, m, CH), 1.96 (1H, m, CH), 1.88 (1H, m, CH), 1.86 (1H, m, CH), 1.84 (1H, *m*, CH), 1.77 (1H, m, CH), 1.73 (1H, m, CH), 1.72 (1H, m, CH) 1.61 (2H, m, CH2), 1.55 (2H, *m*, CH2), 1.53 (1H, m, CH), 1.50 (1H, m, CH), 1.39 (3H, s, CH3), 1.32 (3H, s, CH3), 1.18 (3H, s, CH3), 1.15 (3H, s, CH3), 0.89 (3H, s, CH3), 0.83 (3H, s, CH3); 13C NMR (CD3OD, 125 MHz) δ 167.9, 120.7, 108.2, 86.0, 85.4, 83.6, 81.2, 73.6, 70.7, 67.2, 50.6, 48.7, 43.25, 38.1, 37.2, 36.9, 32.6, 31.8, 29.5, 29.3, 27.3, 25.6, 24.05, 23.8, 22.7, 22.5, 22.5, 17.8, 17.3; ESIMS m/z 559 [M + Na]+ (100), 537 [M + H]+ (36), 518 [M - H2O]+ (3), 541 [M + Na - H2O]+ (12), 501 [M + H - 2H2O]+ (2), 445 (10), 385 [M + H - H2O - C6O3H14]+ (3), 315 (12), 304 (24).
HRESIMS m=z 537.3420 (calculated for C30H49O8, 537.3414).

Appendix D. The Identified Compound No. 4 (20,26-dihydroxyecdyson 20,22-acetonide)

$[R]_D^{25}$ +145 (c 0.005 MeOH); **UV (MeOH)** λmax (log ε) 242 (4.01) nm; **^{1}H NMR (CD3OD, 500 MHz, Chloroform-d)** δ 5.86 (1H, *d*, *J* 2.6 Hz, CH), 3.70 (1H, m, OH), 3.38 (1H, d, J 10.9 Hz, CH), 3.36 (1H, *d*, *J* 11.0 Hz, CH), 2.42 (1H, *dd*, *J* 12.6, 4.0 Hz, CH), 2.33 (1H, *dd*, *J* 9.2, 8.6 Hz, CH), 1.39 (3H, s, CH3), 1.32 (3H, s, CH3=), 1.18 (3H, s, CH3), 1.15 (3H, s, CH3), 0.96 (3H, s, CH3), 0.83 (3H, s, CH3); 13C NMR (CD3OD, 125 MHz) δ 121.8, 85.9, 85.5, 83.6, 73.2, 70.7, 50.5, 49.3, 37.1, 32.55, 29.4, 27.3, 24.4, 23.7, 22.7, 17.8; ESIMS m/z 575 [M + K]+ (14), 560 [M + H + Na]+ (6), 559 [M + Na]+ (5), 542 (100), 521 [M - CH3]+ (23), 519 [M + H - H2O]+ (2), 503 [M - CH3 - H2O]+ (7), 501 [M + H - 2H2O]+ (23), 478 [M - acetone]+ (4), 445 (14), 413 (6), 314 (10), 304 (55).
HRESIMS m=z 537.3418 (calculated for C30H48O8, 537.3414).

Appendix E. The Identified Compound No. 5 (ecdyson 20,22-acetonide)

UV (MeOH) λmax (log ε) 245 (6.01) nm; **^{1}H NMR (CD3OD, 500 MHz, Chloroform-d)** δ 3.70 (1H, m, OH), 3.38 (1H, *d*, *J* 10.9 Hz, CH), 3.36 (1H, *d*, *J* 11.0 Hz, CH), 2.42 (1H, *dd*, *J* 12.6, 4.0 Hz, CH), 2.33 (1H, dd, J 9.2, 8.6 Hz, CH), 1.39 (3H, s, CH), 1.32 (3H, s, CH), 1.18 (3H, s, CH), 1.15 (3H, s, CH), 0.96 (3H, s, CH), 0.83 (3H, s, CH); **13C NMR (CD3OD, 125 MHz)** δ 121.8, 85.9, 85.5, 83.6, 73.2, 70.7, 50.5, 49.3, 37.1, 32.55, 29.4, 27.3, 24.4, 23.7, 22.7, 17.8.
HRESIMS m = z 476.3278 (calculated for C28H44O6, 476.3138).

References

1. Ashour, M.; Wink, M.; Gershenzon, J. Biochemistry of terpenoids: Monoterpenes, sesquiterpenes and diterpenes. *Annu. Plant Rev.* **2010**, *40*, 258–303.
2. Atanasov, A.G.; Waltenberger, B.; Pferschy-Wenzig, E.-M.; Linder, T.; Wawrosch, C.; Uhrin, P. Discovery and resupply of pharmacologically active plant-derived natural products: A review. *Biotechnol. Adv.* **2015**, *33*, 1582–1614. [CrossRef] [PubMed]
3. Babich, O.; Prosekov, A.; Zaushintsena, A.; Sukhikh, A.; Dyshlyuk, L.; Ivanova, S. Identification and quantification of phenolic compounds of Western Siberia *Astragalus danicus* in different regions. *Heliyon* **2019**, *5*, e02245. [CrossRef] [PubMed]
4. Babich, O.; Sukhikh, S.; Prosekov, A.; Asyakina, L.; Ivanova, S. Medicinal Plants to Strengthen Immunity during a Pandemic. *Pharmaceuticals* **2020**, *13*, 313. [CrossRef]
5. Babich, O.; Sukhikh, S.; Pungin, A.; Ivanova, S.; Asyakina, L.; Prosekov, A. Modern Trends in the In Vitro Production and Use of Callus, Suspension Cells and Root Cultures of Medicinal Plants. *Molecules* **2020**, *25*, 5805. [CrossRef]
6. Banerjee, S.; Singh, S.; Ur Rahman, L. Biotransformation studies using hairy root cultures—A review. *Biotechnol. Adv.* **2012**, *30*, 461–468. [CrossRef] [PubMed]
7. Butler, N.M.; Shelley, H.J.; Jiang, J. First generation genome editing in potato using hairy root transformation. *Plant Biotechnol. J.* **2020**, *18*, 2201–2209. [CrossRef] [PubMed]
8. Caputi, L.; Franke, J.; Farrow, S.C.; Chung, K.; Payne, R.M.E.; Nguyen, T.-D.; Dang, T.-T.T.; Carqueijeiro, I.S.T.; Koudounas, K.; de Bernonville, T.D.; et al. Missing enzymes in the biosynthesis of the anticancer drug vinblastine in Madagascar periwinkle. *Science* **2018**, *360*, 1235–1239. [CrossRef] [PubMed]
9. Cardon, F.; Pallisse, R.; Bardor, M.; Caron, A.; Vanier, J.; Ele Ekouna, J.P.; Lerouge, P.; Boitel-Conti, M.; Guillet, M. *Brassica rapa* hairy root based expression system leads to the production of highly homogenous and reproducible profiles of recombinant human alpha-L-iduronidase. *Plant Biotechnol. J.* **2019**, *17*, 505–516. [CrossRef] [PubMed]
10. Carqueijeiro, I.; Langley, C.; Grzech, D.; Koudounas, K.; Papon, N.; O'Connor, S.E.; Courdavault, V. Beyond the semi-synthetic artemisinin: Metabolic engineering of plant-derived anticancer drugs. *Curr. Opin. Biotechnol.* **2020**, *65*, 17–24. [CrossRef] [PubMed]

11. Carrizo, C.N.; Pitta-Alvarez, S.I.; Kogan, M.J.; Giulietti, A.M.; Tomaro, M.L. Occurrence of cadaverine in hairy roots of *Brugmansia candida*. *Phytochemistry* **2001**, *57*, 759–763. [CrossRef]

12. Chahardoli, M.; Fazeli, A.; Ghabooli, M. Recombinant production of bovine Lactoferrin-derived antimicrobial peptide in tobacco hairy roots expression system. *Plant Physiol. Biochem.* **2018**, *123*, 414–421. [CrossRef]

13. Dang, T.T.T.; Franke, J.; Tatsis, E.; O'Connor, S.E. Dual catalytic activity of a cytochrome P450 controls bifurcation at a metabolic branch point of alkaloid biosynthesis in Rauwolfia serpentine. *Angew Chem. Int. Ed. Engl.* **2017**, *56*, 9440–9444. [CrossRef] [PubMed]

14. Dudareva, N.; Klempien, A.; Muhlemann, J.K.; Kaplan, I. Biosynthesis, function and metabolic engineering of plant volatile organic compounds. *New Phytol.* **2013**, *198*, 16–32. [CrossRef] [PubMed]

15. Dyshlyuk, L.; Dmitrieva, A.; Ivanova, S.; Golubcova, Y.; Ostroumov, L. Panax ginseng callus, suspension, and root cultures: Extraction and qualitative analysis. *Foods Raw Mater.* **2020**, *8*, 369–376. [CrossRef]

16. Georgiev, M.I.; Agostini, E.; Ludwig-Müller, J.; Xu, J. Genetically transformed roots: From plant disease to biotechnological resource. *Trends Biotechnol.* **2012**, *30*, 528–537. [CrossRef] [PubMed]

17. Hidalgo, D.; Georgiev, M.; Marchev, A.; Bru-Martínez, R.; Cusido, R.M. Purificación Corchete & Javier Palazon Tailoring tobacco hairy root metabolism for the production of stilbenes. *Sci. Rep.* **2017**, *7*, 17976. [CrossRef]

18. Jeziorek, M.; Sykłowska-Baranek, K.; Pietrosiuk, A. Hairy root cultures for the production of anticancer naphthoquinone compounds. *Curr. Med. Chem.* **2018**, *25*, 4718–4739. [CrossRef] [PubMed]

19. Kai, G.Y.; Xu, H.; Zhou, C.; Liao, P.; Xiao, J.; Luo, X.; You, L.; Zhang, L. Metabolic engineering tanshinone biosynthetic pathway in Salvia miltiorrhiza hairy root cultures. *Metab. Eng.* **2011**, *13*, 319–327. [CrossRef]

20. Kim, Y.J.; Zhang, D.; Yang, D.-C. Biosynthesis and biotechnological production of ginsenosides. *Biotechnol. Adv.* **2015**, *33*, 717–735. [CrossRef]

21. Li, F.S.; Weng, J.K. Demystifying traditional herbal medicine with modern approaches. *Nat. Plants* **2017**, *3*, 17109. [CrossRef] [PubMed]

22. Liao, P.; Hemmerlin, A.; Bach, T.J.; Chye, M.-L. The potential of the mevalonate pathway for enhanced isoprenoid production. *Biotechnol. Adv.* **2016**, *34*, 697–713. [CrossRef] [PubMed]

23. Ma, Y.N.; Xu, D.-B.; Li, L.; Zhang, F.; Fu, X.-Q.; Shen, Q.; Lyu, X.-Y.; Wu, Z.-K.; Pan, Q.-F.; Shi, P.; et al. Jasmonate promotes artemisinin biosynthesis by activating the TCP14-ORA complex in *Artemisia annua*. *Sci. Adv.* **2018**, *4*, eaas9357. [CrossRef]

24. Massa, S.; Paolini, F.; Marino, C.; Franconi, R.; Venuti, A. Bioproduction of a therapeutic vaccine against human Papillomavirus in tomato hairy root cultures. *Front. Plant Sci.* **2019**, *10*, 452. [CrossRef]

25. Mendoza, D.; Pablo, J.; Cuaspud, O.; Arias, M. Phytochemical Screening of Callus and Cell Suspensions Cultures of *Thevetia peruviana*. *Braz. Arch. Biol. Technol.* **2020**, *63*. [CrossRef]

26. Miao, G.; Han, J.; Feng, J.T.; Zhu, C.-S.; Zhang, X. A MDR transporter contributes to the different extracellular production of sesquiterpene pyridine alkaloids between adventitious root and hairy root liquid cultures of *Tripterygium wilfordii* Hook.f. *Plant Mol. Biol.* **2017**, *95*, 51–62. [CrossRef] [PubMed]

27. Nag, S.A.; Qin, J.-J.; Wang, W.; Wang, M.-H.; Wang, H.; Zhang, R. Ginsenosides as anticancer agents: In vitro and in vivo activities, structure–activity relationships, and molecular mechanisms of action. *Front. Pharmacol.* **2012**, *3*, 25. [CrossRef] [PubMed]

28. Normile, D. Asian medicine: The new face of traditional Chinese medicine. *Science* **2003**, *299*, 188–190. [CrossRef] [PubMed]

29. Paek, K.Y.; Hosakatte, N.M.; Hahn, E.-J.; Zhong, J.-J. Large scale culture of ginseng adventitious roots for production of ginsenosides. *Adv. Biochem. Eng. Biotechnol.* **2009**, *113*, 151–176. [PubMed]

30. Latushkina, N.A.; Ivanovsky, A.A.; Timkina, E.Y. Issledovanie himicheskogo sostava i toksicheskih svoystv fitokompleksa, soderzhaschego biologicheski aktivnye veschestva. *Agrarnaya Nauka Evro-Severo-Vostoka [Agric. Sci. Euro North East]* **2017**, *4*, 58–62. (In Russian)

31. Asyakina, L.; Sukhikh, S.; Ivanova, S.; Prosekov, A.; Ulrikh, E.; Chupahin, E.; Babich, O. Determination of the qualitative composition of biologically-active substances of extracts of in vitro callus, cell suspension, and root cultures of the medicinal plant *Rhodiola rosea*. *Biomolecules* **2021**, *11*, 365. [CrossRef] [PubMed]

32. Yang, Y.; Asyakina, L.K.; Babich, O.O.; Dyshlyuk, L.S.; Sukhykh, S.A.; Popov, A.D.; Kostiushina, N.V. Izuchenie fiziko-himicheskih svoystv i biologicheskoy aktivnosti ekstraktov iz vysushennoy biomassy kallusnyh, suspenzionnyh kletok I kornevyh kultur in vitro. *Tehnika I Tehnologiay Pishchevyh Proizvodstv [Tech. Technol. Food Prod.]* **2020**, *50*, 480–492. (In Russian) [CrossRef]

33. Yu, K.W.; Hahn, E.J.; Paek, K.Y. Ginsenoside production by hairy root cultures of *Panax ginseng*: Influence of temperature and light quality. *Biochem. Eng. J.* **2005**, *23*, 53–56. [CrossRef]

34. Zaushintsena, A.V.; Milentyeva, I.; Babich, O.; Noskova, S.Y.; Kiseleva, T.F.; Popova, D.G.; Bakin, I.A.; Lukin, A. Quantitative and qualitative profile of BASextracted from purple echinacea (*Echinacea purpurea* L.) growing in the Kemerovo region: Functional foods application. *Foods Raw Mater.* **2019**, *7*, 84–92. [CrossRef]

35. Zhang, J.H.; Wider, B.; Shang, H.; Li, X.; Ernst, E. Quality of herbal medicines: Challenges and solutions. *Complement. Ther. Med.* **2012**, *20*, 100–106. [CrossRef] [PubMed]

36. Prithviraj, K. Biological activities of flavonoids: An overview. *Int. J. Pharm. Sci. Res.* **2019**, *10*, 1567–1574.

Technical Note

Qualitative and Quantitative Comparison of Liquid–Liquid Phase Extraction Using Ethyl Acetate and Liquid–Solid Phase Extraction Using Poly-Benzyl-Resin for Natural Products

Yannik K. Schneider *, Solveig M. Jørgensen, Jeanette Hammer Andersen and Espen H. Hansen

Marbio, Faculty for Fisheries, Biosciences and Economy, UiT—The Arctic University of Norway, Breivika, N-9037 Tromsø, Norway; solveimj@stud.ntnu.no (S.M.J.); jeanette.h.andersen@uit.no (J.H.A.); espen.hansen@uit.no (E.H.H.)
* Correspondence: yannik.k.schneider@uit.no; Tel.: +47-(0)-77649267

Abstract: A key step in the process of isolating microbial natural products is the preparation of an extract from a culture. This step determines which molecules will be available for detection in the subsequent chemical and biological analysis of a biodiscovery pipeline. In the present study we wanted to document potential differences in performance between liquid–liquid extraction using ethyl acetate and liquid–solid extraction using a poly-benzyl-resin. For the comparison of the two extraction protocols, we spiked a culture of *Flavobacterium* sp. with a diverse selection of natural products of microbial and plant origin to investigate whether the methods were comparable with respect to selectivity. We also investigated the efficiency of the two extraction methods quantitatively, using water spiked with a selection of natural products, and studied the quantitative effect of different pH levels of the aqueous solutions on the extraction yields of the two methods. The same compounds were extracted by the two methods, but the solid-phase extract contained more media components compared with the liquid-phase extract. Quantitatively, the two extraction methods varied in their recovery rates. We conclude that practical aspects could be more important when selecting one of the extraction protocols, as their efficiencies in extracting specific compounds were quite similar.

Keywords: natural products; bacteria; downstream processing; antibiotics; isolation; extraction; secondary metabolites; pharmacognosy; bioprospecting

Citation: Schneider, Y.K.; Jørgensen, S.M.; Andersen, J.H.; Hansen, E.H. Qualitative and Quantitative Comparison of Liquid–Liquid Phase Extraction Using Ethyl Acetate and Liquid–Solid Phase Extraction Using Poly-Benzyl-Resin for Natural Products. *Appl. Sci.* **2021**, *11*, 10241. https://doi.org/10.3390/app112110241

Academic Editors: Guillaume Pierre and Emanuel Vamanu

Received: 31 August 2021
Accepted: 29 October 2021
Published: 1 November 2021

Publisher's Note: MDPI stays neutral with regard to jurisdictional claims in published maps and institutional affiliations.

1. Introduction

Natural products and their derivatives have successfully been developed as medicines that have enabled the treatment of various diseases [1]. Natural products differ from compounds that can be found in synthetic screening libraries by occupying a different chemical space that makes them a valuable source for new chemical scaffolds [2]. Microorganisms in general and bacteria in particular are promising sources for new bioactive natural products that can be developed as drug leads [3,4]. Natural product drug discovery commonly starts with the preparation of an extract, and this extract maybe pre-fractionated or directly investigated for bioactivity using bioassays such as anti-microbial or anti-cancer assays [5,6]. If bioactivity is detected, extracts or fractions are subsequently investigated for the presence of known active metabolites, which will then be removed from the pipeline [7]. This process is called dereplication and usually relies on hyphenated mass spectrometry techniques such as HPLC-MS [8,9]. An extract may also be investigated for novel metabolites without preceding bioactivity testing [10]. Potential new compounds will be isolated and subjected to structural elucidation and bioactivity profiling.

Extraction is a crucial step in the search for new active compounds. If a compound is not successfully extracted from the biomass or fermentation broth, it can obviously not be detected in downstream bioassays or chemical analysis. For bacterial fermentations, many possible extraction techniques and protocols exist, but as it is usually not feasible to apply more than one of them when screening larger collections of bacterial isolates, it is important to be aware of the limitations of the selected protocol. While we have used liquid–solid phase extraction of fermentation broths in previous studies [11,12], there are other relevant techniques such as liquid–liquid partition with ethyl acetate (EtOAc) [13] or extraction of the bacterial mycelium with ethanol in case of actinobacteria [14]. Sample pre-treatment and solvent selection are also dependent on whether intracellular or secreted metabolites are of interest [15].

To assess the suitability of our standard method for extraction of bacterial and fungal fermentations (liquid–solid phase extraction), we conducted a small study to compare it with liquid–liquid partition using EtOAc. We investigated the qualitative difference when extracting a bacterial culture spiked with seven different natural products. So-called spike and recovery tests are an established method for verifying the suitability and detection limits of an analytical method or sample preparation, including extraction procedures [16,17].

In addition, the two methods were compared quantitatively by extraction of four of the natural products from aqueous solution. We also investigated a potential effect of the pH level of the aqueous solution on the extraction yields.

2. Materials and Methods

The chemicals used were of appropriate purity, as indicated by the supplier, and pure water (pH$_2$O) was produced using an in-house MilliQ system. For vacuum filtrations, a Büchner funnel with Whatman No. 3 filter paper (Whatman plc., Maidstone, UK) was used.

2.1. Preparation of the Spiked Culture

A *Flavobacteria* sp. isolate was cultured over 7 d at 10 °C in a 450 mL shaking bottle culture, shaking at 300 rpm. A DVR1 medium was used for cultivation, consisting of 6.0 g malt extract (Sigma-Aldrich, St. Louis, MO, USA), 10.0 g Peptone from casein (Sigma-Aldrich), 6.0 g yeast extract (Sigma-Aldrich), 450 mL filtrated seawater and 450 mL pH$_2$O [11]. The media was sterilized by autoclaving at 120 °C for 30 min under pressure and inoculated using an inoculation loop. After cultivation, the fermentation broth was centrifuged at 4000 rpm for 10 min to remove the cells using a Multifuge 3 S-R equipped with a SORVALL 75,006,445 rotor (Heraeus GmbH & Co. KG, Hanau, Germany).

An aliquot of 300 mL of the supernatant was spiked with eight natural products—namely, rifampicin (R3501, Sigma-Aldrich), vancomycin (Sigma-Aldrich), colchicine (C3915, Sigma-Aldrich), cyclosporine (32425, Fluka®, Honeywell, Charlotte, NC, USA), paclitaxel (86346, Fluka®, Honeywell), ampicillin (A5354, Sigma-Aldrich) and gentamycin (A2712, Merck, Darmstadt, Germany). The concentrations of the respective natural products are given in Table 1 (structures in Figure 1). After the supernatant was spiked, it was thoroughly mixed and split into two samples of 150 mL each. The two samples were processed in parallel; one was extracted using liquid–liquid extraction and one using liquid–solid extraction.

Table 1. Final concentrations of the natural products in the bacterial culture.

Compound	Natural Product Class	Concentration (μg/L)
Rifampicin *	polyketide	2.0
Vancomycin †	glycopeptide	10.0
Colchicine †	alkaloid	5.0
Tetracycline *	polyketide	10.0
Cyclosporine *	cyclic peptide	10.0
Paclitaxel *	terpenoide	2.0
Ampicillin ‡	β-lactam antibiotic	10.0
Gentamycin‡	aminoglycoside	10.0

* dissolved in ethanol; † dissolved in pH$_2$O; ‡ readymade solution.

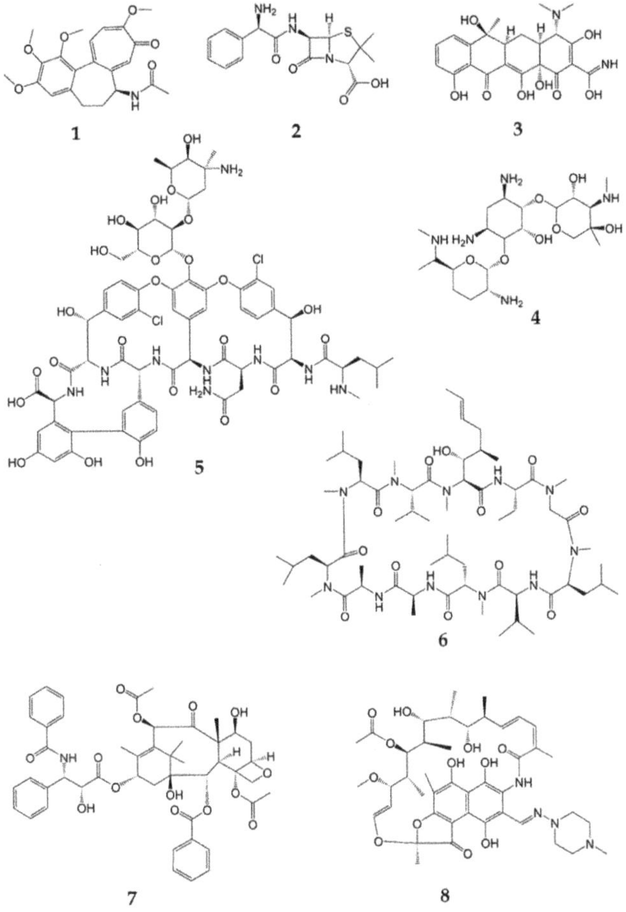

Figure 1. Structures of the natural products used for spiking within this study. (**1**) Colchicine; (**2**) Ampicillin; (**3**) Tetracycline; (**4**) Gentamycin; (**5**) Vancomycin; (**6**) Cyclosporine A; (**7**) Paclitaxel; (**8**) Rifampicin.

2.2. Extraction

2.2.1. Liquid–Liquid Phase Extraction

For the liquid–liquid phase extraction, the 150 mL sample was extracted three times using 75 mL EtOAc (Sigma-Aldrich) and a separation funnel. The extracts were reduced to dryness at 40 °C under reduced pressure, and the yield was determined gravimetrically.

2.2.2. Liquid–Solid Phase Extraction

For the liquid–solid phase extraction, Diaion® HP-20 (SUPELCO, Sigma-Aldrich) resin was used. An amount of 6.75 g resin was soaked in methanol (MeOH, HiPerSolv, VWR, Radnor, PA, USA) for 30 min for activation and transferred to pH_2O to wash out MeOH (~4.5 g resin was used for 100 mL sample or culture). After 20 min of washing in pH_2O, the resin was added to the spiked fermentation broth and shaken for 72 h at 10 °C for extraction. The resin was filtered from the sample using a cheese-cloth filter (1057, Dansk Hjemmeproduktion, Ejstrupholm, Denmark) placed on a Büchner funnel and subsequently washed in pH_2O. Thereafter the resin was eluted twice using 50 mL MeOH. Extraction was done by eluting the resin in MeOH for 1 h and by subsequently removing the resin by vacuum filtration. The extracts were reduced to dryness at 40 °C under reduced pressure, and the yield was determined gravimetrically.

2.3. Preparation and Extraction of Spiked Water for Quantitative Comparison and Investigation of the pH Effect on Extraction Yields

Stock solutions of tetracycline (10 mg/mL aq.), cyclosporine (10 mg/mL in EtOH), colchicine (5 mg/mL in EtOH) and gentamicin (10 mg/mL aq.) were prepared to spike the five water samples. Five 200 mL samples of pH_2O were set to a pH of 4.0, 5.0, 6.0, 7.0 and 8.5, respectively, using 1 M HCl (EMSURE, Merck), 0.3 M NaOH (VWR) and a pH meter (WTW InoLab pH 720, Xylem Inc. Rye Brook, NY, USA) equipped with a SenTix®41 electrode (WTW). The five conditions were spiked with the natural products to reach a final concentration of 10 µg/mL for tetracycline, cyclosporine A and gentamicin and 5 µg/mL for colchicine. Thereafter the five conditions (each 200 mL) were split into two aliquots of 100 mL each for parallel extraction using liquid–liquid phase extraction and liquid–solid phase extraction. For liquid–liquid phase extraction, the 100 mL aliquots were extracted with 100 mL EtOAc using a separation funnel. After shaking the two phases for ~1 min, the funnel was left to rest until the two phases separated properly. The EtOAc fractions were reduced to dryness at 40 °C under reduced pressure. The other aliquot for each water sample for resin or liquid–solid phase extraction was extracted using resin as described above (using ~4.5 g resin/100 mL of sample).

2.4. Analysis of the Extracts
2.4.1. Gravimetrical Analysis

The yields were determined gravimetrically using an analytical balance (Mettler Toledo AB204-S, Mettler Toledo, Columbus, OH, USA).

2.4.2. UHPLC-IMS-MS Analysis

For detection of the compounds, the extracts were re-dissolved in DMSO (Sigma-Aldrich) to a final concentration of 40 mg/mL and diluted 1:4 with 80% MeOH aq. to prepare an injection solution for UHPLC analysis.

For analysis of the presence of the respective natural products, UHPLC-IMS-MS was used. The analytical system consisted of an Acquity I-class UPLC (Waters, Milford, MA, USA) coupled to a PDA detector and a Vion IMS QToF (Waters). The chromatographic separation was performed using an Acquity BEH C18 UPLC column (1.7 µm, 2.1 mm × 100 mm) (Waters). Mobile phases consisted of acetonitrile (HiPerSolv, VWR) for mobile phase B and pH_2O produced by the in-house Milli-Q system as mobile phase A, both containing 0.1% formic acid (*v/v*) (Merck). The gradient was run linearly from 10% to 90% B over 12 min at a flow rate of 0.45 mL/min. Samples were run in ESI+ and ESI− ionization mode. The data were processed and analyzed using UNIFI 1.9.4 (Waters). Stock solutions of the respective natural products were injected as reference for unambiguous identification and to confirm that the compounds were detectable within the analytical setup.

For the quantitative analysis, the extracts were dissolved in 1.0 mL 80% MeOH aq., and a 10-fold dilution with 80% MeOH aq. was prepared for injection. A sample of each of the four stock solutions was diluted to 0.1 mg/mL in 80% MeOH aq. to obtain a reference solution. An amount of 5.0 µL was injected and analyzed using the same UHPLC protocol as above. For the quantification, three injections per sample were made, and the response/ion count of the respective *m/z* signal of the pseudo-molecular ion was used for quantification. The results were analyzed using GraphPad Prism 8.1.0 for linear regression.

3. Results
3.1. Qualitative Comparison

The EtOAc extraction of the spiked bacterial ferment was done three times with the same spiked sample, and the yield was determined for each of the three extractions individually. The yields for the three extractions were 26.9, 11.7 and 15.0 mg, respectively. The combined yield of the extractions was 53.6 mg (dry weight). The liquid–solid extraction yielded 325.7 mg of extract. The first EtOAc extract and the solid-phase extract were analyzed using UHPLC-IMS-MS and investigated for the presence of the spiked natural products. A small quantity of the stock solutions was injected as reference to determine the retention time and mass spectra for the respective natural products. The extracts were investigated for the presence of the spiked compounds by comparison with the retention time and mass spectra of the references. The results for the qualitative comparison of the extraction methods are given in Table 2.

Table 2. Retention times and presence of the natural products in EtOAc and resin extracts.

Compound	Ret. Time (min)	Mw (u)	Present in Extracts: EtOAc	Resin
Rifampicin	6.06	822.9	+	+
Vancomycin	0.97	1449.2	+	+
Tetracycline	2.09	444.4	+	+
Colchicine	3.30	399.4	+	+
Cyclosporine	9.60	1202.6	+	+
Paclitaxel	6.51	853.9	+	+
Ampicillin	1.51	349.4	-	-
Gentamycin	0.46	482.5	-	-

When comparing the chromatograms from the liquid–liquid and liquid–solid extractions, it appeared that many more polar metabolites and media components were extracted using the liquid–solid phase extraction compared with the liquid–solid phase extraction.

3.2. Quantitative Comparison

Water samples with pH values of 4.0, 5.0, 6.0, 7.0 and 8.5 were spiked with natural products, and an aliquot of each sample was extracted using EtOAc as well as resin extraction, as described above. For the quantitative comparison, the response of the protonated pseudo-molecular ions of the respective standard injections was used as reference to calculate the concentrations of the respective spike compounds. For determination of a calibration curve, LOQ and LOD, we injected 0.05, 0.25, 0.5 and 1 µg of cyclosporine, gentamycin, colchicine and tetracycline using the analytical protocol described above. We also injected 0.25 µg and 0.05 µg of ampicillin to determine the LOD of ampicillin that was not detected in the qualitative study. The calibration curves are shown in Figure 2. The LOD for ampicillin was <0.05 µg; LOD for tetracycline, cyclosporine and colchicine was <0.05 µg; LOD for gentamycin was <0.25 µg, and a proper quantification of gentamycin failed (see Figure 2). We also failed to detect gentamycin at 0.05 µg in negative electrospray. The extracts were dissolved in 1.0 mL of 80% MeOH aq., and the final yields were determined by multiplying the concentration with the volume; the final yields are given in Appendix A. Relative yields were calculated as ratios of the initial quantity of spike compound in the 100 mL of aqueous solution to be extracted. The relative yields are given in Figure 3.

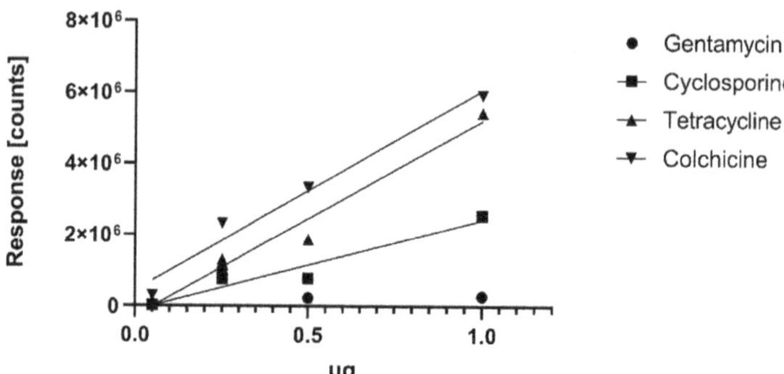

Figure 2. Calibration curves for the quantification of the selected natural products. Single injections were used for the calibration curves. The formulas and R^2 values for the calibration curves were tetracycline y = 5487981x − 112802, 0.97; cyclosporine 253601x − 11282, R^2 = 0.93; colchicine y = 5617051x + 438251, R^2 = 0.97; with y being the response and x being the injected compound quantity in µg. We failed to generate a proper calibration curve for gentamycin.

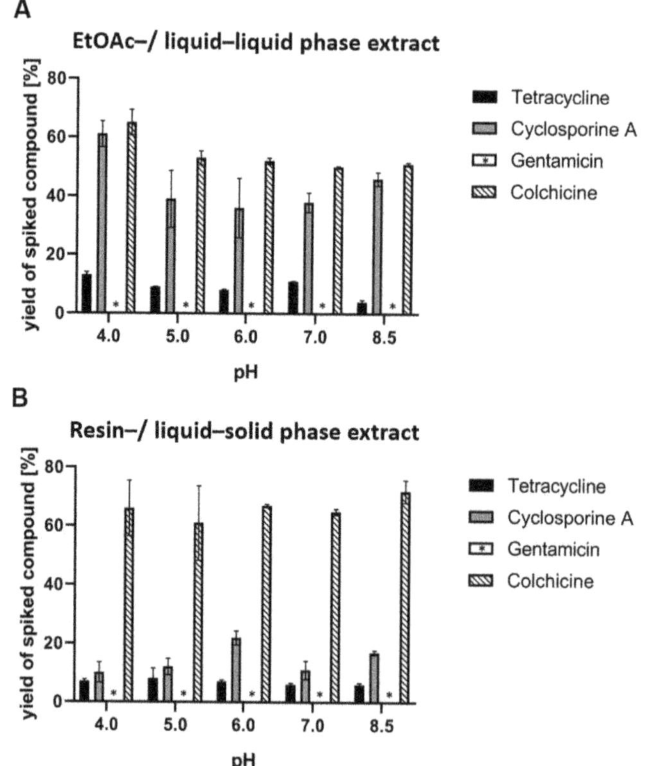

Figure 3. Yield of the natural products in percentage recovery from the spiked solution. A: Results for the liquid–liquid phase extraction. B: Results for the liquid–solid phase extraction. The different pH levels for the aqueous solutions/extraction conditions are given on the *x*-axis. Note that gentamycin (*) was not detected in any of the two extracts; however, the LOD for gentamycin would be around 5.0% of the initially spiked quantity; given the extract was dissolved in 1.0 mL of solvent, for all the other compounds LOD would equal <1.0% of the initially spiked compound quantity.

4. Discussion

The number of different protocols that can be used for extraction of microbial secondary metabolites is high, and they range from classical liquid–liquid extractions to more sophisticated applications, such as ultrasound assisted extraction and supercritical fluid extraction [18,19]. Furthermore, the selection of extraction solvents will have an impact on extracted metabolites, as observed in plant material [20]. However, we chose the two selected techniques in this investigation because they required no specialized equipment that went beyond the common equipment in a chemical laboratory, and they were widely used. When establishing the biodiscovery pipeline for bioactive secondary metabolites in our laboratory, we were not able to find a direct experimental comparison of the extraction methods using a spike and recovery test. To assess the potential difference of extraction efficacy and to make our data available to others were therefore the main objectives of this study.

Qualitative Comparison of the Extraction Methods

The first finding of the qualitative comparison of the extraction methods was the difference in yield when extracting the bacterial cultures. The yield of the resin extract was 325.7 mg, and this was about 6 times higher compared with the EtOAc extract, yielding 53.6 mg. Investigation of the chromatograms of the UHPLC-MS analysis revealed that the resin extract contained a higher share of polar compounds that originated from bacteria and media, as comparison with a media reference for DVR1 media confirmed. For the targeted natural products, the two extraction methods were quite comparable, as we detected the same spiked natural products in the resin and the EtOAc extracts (Table 2). Interestingly, ampicillin and gentamycin were not extracted by any protocols or at least they were not extracted in sufficient quantity for detection. When comparing the retention times in Table 2, it appears that ampicillin and gentamycin had rather short retention times (1.51 and 0.46 min, respectively) and vancomycin eluted between them (0.97 min). We speculate that both extraction methods are rather unsuitable for very polar and hydrophilic compounds. When comparing the structures of the analytes, it appears that vancomycin (**5** in Figure 1) is a relatively large molecule (1449.2 u) and consists of five benzylic rings, while ampicillin (**2**, 349.4 u) and gentamycin (**4**, 482.5 u) have substantially lower molecular weights.

For the quantitative comparison of the two extraction protocols, water was used as matrix to dissolve the natural products for reproducibility reasons since the media com-ponents and bacterial metabolites in a fermentation could affect the extraction (as we observed above that the solid-phase extract contained significantly more of those compounds). We intended to compare the extraction of the natural products in a simple way, from purified water, including the effect of the pH levels of the aqueous solutions on the extraction yield. Some bacteria prefer to grow or grow exclusively at acidic (acidophil bacteria) [21] or alkalic pH levels (alkaliphile bacteria) [22], and we also wanted to investigate if a post-fermentative change in pH by adding acid or alkali could be a strategy to increase yields. For the quantitative study, we selected four natural products (tetracycline, cyclosporine, colchicine and gentamycin), trying to cover a broad range of polarities and molecular weights. We investigated pH levels of 4.0, 5.0, 6.0, 7.0 and 8.5. In Figure 3, the yields of the respective natural products are given as percentage recovery or yield of the initially spiked compound quantity (1.0 mg for tetracycline, cyclosporine and gentamycin and 0.5 mg for colchicine). As expected, after the observations from the preceding qualitative comparison, gentamicin was not detected in the extracts of either extraction technique or at any pH-level. Generally, as visible in Figure 3, there were few variations among the pH levels within the same extraction. Generally, EtOAc extraction had a better recovery of the spiked compounds (except for gentamycin), tetracycline was recovered at about 10% using resin and 15% using EtOAc, colchicine was recovered at levels between 50% and 70%, depending on the condition, for colchicine; the resin extraction performed better. The recovery of cyclosporine was, depending on the condition, at least around 2 times higher in the EtOAc extract samples compared with the resin extracts.

We conclude that the two extraction techniques are qualitatively comparable and that both are less suitable for polar compounds, which was expected given the chemical composition of the poly-benzyl resin. Quantitatively the EtOAc extraction showed higher yields for cyclosporine, while the other two compounds were comparable. One of the prominent advantages of EtOAc extracts is that the extracts contain fewer media-components and polar metabolites, which eases downstream purification of compounds from the extracts. Furthermore, technical considerations can play a decisive role when screening larger sample collections. The presented protocol of resin extraction appears to be rather impracticable for smaller culture volumes (>5 mL), but on the other hand, it can be more convenient when processing larger numbers of fermentations at an intermediate scale. Since we expect many of the microbial secondary metabolites to be excreted to the media in order to exert their biological effect [4], extraction of the media seems to be reasonable. However, extraction of a 10 L culture using EtOAc would demand an equal volume of the organic solvent, while resin extraction would require just gram amounts of resin, but at the expense of extracting more polar molecules. It was important for us to document that the extraction methods were qualitatively comparable for the tested compounds; therefore, other technical considerations can be prioritized when deciding on one of the protocols.

5. Conclusions

The liquid–liquid and liquid–solid state methods are comparable when extracting the selected natural products from a microbial fermentation broth or water solutions at different pH levels. Both methods are less suitable for the extraction of polar metabolites. The liquid–liquid extract contains fewer media components, but this method is less suited when extracting larger cultivations due to high consumption of organic solvents.

Author Contributions: Conceptualization, E.H.H., J.H.A. and Y.K.S.; investigation, S.M.J. and Y.K.S.; writing—original draft preparation, Y.K.S.; writing—review and editing, E.H.H. and J.H.A. All authors have read and agreed to the published version of the manuscript.

Funding: Y.K.S. received funding from the Marie Skłodowska-Curie Action MarPipe, grant agreement GA 721421 H2020-MSCA-ITN-2016, of the European Union, and from UiT–The Arctic University of Norway.

Institutional Review Board Statement: Not applicable.

Informed Consent Statement: Not applicable.

Data Availability Statement: Not applicable.

Acknowledgments: Dana Elizabeth Wright (UiT—The Arctic University of Norway) is gratefully acknowledged for proofreading the manuscript.

Conflicts of Interest: The authors declare no conflict of interest.

Appendix A

Table A1. Table of calculated yields from the quantitative comparison.

Calc. Compound Yields in (µg)	EtOAc Extraction					Resin Extraction				
pH	4.0	5.0	6.0	7.0	8.5	4.0	5.0	6.0	7.0	8.5
Colchicine	327	267	260	251	254	332	303	338	325	359
Cyclosporine	607	389	362	383	461	97	118	220	112	171
Tetracycline	134	88	78	106	44	70	77	69	58	59
Gentamycin	-	-	-	-	-	-	-	-	-	-

"-" = not detected.

References

1. Newman, D.J.; Cragg, G.M. Natural products as sources of new drugs over the nearly four decades from 01/1981 to 09/2019. *J. Nat. Prod.* **2020**, *83*, 770–803. [CrossRef]
2. Rosén, J.; Gottfries, J.; Muresan, S.; Backlund, A.; Oprea, T. Novel Chemical Space Exploration via Natural Products. *J. Med. Chem.* **2009**, *52*, 1953–1962. [CrossRef] [PubMed]
3. Ganesan, A. The impact of natural products upon modern drug discovery. *Curr. Opin. Chem. Biol.* **2008**, *12*, 306–317. [CrossRef] [PubMed]
4. Pham, J.V.; Yilma, M.A.; Feliz, A.; Majid, M.T.; Maffetone, N.; Walker, J.R.; Kim, E.; Cho, H.J.; Reynolds, J.M.; Song, M.C.; et al. A Review of the Microbial Production of Bioactive Natural Products and Biologics. *Front. Microbiol.* **2019**, *10*, 1404. [CrossRef]
5. Wagenaar, M.M. Pre-fractionated Microbial Samples—The Second Generation Natural Products Library at Wyeth. *Molecules* **2008**, *13*, 1406–1426. [CrossRef]
6. Grkovic, T.; Akee, R.K.; Thornburg, C.C.; Trinh, S.K.; Britt, J.R.; Harris, M.J.; Evans, J.R.; Kang, U.; Ensel, S.; Henrich, C.J.; et al. National Cancer Institute (NCI) Program for Natural Products Discovery: Rapid Isolation and Identification of Biologically Active Natural Products from the NCI Prefractionated Library. *ACS Chem. Biol.* **2020**, *15*, 1104–1114. [CrossRef] [PubMed]
7. Cordell, G.A.; Shin, Y.G. Finding the needle in the haystack. The dereplication of natural product extracts. *Pure Appl. Chem.* **1999**, *71*, 1089–1094. [CrossRef]
8. Ito, T.; Masubuchi, M. Dereplication of microbial extracts and related analytical technologies. *J. Antibiot.* **2014**, *67*, 353–360. [CrossRef]
9. Sashidhara, K.V.; Rosaiah, J.N. Various Dereplication Strategies Using LC-MS for Rapid Natural Product Lead Identification and Drug Discovery. *Nat. Prod. Commun.* **2007**, *2*, 193–202. [CrossRef]
10. Zou, Y.; Yin, H.; Kong, D.; Deng, Z.; Lin, S. ATrans-Acting Ketoreductase in Biosynthesis of a Symmetric Polyketide Dimer SIA7248. *Chembiochem Eur. J. Chem. Biol.* **2013**, *14*, 679–683. [CrossRef]
11. Schneider, Y.K.-H.; Hansen, K.Ø.; Isaksson, J.; Ullsten, S.; Hansen, E.H.; Hammer Andersen, J. Anti-Bacterial Effect and Cytotoxicity Assessment of Lipid 430 Isolated from *Algibacter* sp. *Molecules* **2019**, *24*, 3991. [CrossRef] [PubMed]
12. Schneider, Y.; Jenssen, M.; Isaksson, J.; Hansen, K.; Andersen, J.; Hansen, E. Bioactivity of Serratiochelin A, a Siderophore Isolated from a Co-Culture of *Serratia* sp. and *Shewanella* sp. *Microorganisms* **2020**, *8*, 1042. [CrossRef]
13. Hayashida-Soiza, G.; Uchida, A.; Mori, N.; Kuwahara, Y.; Ishida, Y. Purification and characterization of antibacterial substances produced by a marine bacterium *Pseudoalteromonas haloplanktisstrain*. *J. Appl. Microbiol.* **2008**, *105*, 1672–1677. [CrossRef]
14. Wietz, M.; Månsson, M.; Bowman, J.S.; Blom, N.S.; Ng, Y.; Gram, L. Wide Distribution of Closely Related, Antibiotic-Producing Arthrobacter Strains throughout the Arctic Ocean. *Appl. Environ. Microbiol.* **2012**, *78*, 2039–2042. [CrossRef]
15. Pinu, F.R.; Villas-Boas, S.G.; Aggio, R. Analysis of Intracellular Metabolites from Microorganisms: Quenching and Extraction Protocols. *Metabolites* **2017**, *7*, 53. [CrossRef] [PubMed]
16. Sturini, M.; Speltini, A.; Pretali, L.; Fasani, E.; Profumo, A. Solid-phase extraction and HPLC determination of fluoroquinolones in surface waters. *J. Sep. Sci.* **2009**, *32*, 3020–3028. [CrossRef] [PubMed]
17. Xu, J.; Zhu, L.-Y.; Shen, H.; Zhang, H.-M.; Jia, X.-B.; Yan, R.; Li, S.-L.; Xu, H.-X. A critical view on spike recovery for accuracy evaluation of analytical method for medicinal herbs. *J. Pharm. Biomed. Anal.* **2012**, *62*, 210–215. [CrossRef]
18. Grosso, C.; Valentão, P.; Ferreres, F.; Andrade, P.B. Alternative and Efficient Extraction Methods for Marine-Derived Compounds. *Mar. Drugs* **2015**, *13*, 3182–3230. [CrossRef] [PubMed]
19. Baldino, L.; Scognamiglio, M.; Reverchon, E. Supercritical fluid technologies applied to the extraction of compounds of industrial interest from *Cannabis sativa* L. and to their pharmaceutical formulations: A review. *J. Supercrit. Fluids* **2020**, *165*, 104960. [CrossRef]
20. Pintać, D.; Majkić, T.; Torovic, L.; Orčić, D.; Beara, I.; Simin, N.; Mimica-Dukić, N.; Lesjak, M. Solvent selection for efficient extraction of bioactive compounds from grape pomace. *Ind. Crop. Prod.* **2018**, *111*, 379–390. [CrossRef]
21. Hedrich, S.; Schippers, A. Distribution of Acidophilic Microorganisms in Natural and Man-made Acidic Environments. *Curr. Issues Mol. Biol.* **2021**, *40*, 25–48. [CrossRef] [PubMed]
22. Ding, Z.-G.; Li, M.-G.; Zhao, J.-Y.; Ren, J.; Huang, R.; Xie, M.-J.; Cui, X.-L.; Zhu, H.-J.; Wen, M.-L. Naphthospironone A: An Unprecedented and Highly Functionalized Polycyclic Metabolite from an Alkaline Mine Waste Extremophile. *Chem.-A Eur. J.* **2010**, *16*, 3902–3905. [CrossRef] [PubMed]

MDPI
St. Alban-Anlage 66
4052 Basel
Switzerland
Tel. +41 61 683 77 34
Fax +41 61 302 89 18
www.mdpi.com

Applied Sciences Editorial Office
E-mail: applsci@mdpi.com
www.mdpi.com/journal/applsci